The Eclectic Legacy

The Eclectic Legacy

Academic Philosophy
and the Human Sciences
in Nineteenth-Century France

John I. Brooks III

DELAWARE

Newark: University of Delaware Press
London: Associated University Presses

Associated University Presses
440 Forsgate Drive
Cranbury, NJ 08512

Associated University Presses
16 Barter Street
London WC1A 2AH, England

Associated University Presses
P.O. Box 338, Port Credit
Mississauga, Ontario
Canada L5G 4L8

The paper used in this publication meets the requirements of the American National Standard for Permanence of Paper for Printed Library Materials Z39.48-1984.

Library of Congress Cataloging-in-Publication Data

Brooks, John I., 1957–
    The eclectic legacy : academic philosophy and the human sciences in nineteenth-century France / John I. Brooks III.
        p.   cm.
    Includes bibliographical references and index.
    ISBN 0-87413-648-2 (alk. paper)
    1. Philosophy, French—19th century. 2. Philosophy and social sciences—History—19th century. 3. Psychology—France—Philosophy—History—19th century. 4. Sociology—France—Philosophy—History—19th century. I. Title.
B2188.S63B76 1998
194—dc21   97-32351
                                                                                          CIP

101598-4356X8

Contents

Tables

Acknowledgments

THIS project developed over a long period of time, and along the way a number of people provided invaluable support. The work began as a dissertation at the University of Chicago, which I attended under a fellowship funded by the Searle Foundation. Financial support for dissertation research in France came from the Lurcy Foundation, the Social Science Research Council, and the Tocqueville Foundation. Many people read my proposal, gave me advice, and/or guided me through the labyrinth of French libraries and archives during this sojourn in France, including Susanna Barrows, Philippe Besnard, Jean-Claude Chamboredon, Christophe Charle, Jean-Louis Fabiani, François Furet, Victor Karady, Robert Nye, and Pierre Petitmengin. John Carson, James Johnson, and Laura Mason also helped me refine my ideas and direct my research in France. The members of my dissertation committee, Keith Baker, Jan Goldstein, and Robert Richards, guided my work through a long process of definition and focus. Gregory McGuire shared his research and insights on Pierre Janet with me. The participants in the Fishbein Workshop in the History of the Human Sciences, under the leadership of George Stocking, read and commented upon early drafts of the manuscript. The members of my dissertation support group, Sherrie Lyons, Jeff Ramsey, and Marc Swetlitz, lived up to its name, providing unwavering support along with constructive criticism. Other people whose contributions cannot be measured include Nancy Burke and Susan McCord.

I have also received a great deal of help from Teikyo Loretto Heights University, which has contributed to the project in two important ways. First, it has given me insightful colleagues in Nancy Brown, Jim Warnock, and Mark Zellmer, who read drafts of papers presented at several conferences. Equally important, it has generously supported travel expenses to those conferences, where I was able to present my ideas and to get valuable feedback. One such conference was the annual meeting of the Society for French Historical Studies in March 1992, where Phyllis Stock-Morton made helpful comments on a paper on Ribot, and Brian Williams, a copanelist, gave me additional archival references. The hospitality of James Whitman and Alain Fontaine allowed me to follow up these references and to extend my research during a second expedition to France in the summer of 1992, where I benefited from conversations with Erika Apfelbaum and Régine Plas. The research I did that summer resulted in an article that appeared in the *Journal of the History of the Behavioral Sciences* in 1993. Parts of that article have been incorporated into this book, and I would like to thank Barbara Ross for publishing the piece so expeditiously and John Wiley & Sons for permission to reprint selected portions. I should also thank John Wiley & Sons for permission to use portions of another article that appeared in the same journal in 1996.

As the work neared completion, Warren Schmaus provided precise, astute comments on several of the chapters. Rhoda Miller used her extraordinary knowledge of French to review my translations against the originals. Her suggestions have made the quotations both more accurate and more elegant. My wife Ellen read the final draft and saved me from innumerable errors large and small. She also provided much-needed moral support through the process of revision. My greatest debt, however, is to my parents, who first taught me to love the life of the mind by living it themselves, who spared no sacrifice to provide me with every educational opportunity, and who believed in me throughout an uncertain beginning to my academic career. Despite the advice and support of so many people, who often tried to save me from myself, responsibility for the ideas and opinions expressed in this work rests with the author.

The Eclectic Legacy

Introduction

When one gives such a place to chance, one opens the door to the imagination. If chance is indeed found everywhere, it is a *deus ex machina* that one can have intervene at will, each time one has need of it and without any verification being possible.

Chance would have it that all of the academics who have been involved with sociology are professors of philosophy.
—Emile Durkheim, "The Present State
of Sociological Studies in France"

THIS book examines the relationship in nineteenth-century France between academic philosophy—the philosophy taught in lycées and universities—and the kind of scientific psychology and sociology that became institutionalized in those same universities at the end of the century. I was attracted to the topic because, as this quotation suggests, the connection seemed both strong and strongly denied. It seemed strong because most of the early French innovators in psychology and sociology had a lifelong relationship with academic philosophy. They encountered it early in their lives, as students in the French lycées, or secondary schools, where *psychologie* and *morale* formed two important parts of the standard philosophical syllabus—the *programme*. Many went on to specialize in philosophy at the elite Ecole Normale Supérieure, after which they themselves became professors of philosophy in those same lycées. They taught there for years, as they struggled to complete dissertations that would be judged by academic philosophers and published in journals addressed to the same audience. They answered the criticisms leveled by university philosophers, and they recruited the next generation of social scientists from among students of philosophy. Hence it would seem obvious to search among early French social scientists for a legacy bequeathed by academic philosophy, largely defined in the nineteenth century by a school of thought called eclectic spiritualism.

Nevertheless, if the testimony of these early psychologists and sociologists is to be believed, their immersion in philosophical culture had little effect on them. Certainly the pioneering sociologist Emile Durkheim might recall his teacher Emile Boutroux with fondness and respect, and the eminent psychologist Pierre Janet would argue late in life that his

13

uncle, the once-powerful philosopher Paul Janet, had not received his due.[1] In general, though, such individuals depicted their relationship with academic philosophy as one of opposition, struggle, and liberation. They claimed to have freed themselves from the superficial, a priori, and metaphysical approach of their philosophical predecessors and to have founded empirical disciplines modeled on the natural sciences. Moreover, they felt that philosophers both misunderstood and opposed their efforts, and often their complaints reflected real differences of fundamental approach.

Subsequent historians long reinforced this rhetoric of revolution in a variety of ways. Some ignored philosophical antecedents altogether. Thus, such a valuable and well-documented study as Terry N. Clark's *Prophets and Patrons* scarcely mentions academic philosophy, although sociologists often competed for institutional positions with philosophers, and when they did not, they often depended upon the support of philosophers in their struggles with other disciplines.[2] Others discussed philosophical antecedents, but instead of the immediate philosophical milieu, humble and forgotten, great social scientists were elevated to the canon and linked in the chain of Great Books to their remote ancestors. In this fashion, Durkheim was often shown to descend from Montesquieu and Comte, but rarely from Alfred Espinas, and Pierre Janet was related to Maine de Biran but not to his uncle.[3] Finally, most historians simply assumed that some fundamental mutation occurred during the period, and that empirical sciences were indeed constituted. This assumption was so basic and pervasive that it was often simply stated rather than argued.[4]

Philosophy itself has contributed to this appearance of radical change by denigrating or ignoring nineteenth-century French academic philosophers. It is true that these thinkers were not universally esteemed even by their contemporaries, and most of the readily recognizable French names in nineteenth-century philosophy—Saint-Simon, Comte, and Fourier, for example—were outside the university system. Nevertheless, subsequent changes in philosophical fashion helped bury rather than preserve whatever value the academic tradition may have had. Beginning in the 1930s, French philosophers found new idols in German thinkers such as Marx, Nietzsche, Husserl, and Heidegger. Many French philosophers consciously turned away from their native ancestors, whom they considered less rigorous and less original.[5] Jean-Paul Sartre was the archetype of this sea change in French philosophy, but he was only the best known of an entire generation. Partly as a result, names such as Victor Cousin, Théodore Jouffroy, Félix Ravaisson, and Jules Lachelier are no longer cited with the same reverence as their German counterparts. As French philosophers once again rebel against their predecessors, they are rediscovering once taboo philosophers such as Kant, but the reexamination of the French academic tradition has made little progress.[6]

If all of these groups tended to exclude and deny the academic tradition in philosophy, the converse was not true of philosophers closer to the period in question. The histories and surveys of philosophy written by university professors of philosophy from the end of the nineteenth century to the mid-1930s always included scientific psychology and sociology as relevant contributions to philosophy.[7] If their inclusion of these trends was often merely the occasion for reaffirming the bankruptcy of positivism and celebrating the inevitable triumph of spiritualist metaphysics, the fact that a place was found for them at all attests to the philosophers' concession that the "positive" approach was a legitimate philosophical position. Of course, this generosity infuriated the social scientists, who considered themselves perhaps relevant to, but certainly distinct from, philosophy. Furthermore, the assurance with which the philosophers of this period contained the threat of positivism refutes the impression given by some social scientists that philosophy simply disappeared or entered an interminable "crisis" upon the appearance of the social sciences.[8]

Here then are two different pictures of the relation between academic philosophy and the human sciences. According to one, psychology and sociology became essentially distinct from philosophy around the turn of the century, firmly based on methods that had little in common with philosophy; according to the other, these disciplines remained a disloyal opposition within a generously defined arena of philosophical thought, dependent on ultimately philosophical assumptions. As is usually the case, the real story is more complicated, and recent studies have begun to explore the relationship between academic philosophy and the emerging human sciences in France. Most attention has gone to sociology. William Logue, in *From Philosophy to Sociology*, fully recognizes the filiation of sociology and philosophy.[9] Jean-Louis Fabiani, Victor Karady, and W. Paul Vogt have suggested that academic philosophy is the proper context for studying the emergence of the human sciences.[10] A number of historians, philosophers, and sociologists have assessed the philosophical influences on Durkheim.[11] Psychology has received less attention, but Mara Meletti Bertolini has studied philosophy and psychology in one of the early philosophical journals in France, and Jan E. Goldstein has begun to explore the relationship between academic philosophy and psychology.[12] At a more general level, Fritz Ringer, in his *Fields of Knowledge*, has commented upon the connection between philosophy and the human sciences in the larger context of French academic culture during the late nineteenth century.[13] Finally, my own work has explored some case studies of the interactions among philosophy, psychology, and sociology.[14]

This growing body of research both recognizes the importance of academic philosophy in the emergence of the human sciences and permits the problem to be addressed in more general terms. That is what this book attempts to do. I will show that the French academic philosophy created by the group known as eclectic spiritualists contributed to the emergence of psychology and sociology in significant and positive ways—that there was, in other words, an eclectic legacy to the human sciences. The eclectic spiritualists created a philosophical universe of discourse that included and encouraged an empirical approach to human phenomena. They also created an institutional niche in the educational system into which the more specialized practitioners of the human sciences could fit. Psychologists and sociologists used these conceptual and institutional resources to construct their own disciplines, even as they tried to separate themselves from their philosophical roots.

Of course, some of the same features of academic philosophy that could potentially give rise to the human sciences also helped block them. If philosophy already addressed these domains, what need was there for separate disciplines? The eclectic spiritualists who dominated academic philosophy for much of the nineteenth century saw no reason to separate psychology and sociology from philosophy; indeed, they worked hard to exclude those who advocated such separation. The space of eclectic spiritualism could be used to construct human sciences, but only if eclectic spiritualism disappeared. A legacy, after all, is something left after one dies, and eclectic spiritualism had to die if the human sciences were to receive their inheritance.

The death of eclectic spiritualism was as much the work of philosophers as of social scientists. Contradictions within eclecticism and new influences from outside the university conspired to create a rebellion against the doctrine on the part of philosophers, a rebellion that ended the eclectic monopoly in higher education during the early Third Republic. A wide variety of new schools of thought emerged, from more rigorous forms of neospiritualism to more positivistic forms of psychology and sociology. Nevertheless, the common institutions and intellectual universe established by eclectic spiritualism ensured that philosophers and social scientists would continue to talk with and argue with each other even after they had declared themselves independent. These conversations helped redefine both philosophy and the human sciences, and the common space in which they took place is another part of the eclectic legacy to the human sciences.

The two domains, philosophy and human science, did begin to diverge, but at the end of the nineteenth century, which for the purposes of this study can be considered to extend up to 1914, this divergence was still incomplete, conflicted, and contested. This confused state of affairs manifests itself in tensions *within* the texts of the writers in each discipline and in the continuing arguments *between* writers representing different disciplines. It is also demonstrated in the continuing institutional ties between philosophy and the human sciences. Indeed, the whole issue of how many disciplines were competing continued to be a source of controversy. The remainder of the introduction will explain why I have delimited the topic the way I have and the general approach I have taken to the history of philosophy, science, and academic disciplines.

FOCUS

Although this book considers attempts by would-be sociologists and psychologists to free themselves from philosophy and to establish themselves as scientists, it should not be read as a study of the relation between science and philosophy in general. Indeed, I will show that the words "philosophy" and "science" had very specific, changing, and contested meanings during the nineteenth century. Although the two terms were often opposed to each other by authors interested in making a sharp distinction, the meanings of the terms, and hence of the opposition, changed during the course of the century. Most studies of the period have failed to ask what these terms could have meant in specific contexts, and as a result they have either projected onto the past their own conception of this distinction, or they have uncritically adopted the distinction of one of the parties involved, usually that of the social scientists. They therefore assume that the terms meant then what they mean now, and that both parties to the dispute agreed upon this meaning. Even worse, they assume that the practices referred to by these terms, the sets of principles and methods that constituted philosophy or science, actually corresponded to the terms used. In this way, it is assumed that all philosophers were engaged primarily in the analysis of concepts, whereas the social scientists were really engaged in empirical research. This book will show that such assumptions can be misleading.

First of all, nineteenth-century French academic philosophy had peculiar features that distinguish it from nonacademic philosophy in France and from philosophy in other countries. Its highly uniform and institutionalized character had no real counterpart in the English-speaking world. It was taught in all French lycées from a standard syllabus known as the *programme,* upon which state-administered examinations were based. As I show in chapter 1, the syllabus constituted the universe of discourse for academic philosophy—the range of problems, methods, fundamental assumptions, and conceptual parameters that defined the discipline. So basic that it was rarely articulated at the higher level of individual doctrines, it nevertheless constituted the ground on which such variations were built. I would like to bring to light the pervasiveness of the philosophy of the syllabus in the discourse of the early human sciences, because I think that it continued to inform much of the thought and practice of these disciplines long after they claimed their independence. Within this universe of discourse, individual philosophers in both the lycée and the university discussed a variety of particular issues on which there could be considerable disagreement, even during the early days of academic philosophy, when doctrinal orthodoxy was explicitly enforced. It is at this level of explicit, individual doctrines that most historians have looked for influences on the human sciences, but it was neither the only level nor the most important.

The diversity of academic philosophy should not obscure the definite limits on acceptable methods and solutions within the discipline, and these limits differentiated the content of French academic philosophy from its nonacademic counterparts. Nineteenth-century French academic philosophy was created almost single-handedly by Victor Cousin, whose own philosophy, which he called eclectic spiritualism, pervaded the syllabus. The name was not chosen lightly. Cousin's philosophy claimed to be *eclectic* in that it attempted to incorporate and reconcile elements from all philosophical doctrines, however opposed they might be. It was *spiritualist* in that it recognized the existence of thinking substance. And it also claimed to be *scientific,* in that it pretended to be a science of observation. These three features of French academic philosophy—eclecticism, spiritualism, and scientism—had important consequences both for the shape the human sciences took and the way in which academic philosophy reacted to the human sciences.

The *eclecticism* of French academic philosophy was important, first of all, because it meant that a wide range of philosophical doctrines was discussed within the academy, even if only to excerpt or refute them. In some ways, the French eclectics sowed the seeds of their own destruction by translating and commenting on philosophical doctrines which, in the hands of their students, returned to haunt them. For example, the eclectics played a large role in the introduction of Kant into France, even though they disagreed with aspects of his philosophy.[15] Neo-Kantianism later became one of the leading philosophies in France, and eclecticism was one of its primary targets. Furthermore, the eclectic principle that every doctrine contained an element of truth forced eclectics to examine and incorporate even those tendencies to which they were most opposed. In the long run this resulted in a considerable shift in the character of academic philosophy. Finally, the eclectic practice of picking and choosing from different and often opposing systems of thought carried over into the construction of the human sciences, as individuals trained in the eclectic tradition selected the elements they needed for their vision of the human sciences from widely disparate sources. This sort of conceptual *bricolage,* helpful if one was to reconcile the methods of the natural sciences with the insights of philosophy, is one reason why subsequent scholars have found it difficult to categorize French psychologists and sociologists, labeling them as either positivist, Kantian, or idealist, depending on which part of the proverbial elephant they happen to touch. Indeed, perhaps the most systematic thinker of this period—Emile Durkheim—has proved to be the most problematic, precisely because his interpreters have not recognized the eclectic principle behind his sociological system-building.

The *spiritualism* of French academic philosophy was eventually incorporated into the definitions of psychology and sociology, because these remained essentially sciences of the "moral," which in French meant both "ethical" and "spiritual." This was true of the psychology of Pierre Janet, who opposed physiological explanations of psychological phenomena and who extended the domain of the psychological into the subconscious. It was also true of the sociology of Emile Durkheim, who concentrated on the nature of morality and who claimed that his concept of society was not materialist but "hyperspiritualist."

The *scientism* of French academic philosophy is also crucial, even though it has been routinely ignored or dismissed by subsequent philosophers, scientists, and historians.[16] On the one hand, it encouraged the use of empirical methods in psychology and sociology, even though its understanding of those methods differed from those of the more positivistically oriented social scientists. In this way, psychologists and sociologists could say that they were *really* doing what philosophers merely *claimed* to be doing. Moreover, the philosophy of science held an important and growing place in the syllabus, and it was in large part through the philosophy class rather than through the practice of a natural science that most early psychologists and sociologists gained their notion of science. Finally, the rhetoric of science forced academic philosophers to acknowledge the legitimacy of at least part of the positivist enterprise. This was one reason why academic philosophers did not continue to simply reject out of hand the human sciences. Instead, they constructed strategies of recognition, accommodation, containment, and marginalization. Hence, even if the claim to be empirical and scientific was false, it contributed to the changing relationship between academic philosophy and the human sciences.

From the term *philosophy* let us pass to that of *science.* If a spiritualist philosophy could define itself as scientific, then clearly science was capable of many definitions. Throughout the nineteenth century, philosophers and scientists argued about what science was and what a human science should look like. Auguste Comte, founder of positivism and coiner of the term *sociology,* is perhaps the best-known participant in these debates. His advocacy of a science of society independent of metaphysics and free will was so notorious in the nineteenth century that the term *positivism* was quickly taken up and applied to others who in many ways neither followed nor approved of Comte.[17] This raises a problem of terminology. Strictly speaking, positivism is the philosophy of Comte, and in a classic study of *European Positivism in the Nineteenth Century,* Walter M. Simon recommends that the term be restricted to this sense, "because other usages tend to impoverish the language. For an attitude of admiration for the natural sciences and the wish to extend their virtues to other disciplines we have the excellent word 'scientism. . . .'"[18] He chides D. G. Charlton, author of another classic study on *Positivist Thought in France,* for using positivism in the larger and more nebulous sense.[19] Nevertheless, my own usage in this book is closer to Charlton than to Simon. I mean by "positivism" belief in the following: that the natural sciences represent the pinnacle of possible knowledge; that the human sciences should be modeled on the natural sciences; that the principle of determinism applies to human thought and action; and that metaphysics should be separated from science. I have chosen this definition for lack of a better word. "Scientism" is too vague, since I am using it to describe eclectic spiritualism. However, many of the individuals examined in this book disagreed vehemently with Comte, while holding the beliefs I have just described. To restrict positivism to the philosophy of Comte would leave me no convenient way of characterizing these individuals. Moreover, this meaning is historically legitimate, because it was current during the nineteenth century.

Despite my use of the term *positivism* to characterize the advocates of autonomous human sciences, this book is not primarily about Comte. For one thing, he has been the object of many fine studies.[20] For another, he was not the only philosopher in France who sought to put the study of human phenomena on a scientific footing. In addition to the eclectics, many others had similar goals, and even those whose projects resembled Comte more than Cousin could disagree with or adapt the founder of positivism's vision. These competing visions had as much influence as that of Comte, who was after all marginal in his day to the institutions of higher education that assured a wide dissemination of knowledge. While part of the story of the emergence of the human sciences involves the gradual admission of Comte into the French university, an equally important part of that story chronicles the qualifications and revisions of Comte introduced by those who listened to nonpositivist critiques of positivism even as they attempted to create independent human sciences. These debates over the nature of science and the character of the human sciences were not simply irrelevant or inconsequential ventures into metatheory—they mattered. As Warren Schmaus has argued in the case of Durkheim, one's philosophy of science does have consequences for the way in which one practices science.[21] This book will show how different conceptions of science lead to different practices of science and how early social scientists struggled to find conceptions of science that could be used to construct fruitful human sciences.

In any event, this book should show that simple labels such as "positivism" and "spiritualism" are inadequate to capture the complexity of good thinkers, in whichever category one may find it convenient to place them. Labels are important, if for no other reason than that individuals choose to identify themselves by certain labels and to fight attempts by others to apply different labels. Conflicts over the meaning and attribution of philosophical labels are an important part of the history of the human sciences. However, precisely because of these conflicts, the meanings of these labels change over time, and in the process of engaging in arguments, intellectuals appropriate ideas from schools of thought other than the one with which they are primarily identified. Hence, while using generic labels for purposes of identification and historical score-keeping, I also examine the complexities and ambiguities of individual authors. In the process, I hope to put into question the purity of the philosophical categories usually invoked to characterize the early social scientists.[22]

I have chosen to concentrate on the relationship of academic philosophy to psychology and sociology rather than to other possible disciplines for several reasons. First, these disciplines were the last of the substantive domains of philosophy contested by the positivists. The individual mind and society were the last two objects on the positivist agenda, and as such they constituted the final inroads of natural science into the territory of speculative philosophy. For the same reason, they were especially vigorously defended by philosophers. Other disciplines in what are now called the human sciences, such as economics or history, were much less central to the academic self-definition of philosophy.[23] Second, a surprising number of the early psychologists and sociologists began their careers as academic philosophers. Appendix A illustrates this preponderance of philosophers quantitatively. The philosophical background of French social scientists meant that they knew intimately the language of philosophy, and I hope to show that even their opposition to philosophy was shaped by their proximity to it. Third, despite the different objects and contexts of psychology and sociology, philosophy influenced them in remarkably parallel ways. Psychologists and sociologists had to overcome the same kinds of philosophical objections and used the same kinds of arguments. Indeed, finally, psychologists and sociologists often borrowed from each other in their attempts to gain independence from philosophy and to construct their new sciences. These four considerations constitute the unity of the story I am telling.

This combination of psychology and sociology has no ready-made name, and if I have used the term *human sciences,* it is because it is familiar to modern readers. When I use this expression in the following it is only with reference to these two disciplines and not to the wider range of sciences—anthropology, medicine, economics, et cetera—to which it commonly refers. It is also not coextensive with the expression *science de l'homme,* which was used in the nineteenth century to refer to a wide range of disciplines, from medicine to philosophy. Another expression that perhaps more closely captures the connection of these disciplines is *science morale,* playing on the double meaning of *moral*—ethical and spiritual—just referred to. The term *science morale* or *science des phénomènes moraux* was commonly used throughout the nineteenth century to refer to either the study of mind or the study of morals. Although the different meanings have no necessary connection in English, in French their coincidence in one word carried semantic weight. Ethical and spiritual were closely united, and to taint the one invariably brought the accusation of impugning the other. This was why the expression *science morale* was rarely applied to biological disciplines such as medicine or anthropology (which in the nineteenth century implied a biological approach) in France. On the other hand, the *sciences morales* included more than psychology and sociology for the French person of the last century. Political economy, history, political science, and indeed philosophy itself were considered moral sciences. Moreover, the moral sciences connoted an introspective, traditional approach best represented by the Académie des Sciences Morales et Politiques, which was largely hostile to positivism. For this reason, I have in general used the term *human science.* The final term that suggests itself, *social science,* I have rejected because in French psychology is never described as a *science sociale.* This expression refers specifically to sociology, and I have avoided it except to describe the practitioners of these disciplines as *social scientists,* because *human scientist* is not current in English. Unless otherwise specified, any of these terms that appear in this book should be interpreted to refer to psychology and sociology.

Even within the fields of psychology and sociology, I have not treated all those who dealt with subjects that could be construed as falling within these disciplines, but only those who first achieved independent university positions. These academic social scientists should not be considered representative of the entire field. There was a great deal of activity and diversity outside the universities, which a general history of the human sciences would have to address. I have touched on such individuals and writings only to the extent that they compete for recognition in the academy. Generally, such outsiders lost to insiders with a background in academic philosophy. This, however, simply reinforces the thesis that there was a close connection between academic philosophy and the academic human sciences. Moreover, such a focus renders a vast topic manageable, given that very few positions in the human sciences were created during this period. It permits one to go beyond statistical correlations and to investigate the texts of these thinkers in some detail, which is necessary to expose the subtle and deep-seated traces of their philosophical background.

Within this context, I focus on four individuals. Théodule Ribot (1839–1916) taught the first course in experimental psychology at the Sorbonne and held the first chair in that subject at the Collège de France. Alfred Espinas (1844–1922), a friend and classmate of Ribot at the Ecole Normale Supérieure, occupied a chair in "social economy," created in 1893 at the Sorbonne. It is generally considered the first official position in sociology in Paris. Pierre Janet (1859–1947) succeeded Ribot at both the Sorbonne and the Collège de France. And his classmate at "Normale," Emile Durkheim (1858–1917), followed Espinas to the Sorbonne, though in a different position.[24] The backgrounds of these individuals are remarkably similar. All attended the same schools, had the same professors, passed the same examinations, and began their careers in the same way, as professors of philosophy in French lycées. This similarity of formative background underlines the unity of this study despite its interdisciplinary spread and its multigenerational extent. In addition to representing two different disciplines, they also represent two different generations, allowing one to examine changes over time in the way in which the human sciences were conceived. This is important because Espinas and Ribot were not the shadowy, insignificant precursors they are usually depicted to be, when they are depicted at all. On the contrary, they were original thinkers in their own right, crucial pioneers in bringing the human sciences into the French university, and powerful influences on their better-known successors, Janet and Durkheim. Any history of the human sciences in France should give Espinas and Ribot a prominent place.

APPROACH

To define one's focus as the philosophy and human sciences of the university is to choose a criterion that is as much institutional as it is intellectual. Academic philosophy was not simply a set of ideas; it was embedded in a state educational system that ensured it an influence regardless of its intrinsic merits. Academic philosophers could and did use their position to exclude the heterodox, and social scientists could and did appeal to authorities outside philosophy to gain institutional niches. For this reason alone, the relationship between academic philosophy and the human sciences is not simply one of competing ideas. It is also a story of power, of insiders and outsiders, and of struggles for position. Moreover, behind the academic politics lay the more significant political issues of nineteenth-century France, which became attached to issues of philosophy and science in complicated but powerful ways. If I have concentrated on the academic rather than on the larger politics of the disciplines involved, it is because it seems to me that many authors have had a tendency to leap directly from intellectual to political positions, stepping over the middle ground of academic culture.

Among the more recent studies on the history of French philosophy and the human sciences, two works may be taken as examples of the different approaches one can take. William Logue's book *From Philosophy to Sociology* contends that the foundations of French liberalism shifted from philosophy to sociology during the early Third Republic. In making this argument, he examines many of the individuals included in this book. Logue's book is an example of the history-of-ideas approach, concentrating primarily on the arguments of individual authors. I am particularly sympathetic to his attempt to identify a broad middle way between strident positivism and militant Catholicism in what has often been portrayed as a completely polarized society.[25] It seems to me, however, that the concepts of philosophy and science remain unexamined, as well as the reciprocal influence of sociology on philosophy.

Jean-Louis Fabiani's thesis on the field of academic philosophy, published as *The Philosophers of the Republic* (*Les philosophes de la République*), does not directly address the history of the human sciences, although he recognizes that academic philosophy is the primary context for the human sciences in France. Nevertheless, as a study of the academic culture of French philosophy and as a product of the sociological approach of Pierre Bourdieu, it is exemplary. Unlike Logue, Fabiani explains intellectual changes in terms neither of larger political issues nor of internal philosophical problems. Instead, he attributes the increasing diversity and changing nature of French philosophy to transformations in the social conditions of philosophical production that required academic philosophers to create a distinctive personal approach as a means of attaining professional credentials and advancement.[26] A similar approach has been extended to the human sciences by Fritz Ringer, who explicitly acknowledges the influence of Bourdieu.[27]

Each approach has much to say for it, but each is partial. With Logue, I believe that ideas are important and that one cannot reduce intellectual to institutional positions. Only a close reading of the texts can specify the conceptual issues at stake. Although Ringer insists with Fabiani and others of the sociological school that their approach acknowledges the specificity of intellectual discourse, their readings of particular authors do not always penetrate to the subtleties and inconsistencies of the texts cited. At the same time, I agree with them that the philosophical culture of the university introduces features that can be directly explained neither by the intellectual struggle of the individual thinker nor of the political configuration of Third Republic France. As a result, I have tried to integrate the two approaches, emphasizing the intellectual content of the academic debates while recognizing the institutional positions and aspirations of the authors.

In any event, recent scholarship has tended to treat the relationship between intellectual and institutional factors as an empirical matter capable of study and verification in particular cases rather than as an a priori position in the dreaded internalist/externalist debate. This debate pits those committed to the thesis that the progress of science can be explained purely in terms of the "internal" ideas and methods of the discipline against those who argue that all science can be reduced to "external" factors such as the quest for social power and the competition for scarce resources. The externalist position was argued most forcefully by the Edinburgh School, and the internalist standpoint has been recently argued in relation to Durkheim by Schmaus.[28] I have used the term *dreaded* because the debate is one that seems both inevitable and intractable, once one has accepted its premises. One promising way of avoiding this debate is through the growing literature in the rhetoric of the human sciences.[29] A rhetorical approach considers discourse as persuasive action in which agents take their arguments wherever they may be found, adapt them to their own cognitive agendas, and address them to audiences with particular intellectual horizons. Most recent arguments for a rhetorical approach go beyond a "thin" concern with the persuasive strategies of scientific texts to incorporate such "thick" considerations as the disciplinary and institutional structures that construct audiences for scientific works. In the same manner, most of those who adopt the rhetorical stance do not limit themselves to "mere" rhetoric—the figures of speech, appeals to authority, and emotional exhortations that try to persuade through nonrational means—but consider the deployment of evidence, methodological rigor, and proof as parts of persuasion in scientific discourse. My approach embraces this "thick" interpretation of rhetoric.

When the emergence of the human sciences in France is studied from a rhetorical point of view, the opposition between internal and external vanishes. This is not because there is no difference between an intellectual activity such as the construction of a Great Book—Durkheim's *Division of Labor in Society,* for example—and an administrative exercise of institutional power—Durkheim's alleged difficulty in securing an elite university position. Rather, the opposition vanishes because there are a number of intermediate positions that connect the two extremes and elucidate each other. *The Division of Labor in Society* began as a dissertation in philosophy, and it was defended before a committee of philosophers, who had the power to approve, disapprove, or ask for revisions.

Upon their approval depended Durkheim's hopes for professional advancement in academia. Is this an intellectual or an institutional issue? Posed in this way, the question makes no sense, because the situation is obviously both. However, connecting the Great Book with the humble conditions of its production helps us specify the audience for which it was intended and the arguments it would have to address. Analyzing the Great Book as a text helps us understand the complex and creative ways in which Durkheim tried to address his audience while pursuing a research program that his audience was likely to oppose. This book attends to the moments in which the interpenetration of intellectual and institutional considerations is most apparent—dissertation defenses, faculty meetings, administrative evaluations, debates in learned journals, et cetera. In this way I hope to capture the richness of conflict and collaboration in a period of uncertain and incomplete disciplinary differentiation.

A rhetorical approach helps moderate another dreaded controversy in the history of science, the continuist/discontinuist debate. Continuist models of intellectual change, whether of the naive progressivist or sophisticated evolutionary variety, assume that intellectual change generally occurs in small increments, that each concept is rationally related to the concepts it contests, and that each new piece of knowledge is an improvement on the piece it replaces.[30] These models, however, have difficulty explaining the rhetoric of radical and revolutionary change used by the early social scientists. Discontinuist models, on the other hand, help us appreciate the distance that can separate competing explanations and the gaps that can open up between one rational discourse and another.[31] But they have trouble explaining why, if people necessarily talk past each other across chasms of incommensurability, they continue to talk at all. In the case of the French human sciences, philosophers and scientists did indeed continue to talk to and influence each other over the course of many years, even when their antagonisms were most heated.

Both continuist and discontinuist models, therefore, minimize the noise generated by conceptual debate: continuists by assuming the changes to be insignificant and rationally adjudicable, discontinuists by assuming the debate to be fundamentally irrational, since it involves incommensurable perspectives. A rhetorical approach helps account for the fact that individuals committed to the human sciences would nevertheless listen and respond to the arguments of those who did not share their fundamental cognitive aim. In the case at hand, philosophers were for both institutional and intellectual reasons necessarily part of the audience social scientists had to persuade in order to pursue their work. Institutionally, they held power and positions the social scientists needed, and intellectually, they claimed the domains social scientists desired. A rhetorical approach can be extended to include persuasive effects that are not necessarily conscious, effects that work because both author and audience are implicated in the same set of expectations. Hence, such a perspective, broadly construed, can help us understand the extent to which early social scientists still inhabited the philosophical discourse they were trying to escape.

Finally, a rhetorical approach helps adjudicate yet a third dreaded debate, what might be called the reconstructionist/deconstructionist debate. This controversy divides those committed to a rational reconstruction of what an author *really* meant from those who insist upon a postmodern, deconstructive reading that emphasizes the indeterminacy and instability of meaning in a text.[32] It is one contention of this book that basic terms in philosophy and science had contested, overlapping, and sometimes contradictory meanings. Philosophers and social scientists routinely misunderstood each other—sometimes unintentionally, sometimes willfully. They often offered alternative interpretations of the same facts, arguments, and texts. Moreover, precisely because sociologists and psychologists were trying to forge new discourses, they were often uncertain themselves how to couch their intuitions and research results. Their texts reveal inconsistencies, their expressions changed over the course of their careers, and only retrospective reconstructions—often engaged in surreptitiously by the authors themselves—can find a coherence and system that may not have been evident when the works were first written. Indeed, evaluations of the coherence of a text or a research program varied depending on who was evaluating them and the perspective from which they were viewed. In important ways, therefore, it is impossible to reduce a text to one unitary meaning.

On the other hand, precisely because these works were not produced in a vacuum, one can restrict the range of meanings, if not for all time and all readers then at least for a given time and for different specific groups of readers. My aim is not to exploit inconsistencies in texts to reveal an indefinite number of possible readings; instead, I hope to situate the ambiguities of these texts in fields of historically constituted readings. One field is that of the author, whose intentions can, I believe, be circumscribed by the corpus of writings and actions that constitute the author's life. Other fields consist of the various audiences who could have read an author's works—philosophers, colleagues in other disciplines, and competitors in other countries. All of these potential readers had assumptions, terminologies, and methodological expectations that would inform their reading of any text, and the author had very little control over the interpretations that resulted from different readings.

The organization of this book is designed to elucidate the tense relationship between academic philosophy and the human sciences in France, as well as their gradual and incomplete divergence. Chapter 1 describes the institution and discourse of academic philosophy, with its claims to be a science of moral phenomena, on the eve of the Third Republic. Chapters 2 and 3 describe the psychology of Ribot and the sociology of Espinas, respectively. Both contested the right of academic philosophy to call itself a science, and both proposed alternatives taken from the evolutionary, biological, and associationist philosophy of Herbert Spencer. But while Ribot and Espinas wrote of the need to apply the methods of the natural sciences, their actual scientific practice was quite meager and secondhand. In chapter 4, I show how academic philosophy responded to the challenge of evolutionary human sciences. In part this meant criticizing the underlying assumptions of English evolutionary theory, and in part it involved redefining philosophy so that it no longer claimed to be an empirical science. In chapters 5 and 6, I examine the psychology of Janet and the sociology of Durkheim. I argue that these two should be seen as successors to Ribot and Espinas. However, Janet and Durkheim, acknowledging important aspects of the philosophical critiques of evolutionism, changed the direction of their disciplines. Nevertheless, both remained committed to the idea of an empirical approach to moral phenomena, and their own studies included direct research as those of Ribot and Espinas had not. In this way they learned from academic philosophy even as they continued to oppose it.

1

Philosophy as Science:
The Academic Tradition and Its Critics

To unite observation and reason, not to lose sight of the scientific
ideal to which man aspires, and to seek it and find it by the route
of experience, such is the problem of philosophy.
—Victor Cousin, *On the True, the Beautiful, and the Good*

WHEN the early advocates of the human sciences in France denounced
philosophy, they were in one sense continuing a tradition at least as old
as the Renaissance, when advocates of humane letters decried the arid,
empty dialectics of the schoolmen. But they also had in mind something
much more specific—philosophy as it was taught in the lycées and universities of nineteenth-century France. Philosophy in this sense constituted both an institution of considerable weight and a discourse with
an identifiable range of problems, methods, and solutions. This academic
philosophy not only excluded a number of competing philosophies, but
it also claimed the very areas coveted by the human sciences, under the
name of the moral sciences. Furthermore, it also claimed, through the
philosophy of science, the right to define scientificity and to dictate what
could be the object of science and what could not. Finally, it reared the
first generations of those who brought a positivist conception of the
human sciences into the French university. For these reasons, it is necessary to understand precisely what academic philosophy was in
nineteenth-century France. This chapter explores the institution of philosophy, the people who supported it, the intellectual content that defined it, and the critics who opposed it.

Despite the claims of the positivists, the debate between the defenders
and detractors of academic philosophy in France was not simply an
instance of the eternal struggle between philosophy and science, nor an
episode in the transition from the metaphysical to the positive era. Indeed, the history of this debate renders common stereotypes about the
relation between philosophy and science problematic. Academic philosophy in France was dominated during its formative years by the school

of eclectic spiritualism, one of whose distinguishing characteristics was that it claimed to be a science of observation. It claimed to found metaphysical truths by the methods of modern science. Hence, when positivists demanded that the human sciences be established on the model of the natural sciences, they were not simply asking that these disciplines stop being deductive and start being inductive; they were also implicitly challenging the scientificity of eclectic thought. The debate between positivists and spiritualists was as much a rhetorical struggle over the definitions of science and philosophy, and over the right to represent oneself as scientific, as it was an effort to give the human sciences an empirical foundation. Furthermore, the institutionalization of a philosophy that included the subject matter of the human sciences turned the intellectual debate over the nature of these disciplines into a struggle for power between academic insiders and outsiders.

The Institution of Philosophy

During the nineteenth century, in France as elsewhere, philosophy became increasingly professionalized and institutionalized within the state educational system.[1] This had not always been the case. During the Middle Ages, it is true, philosophy was defined by its association with educational institutions—hence the name scholasticism—and philosophy continued to be taught in the lycées and universities up to and beyond the French Revolution. Beginning with the Renaissance, however, and continuing through the Scientific Revolution and the Enlightenment, philosophy became something that could be carried on outside the university. Indeed, the new philosophy—experimental or speculative—often had to be pursued outside the established institutions, since these institutions in many cases resisted the innovations being proposed. Alternative institutions, in the form of scientific and literary academies, provided support for some of these activities, but the type of eighteenth-century *philosophe* was the man of letters, not the university professor. After the French Revolution, it was still possible to philosophize outside the academy. Indeed, many of the greatest contributions to French philosophy came from persons who strictly speaking were amateurs—Maine de Biran, Comte, and Charles Renouvier are good examples. But a powerful tradition of philosophy also arose within the educational system, one that enjoyed the approval and resources of the state. And eventually all varieties of philosophy were elaborated from within the academy.[2]

As philosophers became more and more incorporated into the educational system, that system underwent profound changes as well. Here again the French Revolution was the watershed, for it was the revolutionaries who first attempted on a large scale to wrest control of education from the Catholic Church, where it had resided since the Middle Ages, and to make of education a function of the state. In this they were only partially successful, since the revolutionary state had neither the resources, the personnel, nor the popular support to replace the schools run by the church. Even Napoleon, who gave the educational system, as he gave so much of bureaucratic France, the structure it was to retain throughout the nineteenth century, could not eliminate the Catholic schools and contented himself with placing them under the regulation and supervision of the state system. This latter, however, he built into a powerful and privileged competitor of the Catholic schools by constituting it as an autonomous lay corporation within the state— the Université.[3] The Université was the body of state teachers and administrators as a whole, and should not be confused with any particular university. Indeed, the Université included secondary as well as higher education, and this is the primary reason for the extremely close relation between philosophy in the lycée and philosophy in the university. The Université was given exclusive authority over public education in France, including private schools. This monopoly was to be a source of constant and often bitter strife between church and State, but the monarchs of the Restoration did not relinquish it. They rather fancied the powerful and efficient state apparatus Napoleon had left them. They simply used it more in accordance with the policies of the church, and they gradually encroached upon its administrative autonomy, transforming it from an autonomous corporation into a cabinet ministry more closely assimilated to the rest of the bureaucracy. The idea of the Université remained a powerful one, however, and the decisions of its executive council could not be lightly overturned, especially in matters of curriculum and academic standards.

Philosophy occupied a central place in the French educational system. It was taught in the last year of the lycée curriculum, and hence it culminated one's secondary education. This in itself was enough to give it prestige, since the lycée provided the French with what they considered to be a complete liberal education. Institutions of higher learning offered specialized and often professional education, but nothing equivalent to the American liberal arts program (indeed, the lycée, for reasons that again go back to the late Middle Ages, is ultimately the equivalent of the first two years of an American liberal arts education). Hence the indispensable "general culture" (culture générale) that represented the cultural patrimony of the nation and that identified one as an educated person was dispensed in the lycée. To use a word that would have a powerful influence on the self-identity of philosophers, the philosophy class was the "crowning" (couronnement) of the lycée education.[4] It was supposed to synthesize one's previous learning, to provide the general principles behind what was learned by rote earlier, and to prepare one for the kind of abstract analytical thinking required for higher studies. For this reason, in addition to being the end of secondary education, it was in some ways an introduction to higher education, in a practical as well as epistemological sense. For one thing, philosophy was on the state-administered examination for the baccalaureate (*baccalauréat*), the degree normally sought after completion of the lycée program. And the baccalaureate in turn was needed for admission to any program of higher education. So almost anyone who aspired to further education had to confront philosophy to some extent.

There was one important exception to this reign of lycée philosophy. Students preparing for any of the engineering *grandes écoles* did not take philosophy.[5] Instead, they enrolled in a mathematics class that prepared them for the rigorous entrance examinations administered by those schools, which had little use for philosophy. When the baccalaureate was divided into scientific and literary options in 1852, students wishing to pursue higher degrees in the sciences also opted for the mathematics class. The result was that by and large, scientists and engineers did not take philosophy. The rudimentary education in philosophy included in the mathematics class did not make much impression on them. Philosophy was therefore the crowning only of the humanities, and one can indeed already speak of two cultures—scientific and literary—in France in the nineteenth century.[6] But during the nineteenth century, the humanities, or rather the classics, though increasingly challenged and forced to accommodate the advocates of more science in the lycée, retained their superior prestige, and hence philosophy also retained the high ground in the hierarchy of disciplines, at least at the secondary level.[7]

Some students elected to pursue philosophy as a vocation and therefore entered a university, or to be more precise, a faculty (*faculté*). For most of the nineteenth century, the French faculties—of letters, science, law, and medicine—were each self-governing bodies that maintained only nominal relations with each other, so that the idea of a university as a unified center of learning was not only inapplicable in practice but formally absent. The Université was too broad to provide any particular cohesiveness to higher education. Each faculty was more attached to its particular branch of learning and to its clientele than to the rest of higher education. Moreover, the clientele of the faculties of letters, until the reforms of the Third Republic, was to a large extent the system of secondary education.[8] University professors examined candidates for all degrees in France, including the baccalaureate. This alone took a tremendous amount of time and energy.[9] On the other hand, they had very few students, most of whom were studying to be secondary school professors, since "liberal arts" education was provided in the lycée, and teaching was one of the few careers a degree in the humanities could offer.[10] This orientation of the university toward professional training and the secondary school system was no accident. Aside from the fact that, as earlier indicated, the lycée was the center of the educational system, the creator of the faculty system—Napoleon—did not want the free and critical thinking to which a research-oriented university might lead. French universities were for this reason very different from their German counterparts, and the rising prestige of the German universities in the nineteenth century provoked calls for reform within France. These calls became increasingly urgent after the French defeat in the Franco-Prussian War of 1870–71, which was widely attributed to French inferiority in education.[11] One final effect of the close connection between secondary and higher education was that the subject matter of university courses tended to repeat that of the lycée philosophy class, at a higher level and in greater detail, but remaining within the framework of problems provided by the lycée *programme,* or "official syllabus."

A student who showed great promise of intellectual achievement in philosophy and who desired a good career in education, secondary or higher, would attempt to avoid the moribund universities at all cost and instead seek admission to the Ecole Normale Supérieure, a *grande école* established during the Revolution to train secondary schoolteachers. Aspiring students usually spent at least one year after the baccalaureate in a special class—the *khâgne*—preparing for the entrance tests. Many aspiring students failed on their first attempt, and overall only a miniscule percentage of applicants was eventually admitted.[12] For the chosen few, however, an outstanding education at state expense and a secure career were assured. Unlike the universities, which offered lectures but no supervision of study, the Ecole Normale provided a carefully organized and highly disciplined education.[13] Indeed, like most lycées but again unlike the universities, it was a boarding school, with a Spartan regimen reflecting the military values of its renovator, Napoleon. Students were evaluated annually, not only by their own professors but also by a member of the Paris Faculty of Letters. Upon completion of their study, *normaliens* were obliged to give ten years of service to the state, usually as lycée professors. Because of its prestige, however, students were attracted who had no intention of remaining in the lycée for the rest of their careers, and many of the greatest names in both the humanities and the sciences in nineteenth-century France passed through the Ecole Normale.

After studying at a university, usually for two years, one could begin to approach the series of higher degrees and certificates offered by the state. The least among these was the *licence,* which certified that one had successfully completed the course of study, but which in theory did not permit one to teach in the secondary school system. Admission to professorship in the lycée was restricted to those who had passed the rigorous *agrégation,* administered in the name of the state by a committee of university professors.[14] In practice, there were never enough *agrégés* to fill all the lycée positions in any subject, so that many positions were filled by *licenciés,* but the best posts, especially those in Paris, were occupied by *agrégés.*[15] Normaliens dominated the *agrégation,* so much so that for anyone from a university—especially a provincial one—to succeed was a matter of general astonishment.[16]

University chairs were open only to those with the doctorate, which until 1902 required two dissertations, one in Latin and one in French. These were evaluated by, and one's public defense was made to, a dissertation "committee" or *jury* made up of university professors. The usual method of obtaining a university chair was first to teach in the lycées and distinguish oneself there, yet another instance of the close ties between secondary and higher education. Then one might seek a position as *suppléant,* or "substitute lecturer," for a titular professor (a *titulaire*) on leave. Such positions were common and could last for years, since a *titulaire* acquired a virtually proprietary right to his chair, and could retain both title and salary without actually lecturing, provided he found a substitute who would be paid out of the total salary attached to the chair.[17] When a chair became vacant, a new professor was often appointed first as a *chargé de cours* before being *titularisé*—"given tenure."

Besides obtaining a teaching position, an individual with a degree in philosophy might end up in educational administration. Since all public education in France was centralized under the Ministry of Public Instruction (Ministère de l'Instruction Publique), a position in the central administration was a great responsibility that gave in return considerable power, prestige, and remuneration. As the public education system grew in size and importance during the nineteenth century, administrative positions grew proportionately. *Agrégés* in philosophy, representing the best minds in the system, enjoyed considerable success in administration. At the end of the nineteenth century, the directors of both secondary and higher education were *normaliens* and former professors of philosophy.[18] These were the highest civil service positions within the Ministry, and though technically they were subordinate to the minister, in practice they had great influence. The minister was usually a politician with little expertise in educational matters, whose tenure furthermore was often brief.[19] The director was a career educator with an intimate knowledge of the educational institution, representing the weight of both expertise and continuity.

Educational policy was established by the Council of Public Instruction (Conseil de l'Instruction Publique), a body of varying size, composition, manner of appointment, and appellation during the nineteenth century. At times, such as during the July Monarchy, it was small and consisted of the heads of the various branches of the educational system.[20] In these circumstances an individual member could have a great influence. During the Third Republic, it was a large body and elective, which diminished the power of any individual. Yet at all times it was the supreme legislative body of the Université, deciding policies on curriculum, degrees, staffing, and administration, within limits established by law. Membership was prestigious and also provided professors with an eye on an administrative career an opportunity to display their skills.

More relevant to philosophy, and to professors of philosophy, was the Inspection General, an office also created by Napoleon.[21] The inspectors general were hired by the Ministry to visit every lycée in the country and to monitor the performance of professors and administrators. As former professors themselves, chosen for their excellence in teaching and experience in a particular subject, they were knowledgeable and severe judges, upon whose reports the future of the lycée professor in large measure depended.

Below the national level, there were further levels of bureaucracy. The Université was divided into academies (*académies*), regional districts each with its own rector (*recteur*), council, and inspectors. Each academy was supposed to include a faculty of sciences and one of letters as well, to ensure the academic integrity and the staffing of the baccalaureate juries of that area. Finally, at the lowest level, each institution—lycée or university—had its own rector and administrative staff.

In addition to the close integration of secondary with higher education, what stands out in this system is the extraordinary degree of control that could be exercised by senior members of the hierarchy over their subordinates. As teachers, mentors, examiners, and inspectors, those established within the system could determine the fate of an aspiring individual at many stages of his education and career. It was important to cultivate mentors and to solicit superiors to advance one's career, although this could not substitute for poor performance.[22] But performance itself would be judged by the elite and hence would tend to reproduce what the elite considered good. This does not mean that professors of philosophy were supposed to merely repeat the dogmas of their superiors. On the contrary, the mark of a true philosopher, even one destined for the lycée, was to think independently and originally. But this individual distinction was held within carefully defined limits, and everything else in the system favored conformity over originality. In short, the institutional structure of French academic philosophy had a built-in tendency to uniformity.

PERSONNEL

Whether the available means of control would be used, and whether the tendency to uniformity realized, would depend to some extent on the individuals in the system. Academic philosophy had in its formative years after the Restoration a chief who was determined to use the means at his disposal to the fullest. Victor Cousin (1792–1867) rose from humble origins to dominate the French academic philosophy of his day.[23] He gave academic philosophy its exceptional coherence both as an institution and as a doctrine by his vigorous and authoritarian leadership. He established his fame by his lectures on philosophy at the Sorbonne, where he was appointed *suppléant* to the anti-Ideologue Royer-Collard in 1815, at the tender age of twenty-three. His lectures were a sensation and soon attracted both fashionable women and ardent youth. Although he was an outspoken supporter of the charter of Louis XVIII, he considered himself and was considered by others a liberal because he defended constitutional government. This proved too much during the reaction that followed the assassination of the Duc de Berry in 1820, and he was forced from his chair in 1822. His popularity and reputation were only reinforced by this injustice, and when he was reinstated in 1828, he took the podium as a martyr to liberty.

The July Revolution of 1830 brought into power a number of Cousin's friends and supporters, giving him an opportunity to spread his influence over academic philosophy. Despite his previous support of the legitimists, he leaped at the chance. He retained his chair in philosophy at the Sorbonne, and in fact was finally made *titulaire,* although his administrative duties obliged him to seek a *suppléant,* which he continued to do for over twenty years rather than give up his claim to the post. But he was also, in 1830, appointed member of the Royal Council of Public Instruction, which at this time consisted of only eight high-ranking scholars and educators. He worked closely with his friend François Guizot, the historian and government minister under the July Monarchy, on matters relating to education, most notably the law of 1833 establishing a state system of primary instruction. In short order he was elected to the Académie Française and the Académie des Sciences Morales et Politiques. He was also made a peer, which entitled him to a seat in the Senate and which gave him direct access to the political arena. He was the director of the Ecole Normale Supérieure from 1835 to 1840 and was closely associated with the school both before and after his directorship. He was president of the *jury d'agrégation* and a member of the *jury de doctorat* for over twenty-five years. Ironically, for all the power he accumulated, he was only the minister of Public Instruction for a mere eight months.

But the positions he retained for longer periods were quite enough to ensure his stamp on academic philosophy. His professorship gave him control over one of only three or four university chairs in philosophy in Paris and entitled him to sit on the committees of the *agrégation* and doctorate. As the director of the *jury de l'agrégation,* he effectively decided the career of every aspirant to secondary education, and by sitting on the *jury de doctorat,* he had the same influence over those who sought university careers. As the only philosopher on the Royal Council of Public Instruction, Cousin effectively represented philosophy on the body that established academic policy. His opinion in matters of policy concerning philosophy was usually final. For example, in 1830 he was instrumental in getting the official language in the philosophy class changed from Latin to French, and the syllabus of mandatory studies revised in 1832 reflects his own philosophy. By controlling the syllabus, Cousin effectively constrained both what students had to learn and what professors were permitted to teach. His membership in the Académies greatly enhanced his prestige and made him a person to be reckoned with by anyone aspiring to election to those organizations—especially philosophers, since other members would usually defer to Cousin's judgment of the candidates he knew best. He used his seat in the Senate to defend philosophy on more than one occasion against the attacks of Catholics who demanded the subordination of morals and metaphysics to religion. Finally, as the director of the Ecole Normale, he had under his supervision and instruction the best and brightest future professors of philosophy.

Cousin used this power to further his vision of what philosophy should be, both as an intellectual discipline and as a social institution. As an intellectual discipline, philosophy both trained the mind and revealed incontestable truths.[24] Both of these functions made it an essential part of a complete education, and Cousin was instrumental in developing and deploying the justifications of philosophy just mentioned, making it the crowning achievement of the educational system. He also developed the content and scope of the philosophy courses to be taught, through his own teaching and through his control of the syllabus. And he trained the personnel who taught this philosophy both in the lycée and in the university. Through his efforts academic philosophy became a well-defined discipline and a firmly entrenched institution in French education.

Cousin also held definite views about the role of philosophy as a social institution and political ideology: he wanted it to help unify and reconstruct France after the division and destruction of the French Revolution, while at the same time preserving what he considered to be good in the revolutionary legacy.[25] Although the reconstruction of France was a common enough aspiration in the early nineteenth century, Cousin's faith in his philosophy as the means to this regeneration implied a particular vision of society. It implied first of all that philosophy had a central place in society. This may have been wishful thinking, but it meant that society should be based on reason. This for Cousin implied constitutionalism and the rule of law, rather than either divine right and arbitrary rule or democracy and the rule of the majority. Since the *ultra* party represented the former and the Jacobins the latter, according to Cousin, his insistence on the centrality of philosophy was also an argument for the doctrinaire liberalism he shared with Guizot. Furthermore, if philosophy is at the center, then religion is not, at least not in the way the Catholic Church intended it to be. Cousin's advocacy of philosophy as an intellectual and moral bond resulted from his conviction that French society must recognize its religious plurality and no longer base the nation on religious affiliation. This did not mean that Cousin was antireligious. On the contrary, he repeatedly emphasized that philosophy demonstrated the truth and necessity of revealed religion, to the extent possible within the bounds of reason. But he maintained that neither philosophy nor the state should be under the control of any particular religious authority. He also thought that philosophy unaided by religion could discover moral truths that could hold society together, and that philosophy could have a great influence on the morals of the nation's youth. These positions inspired hostility from extreme Catholic circles that dogged him throughout his career.[26]

Cousin enjoyed great power during the July Monarchy, and he fell with it. He abhorred both the Second Republic, which eliminated the monarchy, and the Second Empire, which ignored the rule of law. The February Revolution of 1848 dissolved the Royal Council of Public Instruction (although it was later reconstituted without the "Royal"), and Cousin retired discreetly after the coup d'état of Louis Napoleon rather than serve under a government hostile to his philosophy and politics. His direct influence on French philosophy was at an end. Nevertheless, his work lived on, both in the institution he had forged and in the people he had trained. The next generation of academic philosophers, who were to occupy the positions of philosophical authority from the 1850s to the end of the century and who taught the first generation of academic social scientists, were all deeply affected by the legacy of Victor Cousin.

These academic philosophers constituted a substantial group, since nearly every lycée in the country had a philosophy class.[27] Nevertheless, the summit of the hierarchy was extremely small, dominated by the few positions in the institutions of higher education in Paris: the Faculty of Letters, with two or three chairs of philosophy; the Ecole Normale Supérieure, with two; and the Collège de France, with two or three. In addition, at least one inspector general was usually a philosopher. The juries of the *agrégation* and the *doctorat* were almost always selected from the holders of these positions, since these competitions were national in scope. Taken together, the holders of these positions constituted a group of elite academic philosophers who largely controlled the careers of the next generation of aspiring academic philosophers. A look at the careers of some of these academic philosophers illustrates the dynamics of the institution Cousin created.

Not all the philosophers with academic appointments in the latter half of the nineteenth century were students or creations of Cousin. Adolphe Franck (1809–93) may not have encountered Cousin until after he passed the *agrégation* in about 1832 or 1833.[28] Franck was a product of the provinces, bypassing the Sorbonne and the Ecole Normale, where Cousin already held sway, and taking his doctorate before Cousin reached his apogee. His career followed the pattern that was to become standard. He began as a professor of philosophy in a provincial collège,[29] but by 1840 he was teaching at the Collège Charlemagne, one of the prestigious Parisian secondary schools. He was elected to the Académie des Sciences Morales et Politiques in 1844, even before he began to lecture at the university level. After teaching a course in social philosophy at the Sorbonne in 1847, he was *suppléant* to Barthélemy Saint-Hilaire, a professor of natural law and international law at the Collège de France, from 1849 to 1852, succeeding Saint-Hilaire in 1854, first as *chargé de cours* and from 1856 as *titulaire*. He held this position until he retired in 1887. As the title of his chair indicates, he was primarily interested in the philosophy of law, but only as it related to moral philosophy. He was an active member of the juries throughout his long career. And despite his independence from Cousin, he agreed with much of the eclectic project, to the point that the *Dictionary of the Philosophical Sciences (Dictionnaire des sciences philosophiques)* he edited with numerous collaborators can be called a collective manifesto of eclectic spiritualism.[30]

Others, like Paul Janet (1823–99), were direct pupils of Cousin.[31] Unlike Franck, Janet was Parisian by birth and education, attending the Lycée Louis-le-Grand and the Ecole Normale Supérieure. After passing the *agrégation* in 1844, he served as secretary to Cousin for a year before beginning his professorial career at the Collège Royal at Bourges. In 1848 he received his doctorate, a second *agrégation* in letters, and a post in the faculty of Strasbourg. In 1856, he accepted a chair of philosophy at the Lycée Louis-le-Grand, a fairly common career move that indicates the relative status of provincial faculties and Parisian lycées. He became a professor of the history of philosophy at the Sorbonne in 1864, and in 1887 he succeeded Elme Caro in the chair of philosophy at the Sorbonne. From 1889 to 1896 he was a member of the Higher Council of Public Instruction (Conseil Supérieur de l'Instruction Publique). Throughout his career he was a constant member of the juries of the *agrégation* and the *doctorat*. He was considered to be the successor of Cousin as leader of the eclectic school, and he played a large role in determining the syllabus during the late Second Empire and early Third Republic.[32]

Finally, some suffered from the disapproval of Cousin but managed to find other patrons to enable them to participate in academic philosophy. Félix Ravaisson-Mollien (1813–1900) studied under the German Idealist Schelling before returning to Paris to write a prize essay for the Académie des Sciences Morales et Politiques on *The Metaphysics of Aristotle* (1837) and a dissertation for the Sorbonne on *Habit* (1838).[33] His originality and opposition to eclecticism were already apparent in these youthful works, as was his talent. He was given a position on the faculty of letters at Rennes, in Brittany, a marginal appointment that he attributed to the disfavor of Cousin. He left academic philosophy in 1840 for a position as inspector general of libraries. He gained the attention of Napoleon III and as a result was made inspector general of higher education in 1852. He wrote a famous report on *Philosophy in France during the Nineteenth Century (La philosophie en France au XIXe siècle)* at the request of the imperial government for the Universal Exposition of 1868. In 1870, while retaining his position as inspector general, he was also named curator of classical antiquities at the Louvre. He was for many years president of the *jury de l'agrégation,* retiring only in 1887. His career seems to have been derailed from teaching by the opposition of Cousin, but he eventually managed to reclaim a position of power within academic philosophy.

Although, as the examples of Franck and Ravaisson make clear, it was possible to reach a high position in academic philosophy without having attended elite schools or working one's way up from the lycée, Janet's career was more typical. The tables in Appendix A give an overview of the education, careers, and philosophical orientions of elite academic philosophers. Over 80 percent of the elite academic philosophers of Janet's generation (born between 1811 and 1830) attended the Ecole Normale, and over 90 percent began as humble professors of philosophy in a provincial lycée (tables 1 and 2). On the other hand, only about half of Janet's generation passed through provincial universities, nearly the same percentage as those who were professors at a Parisian lycée (table 2). As the examples indicate, ambitious academics would readily relinquish a chair at a provincial faculty for a chair at a lycée in Paris. Proximity to Paris was more important than being in higher education, and the Parisian lycées, as the schools of the elite, were more prestigious than provincial faculties, where professors lectured either to dilettantes or to thin air. Despite the meteoric rise of Cousin, the average academic of Janet's generation labored in the lycées and provincial faculties for over twenty years before attaining an elite position (table 3). And once they got there, they held onto those positions for a long time—nearly twenty years. Given that several of Janet's contemporaries were not appointed until the early Third Republic, his generation remained a presence in academic philosophy almost until the end of the century.

Nearly all of them were disciples of Cousin as well. Almost all elite academic philosophers of Janet's generation were eclectic spiritualists (table 4). Ravaisson was an exception, in that he managed to oppose Cousin and still gain a position in elite academic philosophy. Even then, he attained it through the back door, as it were. Others were not as fortunate; they were forced out of academic philosophy, because their views were considered too radical for a public employee. Given the power of the elite at every stage of one's career, it was very difficult to oppose their views and still aspire to join that elite. This power was not directly related to intellectual prestige or philosophical significance. Many professors achieved elite positions without overly impressing either their peers outside the academy or posterity. Only slightly more than a third of Janet's contemporaries in the highest echelons of academic philosophy were considered notable by the authors of contemporary and subsequent histories of philosophy (table 5). Nevertheless, the content of Cousin's philosophy determined what could be allowed and what must be excluded from the academy. For that reason, in the next section, I will examine the doctrines that constituted the mainstream of academic philosophy.

THE SYLLABUS

The hierarchical and integrated character of academic philosophy combined with the small number of controlling personalities to encourage an extraordinary degree of doctrinal uniformity. Since the syllabus of the philosophy class was established by the same individuals who taught philosophy in the universities and evaluated future lycée professors, all parts of the institution reinforced the homogeneity of the content. The institutional parameters alone, however, do not specify the intellectual content of the philosophy. The political aims of Cousin and the social demand of the ruling classes dictated a doctrine of the *juste milieu*. But it would be a mistake to reduce academic philosophy to its ideological function. The claim to pursue true knowledge and the tradition of philosophical inquiry formed a context in which academic philosophers created a systematic intellectual response that must be analyzed at least partly in intellectual terms.

It is possible to reconstruct a universe of discourse for academic philosophy, since it was literally a matter of public record. School philosophy in nineteenth-century France was dominated by the *programme* or "syllabus," a list of topics and questions published by the Ministry of Public Instruction from which the examination for the baccalaureate was constructed.[34] Translations of some of the syllabi appear in Appendix B. Theoretically independent of the philosophy class, in practice the syllabus determined what was taught there, since the success of both pupil and teacher was measured by performance on the baccalaureate. Educators often complained about the negative effects of the baccalaureate on the teaching of philosophy, but the system was never significantly altered, and for better or worse the syllabus retained its power throughout the nineteenth century.[35]

The syllabus was created with the Université, but it achieved its definitive form under Cousin, who added to the traditional subjects—logic, ethics, metaphysics, and the history of philosophy—a new division—psychology. These five areas constituted the domain of philosophy recognized by the Université, and although the order, denomination, and relative weight of the parts varied somewhat, overall the syllabus changed remarkably little from 1832 to 1902, as a glance at Appendix B demonstrates.[36] These areas may seem relatively unproblematic, but their mere incorporation already begins to define a particular philosophy among other possible philosophies.[37] The inclusion of metaphysics, for example, indicates a potential conflict with other philosophies that do not recognize it (e.g., positivism and certain forms of Kantianism). Moreover, the specific topics within each division also have definite implications; to ask for a theory of the faculties of the soul implies that *faculty* and *soul* are meaningful terms.

There are limits to what can be learned from the syllabus, however, since it is by definition a list of topics and questions rather than answers. Indeed, a case could be made that the openness of the questions indicated a liberality of intention.[38] If there was a range of acceptable answers, however, there were also limits to what would be tolerated, and it would be a mistake—for the examinee and the historian—to suppose that any well-justified answer would be accepted by a baccalaureate jury. To discover these limits, however, it is necessary to go beyond the syllabus. One obvious source of potential answers is Cousin himself, since he made the syllabus what it was to remain, for most of the century. His philosophy had its own developments, which are not without significance, but for the purposes of discovering his mature thoughts on what philosophy should be, the later editions of his book *On the True, the Beautiful, and the Good* (*Du vrai, du beau, et du bien*) are the best source.[39] Other sources exist, however, which are closely associated with the Cousinian project but not by his hand. The standardization of the baccalaureate and the anxiety of the students produced a market for manuals of philosophy, usually written by lycée professors as textbooks for their classes. These were turned out by the score during the nineteenth century, and they provided answers guaranteed to get students past the judges of the baccalaureate. Two representative manuals are Charles Jourdain's *Notions of Philosophy* (*Notions de philosophie*), first published in 1852 but based on his lycée lectures during the July Monarchy, and Janet's *Elementary Treatise of Philosophy (Traité élémentaire de philosophie)*, which appeared in 1879. Franck's *Dictionary of the Philosophical Sciences*, first published in 1843 and reissued in 1875, provides a collective and literally definitive treatment of the questions recognized by philosophers who sympathized with Cousin's project. Using these sources in conjunction with the syllabus, it is possible to delineate the contours of the philosophy taught in France.

General Features

Cousin used two terms to refer to his own philosophy—*eclecticism* and *spiritualism*—and their combination—*eclectic spiritualism*—is a common as well as accurate characterization of his school. The two terms refer to a method and an object respectively. Eclecticism is the earlier term, first used in Cousin's Sorbonne lectures of 1818.[40] It refers to his belief that all systems of philosophy contain a portion of truth and that the task of the philosopher is to distill and combine the elements of veracity from each school. The other defining epithet of this philosophy, "spiritualism," emerged only in the 1830s, after Cousin had studied and edited the works of the philosopher Maine de Biran, and after Catholic critics began to denounce what they considered to be traces of pantheism in his early works.[41] Spiritualism meant that both God and the human soul were real and autonomous immaterial entities, distinct from matter and from each other. The best way to see how eclecticism and spiritualism informed academic philosophy is to go through the parts of the syllabus, following its order and logic.

Introduction

The syllabus began by defining philosophy and placing it within the universe of human knowledge. This introductory section was never counted as a separate division of philosophy on a par with psychology or metaphysics, yet it was one of the most important and controversial parts of the enterprise of academic philosophy, for the eclectics invariably claimed that philosophy is a science. The first question of Cousin's syllabus of 1832 concerns the relation of philosophy with "the other sciences," and the syllabus of 1874 also includes this implicit claim to scientific status. Franck's *Dictionary* was of the philosophical sciences, and his article on "Philosophy" also treats the discipline as a science. Since one of the issues in the rise of the human sciences is held to be the opposition of a "philosophical" to a "scientific" approach, it is important to understand precisely the sense in which the eclectics saw philosophy and science as the same thing, as well as what distinguished the *sciences naturelles* from the *sciences morales,* of which philosophy was a part.

On one level, the eclectics were simply preserving a use of the term *science* that was quite common before it became synonymous with the "natural sciences." Science, says Franck in the article "Philosophy," is "true knowledge," characterized by "unity and certitude." Perfect science would be "absolute and universal," but this kind of knowledge is possible only for the mind of God.[42] Nevertheless, perfect science acts as an ideal for human knowledge, which may be placed on a scale from disconnected and doubtful to unified and certain. This definition of knowledge is not trivial, for it implies that there is only one kind of knowledge, one that includes both metaphysics and the natural sciences. The only differences are of degree.

But there is a stronger sense in which the eclectics claimed to be doing science, one peculiar to that school and one that led them to compare philosophy with physics. Philosophy, according to the eclectics, is a science of observation and experimentation. Cousin thought it one of his chief merits to have made psychology the base of philosophy, and experience the base of psychology. The phenomena and principles upon which philosophy is based should be derived from carefully controlled experiences. This, according to him, was the true method of modern philosophy, which Descartes had pioneered, which rationalists such as Spinoza and Leibniz had forgotten, and which Locke had had the good sense to retain.[43] References to the *méthode expérimentale* and to the *méthode d'observation* abound in Cousin's works, as well as comparisons of philosophy with the physical sciences.[44] This did not mean that Cousin embraced empiricism, for he thought that the natural sciences were guided by reason as well as by the senses. Characteristically, he claimed to take the best from both the empiricist and the rationalist traditions. But this claim did open eclecticism to both revision and refutation, since at least in principle observations could be improved and theories disproved. This insistence on the methodological affinity of philosophy and the natural sciences remained in eclectic writings to the end of the century.[45]

If philosophy does not differ from the natural sciences either in its cognitive aim—certain knowledge—or in its method—observation guided by reason—it does differ in its object. This is made clear by the classifications of the sciences that were a part of this introductory section of the syllabus. These classifications were generally based on the object studied, consistent with the premise that philosophy and science do not differ methodologically. Indeed, eclectics generally rejected the classification of Bacon and the encyclopedists, based on the human faculties of memory, reason, and imagination, arguing that all the faculties are used in all the sciences. They preferred the classification of Ampère, the physicist who was also a friend of Maine de Biran, an early spiritualist. Ampère made the first division of his system a distinction between the *sciences de la matière* and the *sciences noologiques,* or sciences of the spiritual.[46] Within the *sciences noologiques,* or *sciences morales,* as they were more commonly designated, philosophers distinguished between those disciplines that studied the spirit through its external manifestations— the *sciences morales extérieures*—and those that studied it directly, through the inner examination of conscience—the *sciences morales intérieures.*[47] The former, which included history, linguistics, jurisprudence, and political economy, inferred the capacities and motivations of moral agents through their words, acts, and institutions. The latter investigated the mind through one's firsthand experience of consciousness. This included but was not limited to psychology; indeed, the *sciences morales intérieures* constituted the domain of philosophy.[48]

The classification of the sciences, culminating in the definition of philosophy, was not simply an effort to link philosophy with the natural sciences. It also provided, by the process of elimination, a way of constituting the unity of a discipline that contained disparate objects. At first glance it is not clear how the various parts of the syllabus fit into a single discipline—why, for example, subjects as disparate as psychology and metaphysics were a part of the same enterprise, or why these and only these topics constituted philosophy. There was, however, a unity and logic to the syllabus of Cousin, constituted in the first instance by the necessary connection between metaphysics and psychology. The *sciences morales intérieures* included the soul as a spiritual substance, not just psychological phenomena, and God as a spiritual being, not just humans as rational actors. Absolute knowledge of the first principles of being, and of the human mind, our access to these principles, are therefore the inseparable twin objects of philosophy.[49] Logic, the science of the proper conduct of the human mind, and morals, the study of the proper conduct of the human will, constitute the passage between these two goals.[50] And since the history of philosophy provides attempts to address these questions, as well as examples of the human mind at work, it too is an integral part of philosophy. The parts of the syllabus therefore comprised a unified discipline for the eclectics, no part of which could be removed without mutilating philosophy itself.

To define philosophy in this way, however, makes it a science unlike any other. Ultimate truth and its conditions would seem to embrace all the sciences, and indeed the eclectics claimed that philosophy was the science that unified, synthesized, and generalized the results of the particular sciences. Furthermore, as the science of thought, philosophy explained the principles and methods of the other sciences. Eclectic philosophy therefore pretended to be the "science of science." Philosophy was not just another particular science; it encompassed and surpassed the other sciences, and this dual function rendered its position ambiguous within the hierarchy of the sciences.

Philosophy is also unlike the other sciences in that its object cannot be observed by means of the corporeal senses.[51] Entities like God, moral truth, and reason escape the definition of phenomena to which the natural sciences limit themselves, because they are either beyond the phenomenal world, as underlying or transcendent, or within the mind and hence inaccessible to outside observers. The objects of philosophy are therefore not subject to many of the ordinary procedures of the natural sciences, such as measure, intersubjective verification, and replication. According to Cousin, this peculiar character does not invalidate the claim that philosophy is a science of observation. It does, however, mean that philosophical observation is a special type, which Cousin called "reflection." The data of philosophy are one's own spontaneous states of consciousness, which through reflection become an object of inquiry. Here again Cousin took the *cogito* as the model and defined philosophy as the "analysis of thought."[52] This unique relation of subject to object creates difficulties, not the least of which is the possibility that spontaneity and reflection will be confused, but it also assures a direct access to the object denied other sciences. It turns out, then, that philosophy does have a method peculiar to it, after all. Nevertheless, this peculiar method is a function of the object, not an a priori "philosophical" approach that would differ essentially from an empirical "scientific" approach.

Psychology

Before Cousin, psychology was not part of the syllabus. Cousin made it the foundation of the syllabus. He held that one could only begin to speculate about God and the world after achieving a firm knowledge of the capabilities and limits of the human mind. This knowledge was to be gained by the methods of observation, and the eclectics' insistence on the necessity of observation in psychology made the *méthode psychologique* a distinguishing trait of their school. They occasionally contrasted psychology with logic, morals, and esthetics, which were more ideal and normative in character, and in which deductive reasoning played a larger role.[53] Psychology began as a science of phenomena, and in this respect eclecticism did not differ from British empiricism. The syllabus always began by asking for psychological facts or phenomena, from which inductions were to be made.

The peculiarity of psychological facts, and hence of psychological observation, was that they were not accessible to the external senses. Psychological facts such as reasoning or feeling could not be directly observed in others, only within oneself in the *sens intime* or *conscience,* defined as "that particular power, distinct from the [external] senses, by which we become the witnesses of our interior life."[54] Hence psychology was the science of "the soul and its faculties studied solely by means of consciousness."[55] This implies that psychology is a science of consciousness; the psychological is coextensive with the conscious.[56] But it also provided a means of distinguishing psychology from physiology, since physiology studied the functions of life through the external senses. The interiority and specificity of psychological phenomena were used as an argument against thinkers such as Cabanis, Broussais, and Comte, who argued that mental phenomena should be studied as functions of the brain.

Eclecticism differed from the empiricism of Hume or Condillac, not in the distinction between psychology and physiology, but in the kinds of facts adduced from introspection and in the inferences drawn from them. The eclectics distinguished three different kinds of mental phenomena—sensory, intellectual, and voluntary.[57] Self-observation showed that these phenomena cannot be derived from the senses. In addition to the a priori principles discovered through observation and analysis, the mind's operations are not determined by the environment, nor are they reducible to association. The phenomenon of attention, which Laromiguière and Maine de Biran had made the basis of their break with ideology, demonstrated the activity and independence of the mind in its interaction with the stimuli of the senses.[58] Moreover, the eclectics claimed that these irreducible kinds of phenomena proved the existence of corresponding mental causes—powers—of the faculties of the mind.[59]

The will was also central to eclectic psychology and here again their analysis differed from that of classic empiricism. Attention forms a bridge between the intellect and the will, since the mind can know only that at which the will directs it. The eclectics maintained that attention cannot be explained by a purely passive theory of sensory stimuli; it must be the result of an independent and active human faculty.[60] But if this is true of the simplest of determinations of the will, it is a fortiori true of higher acts. This theory of the autonomy of the will provided eclectics with one proof of the existence of spiritual substance. In voluntary activity one has an immediate intuition of oneself as the determining cause of one's actions. But anything that is a cause must exist, and hence the soul as willful activity exists.

Finally, despite the tripartite nature of mental phenomena and faculties, consciousness is always able to distinguish between what belongs to it and what does not. The sense of self, the *moi,* is an undeniable fact of mental life, and constitutes the *personne,* the psychological and moral individual. But more than a simple fact, the unity and identity of the self was used by eclectics as another argument for the existence of spiritual substance. Sensations come and go, the body incessantly changes, and yet the soul retains a sense of identity. This can only be the case if the soul is a simple substance distinct from the body.[61] Hence, the syllabus in psychology, which began with an epistemological distinction between the phenomena of psychology and physiology, ends with an ontological distinction between body and soul.

Logic

If psychology is the study of human mental capacity, logic is the study of the proper conduct of human intellectual capacity. The eclectics defined logic indifferently as the science of the laws of thought and as the science of truth.[62] Such indifference is deliberate; the eclectics, like Descartes and Hegel, held that the real is rational and the rational real. They did not take this to imply that one can deduce the world from the idea of God or Spirit; logic had to be supplemented by the method of observation, and it is significant that the study of the first principles of reason is a part of psychology, not logic. But they did think that the laws governing human thought were also those governing divine thought and Creation. The human mind was limited not by the nature of reason but by the frailty of our concentration and grasp along with our ignorance of the ways in which God chose to arrange the universe.[63] Hence the proper use of reason was not simply a study of the conditions of validity; it could lead to absolute truth, to the extent possible within human limitations. For this reason, the section of the syllabus on logic included a question on certitude, and eclectic textbooks invariably included a refutation of skepticism.

Consistent with his views on the relation of reason and observation, Cousin supplemented syllogistic logic with "method," including scientific method, in the syllabus of 1832.[64] In other words, he introduced the philosophy of science into the curriculum. Since Cousin and the early eclectics liked to compare philosophy with the natural sciences, this innovation provides an opportunity to examine the understanding of the natural sciences that made this comparison possible. It has already been noted that students of philosophy received very little scientific training, and this was especially true of the early eclectics, who were products of a secondary educational system that was still almost exclusively literary. The philosophy of science propounded by the early eclectics was therefore derived neither from the practice of science nor from contemporary practicing scientists but primarily from a literature in the philosophy of science stretching back to Bacon and Descartes. Indeed, the eclectic philosophy of science tried to be a synthesis of Bacon and Descartes, acknowledging the need for experience but insisting on the a priori character of reason. While it may be true that every philosophy of science, extremes excepted, is a reconciliation of reason and experience, it does not follow that all philosophies of science are the same. Eclectics exhibited a suspicion of hypotheses, an incomprehension of the mathematical sciences, and an inability to reconcile the deductive with the inductive sciences that left them open to criticism from both positivists and subsequent spiritualists.

The eclectics, it has been noted, were enthusiastic supporters of the inductive method. They acknowledged that general ideas about the natural world, and even about the moral world, had to be built up from the comparison of particular experiences. As a consequence, they were quite content to reproduce Bacon's table of facts and rules of induction. But they denied that this process was reducible to the mechanical association of ideas, as eighteenth-century empiricism had argued. The scientific mind begins with observation and comparison, but it must call to the aid of experience "on the one hand reason and its necessary principles, on the other faith in the stability of the laws of nature" to extend particular findings to all similar objects and all possible time and space.[65] Reason, its necessary principles, and faith in the stability of the laws of nature are not themselves derived from experience. Somewhat paradoxically, Cousin claimed that the a priori character of the principles of reason was itself a matter of experience. We know we have these principles, but we also know both that we did not make them up, and that we cannot find anything in experience that would suggest them. "We can therefore affirm that the existence of universal and necessary principles rests on the testimony of observation, and indeed of the most immediate and the most reliable observation, that of consciousness."[66]

Cousin distinguished two types of abstraction, one in which comparison leads to the abstraction of a common character, and one in which the mind abstracts the universal immediately, from a single example. The former applies to the general ideas of the natural sciences, but the latter occurs when universal and necessary principles are the object of abstraction.[67] By this distinction Cousin hoped to give both experience and reason their due in the construction of natural sciences. Indeed, by claiming that substance and cause were necessary elements of science, he hoped to establish that metaphysical concepts were inextricably linked to natural science.

If the eclectics denounced mere empiricism, and positivism, which they equated with mere empiricism, they denounced with equal vigor mere speculation, reason unaided by the senses. Martin railed against those who proceed from a demonstration of the identity of the subjective and the objective to "construct the universe to suit their fantasy," and Cousin denounced those who forgot that the conditions of science lie in experience.[68] Thus, although Cousin declared himself in general for Plato and Descartes against Aristotle and Locke,[69] because the former believed in the existence of a higher realm of absolute ideas, he also maintained that "experience saves philosophy from hypothesis, from abstraction, from the exclusively deductive method, that is to say from the geometrical method."[70]

The use of the term *hypothesis* in a pejorative sense here is neither accidental nor isolated. Cousin habitually equated it with speculation, and asserted that it had no part in either philosophy or the other sciences.[71] This distrust of hypotheses stems from the desire to distance his philosophy from the excesses of seventeenth-century rationalism and nineteenth-century idealism. It is inherited ultimately from the eighteenth-century philosophy of science that took Newton literally—*hypotheses non fingo* —and used that slogan as a battle cry against system-building schoolmen and Cartesians. All general laws should be deduced from the phenomena. Cousin would of course except the universal and necessary principles of reason, but these are not hypothetical. Hypotheses are subjective, made up; the principles of reason are objective, impersonal, and absolutely true. Hence it is quite consistent for him to emphasize the role of a priori concepts in science and yet deny the need for hypotheses. Coupled with the derogatory reference to the exclusively deductive method, this disdain for hypotheses makes a hypothetico-deductive model of science quite inconceivable.

The mention of the geometrical method, repeated throughout Cousin's work, points up another feature of eclectic philosophy of science—its essentially qualitative character. There is very little philosophy of mathematics among the early eclectics, and very little discussion of the role of mathematics in the natural sciences. This is not only because these philosophers did not know their math; it is also because they did not consider mathematics essential to the definition of the natural sciences. Mathematics was helpful as an aid to precision and rigor, but it was not constitutive of the natural sciences.[72] None of the natural sciences was defined by its mathematical approach; instead they were characterized as inductive, experimental, or observational. Indeed, the mathematical sciences were peculiar because of their exclusively deductive character, and they could not be used as models for the inductive sciences, unless they were used as ideals of the completed science. This made the mixed sciences—such as rational mechanics—somewhat anomalous in the classification of the sciences. Despite the repeated insistence that all sciences employ both reason and observation, deduction and induction, in fact there was a gulf in eclectic philosophy of science between the deductive and the inductive sciences, a gulf on opposite sides of which stood physics and mathematics.

The problematic relation of induction to deduction extended to the philosophy of the moral sciences. On the one hand, eclectics held that the moral world should be a science of observation, and they denounced a priori psychology, morals, even metaphysics. Martin defined the moral sciences as "the inductive sciences that concern intelligent beings."[73] On the other hand, the rules of logic and morals were held to be prescriptive in character, and the principles of metaphysics, as the laws of being rather than appearance, could by definition not be induced from experience. This led to distinctions and disagreements. The *sciences morales extérieures*—history, political science, and the like—together with psychology were generally held to be inductive in the same sense as the natural sciences. Morals and logic were viewed as mixed: one discovered the principles of these sciences through observation, but one deduced the consequences of these first principles in the manner of the deductive sciences. Metaphysics was held by some to be strictly rational and hence equivalent to mathematics, by others to be a mixed science in which one's experience of the world provides the model from which principles are constructed.[74]

None of this implies an essential methodological distinction between the natural and the moral sciences. The major difference between these sciences was not methodological but ontological, in the nature of their objects and hence of the laws that governed them. Eclectics believed firmly that spiritual beings have free will, and as a result escape from the strict determinism of the physical world. On the other hand, this freedom is regulated by other laws, unique to the spiritual world, which act as rules of liberty. Moral law is regulative rather than determining.[75] But these laws are objective, universal, and necessary nonetheless, and in this sense they do not differ at all from the laws of nature.

Ethics

If the spontaneous activity of the will was the culmination of eclectic psychology, it was the foundation of eclectic ethics. Eclectic moral philosophy depended upon the existence of a morally free and responsible individual. Proofs of free will were a part of psychology, and indeed one of the chief arguments was precisely the need for free will if moral principles and sanctions were to be meaningful.[76] Ethics was therefore the search for the principles that should regulate free action.[77] But such principles define the good, and therefore ethics is the study of the good.

Moral principles are different from the laws of the physical world; nevertheless, the eclectics maintained that morals is at least in part a science of observation. The syllabus asked for a description of "moral phenomena," and Cousin reserved some of his most explicitly experimentalist rhetoric for his discussion of morals.[78] Ethics begins with a description and analysis of moral practice, not with the ideal moral agent.[79] As it turns out, observation reveals that human beings do in fact universally and instinctively make moral judgments of good and bad, right and wrong. The phenomenon of a moral conscience is an undeniable part of human nature.

Philosophers have differed more over the nature and explanation of moral conscience than over its existence. The eclectics generally reduced systems of moral theory to three types, based on the guiding principle of each: self-interest and pleasure, of which utilitarianism was the type; sentiment and sympathy, as in Adam Smith's *Theory of the Moral Sentiments;* or duty, as in Kant's categorical imperative. Characteristically, eclectics acknowledged a grain of truth in each system. People act out of all these motives. The mistake of the theories of interest and sympathy was first of all to think that all actions could be reduced to these principles, and second to hold that interest or sympathy is what makes action moral. The morality of an action is actually determined by its conformity to moral law, and hence only moral law and the concept of duty can truly form the basis of morality. Eclectic ethics was essentially an ethics of duty, and Kant was the moral philosopher most often invoked. Nevertheless, eclectics tried to mitigate the austerity of Kantian ethics by integrating sentiment and interest. Duty determines the goal, but sentiment aids and pleasure rewards moral action.[80] Not only does duty determine morality, but moral law, and therefore moral duty, is universal and necessary. Moral law is exactly like natural law in this sense. This means that the diversity of custom among various peoples must be due to different understandings of moral law, not to the existence of essentially different moral laws.

The syllabus divided practical ethics into several parts, according to the object of the moral tie: oneself, the family, the State, humanity. Two extremes were sometimes included, sometimes not: animals, as sentient but not rational agents, and God, as the ultimate being toward whom we have obligations.[81] Although obligations of the different moral relationships varied, all were couched in general terms. There was one set of duties to oneself, one for the family, et cetera. Since moral principles were abstract and universal by their nature, variations in the moral practices of different cultures and ages had to be explained either as the contingent consequence of ignorance and error, or as the necessary process of moral development in the human race.

Just as the culmination of psychology was the demonstration of the spirituality of the soul, so the culmination of morals was the demonstration of the immortality of the soul. Indeed, the existence of moral law was the strongest proof of the immortality of the soul, since it is obvious that the good are not always rewarded and the bad always punished in this life.[82] The incompleteness of morality without a divine sanction was one reason why Cousin originally placed the proof of the existence of God at the end of the section of the syllabus devoted to ethics.[83] But even after He was moved to a separate part of the syllabus, eclectics insisted that a science of morals could not be indifferent to the question of the existence of God and the immortal soul.

Metaphysics or Theodicy

Every section of the eclectic syllabus includes some metaphysical proposition. Psychology ends with a demonstration of the existence of the spiritual soul; logic affirms the necessity of substance, cause, and absolute truth; ethics depends on the immortality of the soul and the existence of God. Eclecticism began with phenomena, but it was not content to restrict itself to the phenomena, or even to the examination of the conditions of knowledge. The eclectics were committed to the possibility of knowing things-in-themselves, being qua being, and hence to the validity of metaphysics defined as first philosophy, "the science of first principles and first causes."[84] Cousin called Kant a skeptic because he denied that we could know anything beyond phenomena and that our judgments would hold true of rational beings with a different mental makeup.[85] To counter this skepticism, Cousin developed a theory of "impersonal reason," which held that our reason is neither subjective nor arbitrary; on the contrary, "just as sensation puts me in relation with the physical world, so another faculty puts me in communication with truths that depend neither on the world nor on me, and this faculty is reason."[86] The impersonality of reason allowed Cousin to pass from psychology to ontology, leaving phenomenalist and critical philosophies far behind.

Given this decided commitment to metaphysics, it is somewhat curious that the syllabus of 1832 did not include metaphysics as a separate section. Janet invokes this as evidence of Cousin's radical belief in the *méthode psychologique*, although it is just as likely that he did not want to offend the clerical party by encroaching more than necessary on their domain.[87] The brief section at the end of ethics on proofs of the existence of God and of the immortal soul eventually began to expand, more under the pressure of liberal philosophers than of the church, which would rather have seen metaphysics suppressed altogether than taught by laymen.[88] Only after Cousin's downfall, under the Second Empire, did it become a separate section. It was originally called theodicy, which reflects the primary purpose of a metaphysics for the lycée. The Third Republic retained the separate section and the proofs of God's existence, but gave them the slightly more laical name of metaphysics. By the 1870s, schoolchildren had to master the definition of substance and accident, the concept of the infinite, and the difference between potency and act, although in practice and on the syllabus metaphysics was the least developed of the divisions of philosophy.

In addition to the particular metaphysical issues addressed in the other parts of the syllabus and the demonstration of the existence of God, the overarching concern of the eclectics was to affirm a dualism of spirit and matter that avoided the opposite but equally repugnant reductions of materialism and pantheism. This latter involved Cousin in charges of disingenuousness, since his early philosophy was heavily influenced by his association with Hegel and Schelling. But it remains true that the public face Cousin wanted to place on philosophy and the one most eclectics supported was dualist.

History of Philosophy

It is not surprising that a philosophy calling itself eclectic should give a generous place to the history of philosophy. The subject was part of the syllabus before Cousin, but his insistence that the history of philosophy was an integral part of philosophy itself increased the importance of this section. If there is some truth in every system of philosophy, then one way of finding truth is to seek it in the past products of philosophical thought.[89] He even went so far as to say that all philosophical truth had been stated at one time or another; the task of the philosopher was to extract the truth and to synthesize it into a coherent whole. Cousin practiced what he preached by producing a number of new editions of classical philosophers, delving into archives for a series of historical studies of seventeenth-century thought and letters, and encouraging the academic study of past thinkers. Even his critics have had to acknowledge that he made the history of philosophy an academic discipline.[90]

Cousin was not content simply to advocate and practice the study of individual philosophers, problems, and schools. He thought that the history of philosophy showed both order and progress, and he developed a theory to account for it. In its most elaborate form, this theory held that all systems necessarily conformed to one of four types: sensualism, idealism, skepticism, or mysticism. Thinkers first put their trust in the senses and tried to derive all knowledge from the things revealed by the senses. When this failed, they went to the opposite extreme and tried to derive everything from ideas and reason. Unsuccessful in this as well, philosophers despaired of finding truth at all, and became skeptics. But unable to suspend judgment for long, they fled into mysticism, the belief in an access to truth without the mediation of either reason or the senses. Not many eclectic philosophers followed Cousin in the details of this classification. Usually, some variation of the struggle between empiricism and rationalism, between materialism and spiritualism, provided the framework for telling the story of philosophy.

The eclectics did not see this struggle and fluctuation among systems, whatever the number, as static. Philosophers learn from the past; they do not simply repeat it. If the general outlines have remained the same, the problems raised, concepts used, distinctions made, and solutions offered have all increased in clarity, subtlety, and accuracy.[91] Eclectics therefore denied that philosophy could be distinguished from the natural sciences on the ground that the former never changed while the latter progressed, and they used the history of philosophy as proof. Hence the same discipline that sought all truth in the past also provided a justification for the future of philosophy.

CRITICS AND THE LIMITS OF ECLECTICISM

Such was the doctrine that determined the syllabus in French philosophy through the end of the nineteenth century. The persistence of eclecticism can be attributed in part to the sheer institutional inertia of the French state apparatus and to the longevity of the persons holding the chief positions. But for such a syllabus to persist for seventy-five years through the political and social upheavals that France suffered, it must have answered some intellectual and social need. To some extent eclecticism did succeed in being the via media, the reconciler of antinomies envisioned by Cousin. The very comprehensiveness of its universe of discourse ensured that there would be something for everyone: independent of theology yet respectful of religion, based on observation yet affirming metaphysical entities, founding individual rights yet recognizing the need for legitimate authority, et cetera.

Perhaps more important, as the only philosophy that explicitly recognized the partial truth of all systems and tried to accommodate them, eclecticism had an important advantage over its more exclusive opponents, particularly Catholicism and positivism. However much these two schools disliked eclecticism, they disliked each other even more. In general, it was unlikely that they would agree on how to replace eclecticism, and each would prefer the status quo to the victory of an even more dogmatic enemy. But even if they did manage to compromise with each other, then the result would by definition be eclectic; innovations introduced piecemeal to placate either side would only reinforce the eclectic character of the syllabus. Either way, eclecticism won, and the result was a philosophical stalemate in which this school, *faute de mieux,* continued a dominance far in excess of its representation in the field of philosophical discourse.

But this dominance could not hide the fact that eclectic spiritualism was criticized from all sides, to the point that Janet, in a series of articles published in the *Revue des deux mondes* in 1864, spoke of a crisis in spiritualist philosophy.[92] Nor could eclectics simply ignore these attacks, if for no other reason than that, despite their strong institutional position, they were vulnerable on a number of fronts. If Cousin's power was real, it was also limited. He could not, for example, take away the chair of a *titulaire*. Pierre Laromiguière, who carried on the tradition of the Ideologues, continued to occupy his chair unmolested, although Cousin made life difficult for his *suppléant* and followers. Cousin ensured that one of his own former pupils, Théodore Jouffroy, obtained Laromiguière's chair when his death in 1837 finally made it vacant, despite the claims of Laromiguière's own longtime *suppléant*.[93] Once Jouffroy was settled in Laromiguière's position, however, he began to manifest what Cousin considered an excessively skeptical attitude toward the limits of human knowledge, but there was little the former master could do. Even among the lycée teachers who were most beholden to him, his autocratic style evoked a silent opposition that finally voiced itself in the pointedly entitled journal *La liberté de penser,* founded in 1847 by Jules Simon, Janet, and others as an organ outside the control of their jealous master.[94]

If Cousin's reach over professors of philosophy was to some extent restricted, his influence on other parts of the educational bureaucracy was even more so. Hippolyte Taine and Ernest Renan, two of eclecticism's most outspoken and respected critics, were both professors at other Parisian institutions, Taine at the Ecole des Beaux Arts, Renan at the Collège de France.[95] Félix Ravaisson-Mollien and A. A. Cournot, who both published major studies critical of academic philosophy, were inspectors general. Taine and Ravaisson had both been excluded from academic philosophy early in their careers, and both came back to haunt their persecutor from positions outside his reach.

Above all, Cousin and his successors had very little control over the larger political developments that might affect institutional philosophy. If Cousin had sufficient personal prestige to influence legislation and to resist legislative attempts to dismantle philosophy during the July Monarchy, the Revolution of 1848 and the coup d'état of 1851 completely overwhelmed the resources of the philosophical establishment. Louis Napoleon, both to placate the church and to rid himself of what he considered a subversive element, decimated the ranks of professors and truncated the philosophy class. Many professors were simply dismissed; others resigned rather than take the oath of allegiance to the new emperor. Philosophy was restricted to logic in the lycée, and the *agrégation* in philosophy was eliminated. Although these parts of the institution were restored in 1863, the experience made it clear that determined political pressure could radically alter the conditions of academic philosophy.

But for these opportunities for institutional encroachment to be legitimated, they needed reasons. Reasons were, in fact, easily found to impugn both the validity of eclecticism as an intellectual enterprise and its effectiveness as an instrument of moral regeneration. Critics to voice these reasons were also easily found, since a doctrine that borrowed from all schools was bound to offend all schools. Doubters asked whether dogmatic eclecticism was a coherent concept, since extracting the truth from different systems implied a criterion of selection. If one had such a criterion, one would not need the other systems, and if one did not, then how would one discern the true?[96] Others asked whether one could define a philosophy by both a method and a doctrine. If it is defined by its method, whether eclecticism or the *méthode psychologique,* then it ought to admit the possibility that its current positions might be disproved by further application of the same method. If it is simply defined by its content—spiritualism—then why bother giving the pretense of methodological openness?

One feature of Cousinian eclecticism, however, united critics of all schools in universal condemnation: its pretension to demonstrate the truths of spiritualism using the methods of the natural sciences. The mélange of metaphysics and experimentalism disturbed thinkers at both extremes of the philosophical spectrum, who questioned whether philosophy should consist of a science of observation, at least in the manner of the natural sciences. Some thought it should not be, and attempted to find other foundations for the study of metaphysics. This attempt led to the more intransigent forms of idealism and spiritualism. Others thought that at least part of philosophy should consist of natural sciences, but that Cousin had not made them natural sciences. This line of thinking led to various positivistic conceptions of the moral sciences. Both directions worked to pull apart the reconciliation eclectics thought they had achieved between spiritualist metaphysics and modern science.

Spiritualism

Among those who criticized the eclectic project from the standpoint of spiritualism, none was more implacable than the Catholic Church. However much Cousin tried to reassure Catholics that nothing in his doctrine could offend them, he also maintained that philosophy should be independent of theology. The church held that philosophy—especially one including moral precepts and proofs of the existence of God— could be properly studied and taught only from within the faith, and therefore it was bound to oppose Cousin in principle, even if none of his particular propositions concerning the Divinity proved heretical. Reason without the guidance of faith, much less reason in the service of the methods of the natural sciences, would inevitably lead to error in matters of faith. But Catholic writers did find fault with Cousin's propositions. In particular, they interpreted some of his early statements concerning God and the world, made under recent contact with Schelling and Hegel, as pantheistic, and the charge of pantheism echoed throughout the 1830s and 1840s.

Behind the intellectual differences between eclecticism and Catholicism lay a larger issue—freedom of instruction. Once the end of the Reaction of 1822–28 had restricted the influence of the church in the Université, Catholics sought to escape state control by seeking the end of the monopoly of the Université over educational institutions, thereby allowing Catholic schools greater independence. Part of their strategy included denouncing the errors of eclectic philosophy and the dire consequences of allowing children to be exposed to them. For his part, Cousin defended the Université, viewing education as a public function to be exercised only with the permission and under the supervision of the state. The attacks of the church reached a fever pitch in the early 1840s, and in 1844 a law was introduced to grant freedom of instruction. Cousin opposed it in the Senate, and the law was defeated, but the Loi Falloux of 1850 accomplished much of what Catholics had demanded.

Catholics were not the only spiritualists who criticized Cousin. Félix Ravaisson agreed with Cousin that philosophy should be independent of theology, but he objected to the methods Cousin employed. In his widely read report on *Philosophy in France during the Nineteenth Century*, produced for the Ministry of Public Instruction in 1868, he criticized the *méthode psychologique* for sharply distinguishing between interior and exterior phenomena, between conceptions and perceptions, and for treating both according to the precepts of Bacon, as phenomena to be observed and to be made the basis of inductive generalizations.[97] By distinguishing radically between conception and perception, and by making perception correspond to phenomena while ideas correspond to metaphysical entities, Cousin robbed the ideal of its concreteness and vitality, while robbing perception of its ability to evoke the ideal. By making mental events phenomena like those of the other sciences, he artificially externalized that which is closest to us and made it impossible to know anything in more than a superficial way. Ravaisson proposed instead to get rid of the distinction between idea and sensation, as well as the characterization of mental life as "phenomena" to be studied by the methods of the natural sciences.

> The true psychological method, at least that which extends to what is called rational psychology or metaphysics, must therefore not, it seems, be defined as that which, from so-called internal phenomena or phenomena of consciousness, identifies their causes through induction, but rather as that which allows us, in everything of which we are conscious and which, viewed from the outside, is in some sense phenomenal and natural, to discern our act, which alone must be called strictly internal, and which indeed, beyond any condition of extension and even of duration, is supernatural or metaphysical

in its essence. The true psychological method is that which, from the fact of some sensation or perception, distinguishes by an entirely unique operation, that which completes it in making it ours, and which is none other than our self. (ibid., 28)

Ravaisson called the operation by which we discern our act in all consciousness *réflexion*, using the same term as Cousin, but giving it a significantly different meaning. For by reflection we have immediate access to ourself as substance and cause, not simply to a phenomenon like any other. This view of reflection, which Ravaisson attributed to Maine de Biran, was not entirely lost on Cousin and was formally adopted by several of his pupils.[98] Nevertheless, eclecticism remained wedded to the Baconian method in psychology. But the scientific method can only give us knowledge of phenomena, whether physical or mental; it cannot grasp the metaphysical ground of phenomena. Ravaisson thought that since the restoration of the *agrégation* in 1863 philosophy had been gradually abandoning this phenomenalist approach and adopting what he termed a "spiritualist realism or positivism, having as its generating principle the mind's inward awareness of an existence from which it recognizes that all other existence derives and depends, and which is none other than its action."[99]

Positivism

Ravaisson's transformation of reflection would indeed revolutionize spiritualism in the years to come, but what seemed most prescient to his successors was that he made his prediction at a time when spiritualist philosophies were on the defensive. Janet's piece mentioned neither Ravaisson nor the Catholics: its theme was that the "spiritualist idea" itself was in danger from the rise of philosophies that denied the independence of thinking substance. Ominously, these philosophies claimed like eclecticism to base themselves on the natural sciences.[100]

The *Refutation of Eclecticism* (*Réfutation de l'éclectisme*, 1838) of Pierre Leroux may serve as a convenient transition from the spiritualist to the positivist critiques, combining as it does elements of both. Leroux was a Polytechnicien who converted to Saint-Simonianism at the height of its religious phase.[101] Although he soon distanced himself from the orthodox Saint-Simonians, his association left him with the conviction that traditional Christianity was outdated and that society needed to be regenerated on the basis of modern science and a new philosophy that recognized Humanity rather than God as the supreme reality. Nevertheless, Leroux did not think that philosophy should be a science. On the contrary, he maintained that philosophy and religion are essentially one; the task of modern philosophy, he held, is to become the new religion.[102]

Hence he pilloried Cousin both for trying to maintain the distinction between philosophy and religion and for continuing to support the moribund religion of Catholicism. Moreover, if philosophy is religion, it is not science, at least not like the natural sciences. Leroux waxed indignant over the assimilation of philosophy to physics:

> Philosophy a science like *physics!* a science directly objective like physics, a science of observation and of experimentation like physics and industry! . . . But in physics, the object is outside us; it belongs to an order of life incommunicable to us. In philosophy, on the contrary, it is above all a question of the *self* and the us, of human life either individual or collective. (ibid., 360)

Leroux thought that philosophy, as the expression of life, had to receive its inspiration from the heart, like poetry. Philosophy was therefore as much an art as a science (ibid., 365).

This comprehensive view of the nature and role of philosophy constituted another reason why Leroux opposed eclecticism. He thought that the philosopher should play a leading role in the regeneration of society, and he accused Cousin of making philosophy a study apart from society, insulating it within the walls of the Ecole Normale, where "the attempt was made to cultivate languages, literature, and philosophical matters for themselves, independently of political and social life. They tried to train rhetoricians and dialecticians just as the Ecole Polytechnique trained engineers or artillery officers" (ibid., 299). The result was that "a psychologist, and especially a psychologist of our time, is a man who has neither tradition nor goal; in this he resembles a chemist or a physicist" (ibid., 300). This was a legacy of the Napoleonic origins of the Ecole Normale and reduced philosophers to the role of obedient servants of the status quo.

Another errant pupil of Saint-Simon, Auguste Comte, also condemned Cousin's *méthode psychologique*. In the forty-fifth lesson of the *Course of Positive Philosophy (Cours de philosophie positive),* entitled "General Considerations on the Positive Study of the Intellectual and Moral, or Cerebral, Functions," Comte explained why he considered the study of mental life to be the highest part of biology rather than the material of an independent science. One set of objections was methodological: Comte thought that interior observation was inherently impossible, that one cannot watch oneself think.[103] Even if one could, he continued, such a method would restrict psychology to the adult, sane man, ignoring children, animals, and the insane. Thought was the function of the organ "brain," according to Comte, and since the positive study of vital functions consisted in relating them to the structures that produced them, psychology could only be studied as a part of neurophysiology. The other way to observe intellectual acts was to observe them, not in one-self, but in others through their actions. This would make psychology part of natural history. And since both natural history and physiology are parts of biology, psychology should take its place at the summit of biology, as the study of the most complex biological organism. Unless one connected intellectual phenomena with either their conditions or their effects, one would be hopelessly mired in an "unintelligible logom-achy, in which purely nominal entities are incessantly substituted for real phenomena, following the fundamental character of any metaphysical conception" (ibid., 1:855). Comte did not distinguish between eclectics and ideologues in his condemnation of an independent psychology.

Another set of objections to psychology was substantive rather than methodological. Comte denounced the tendency of psychologists to de-fine humanity in terms of intellect when it was obvious that the emotions played a greater role in determining behavior. He attributed this ten-dency to two philosophico-theological motivations. One was the "vain" desire to effect a clear-cut demarcation between humans and animals. The other was the need to preserve the unity of the self, "in order for it to correspond to the rigorous unity of the soul" (ibid., 1:857). With-out this need, there would be no obstacle to recognizing what the phre-nologists had established, that the mind was made up of distinct and independent powers, and that the self was simply the feeling of the "universal consensus of the whole of the organism" (ibid., 1:858). On the other hand, they would recognize that these powers are not meta-physical entities but simply the functions of various parts of the brain.

With these criticisms, Comte hoped not merely to defeat psychology but to raze the walls, disperse the inhabitants, and sow salt among the ruins to prevent it from ever reappearing in its traditional form. Psychology, of course, was only one small part of Comte's enemy. Meta-physical thinking was the greater enemy, and he treated it with the same contempt, although he recognized its necessity as a stage in human development. In the formulation of his famous theory of the three states, he defined the metaphysical state as that in which

> supernatural agents are replaced by abstract forces, veritable entities (personi-fied abstractions) inherent in the diverse beings of the world, and conceived as capable of engendering by themselves all observed phenomena, of which the explanation then consists of assigning for each the corresponding entity. (ibid., 1:21)

Metaphysical thinking also sought absolute knowledge of first and final causes. The positive mode of thought eliminated this twin tendency to postulate entities beyond or beneath phenomena and to strive for abso-lute knowledge. In the positive state,

the human mind, recognizing the impossibility of obtaining absolute notions, renounces the search for the origin and destination of the universe,
and the knowledge of the intimate causes of phenomena, to attach itself
solely to the discovery, through the well-combined use of reason and observation, of their actual laws, that is their invariable relations of succession and
similarity. (ibid., 1:21–22)

This was the character attained successively by the various natural sciences, and Comte hoped to extend it to philosophy and to the other
moral sciences.

It is clear that such an extension would pose a threat to eclecticism,
but the nature of this threat is often misunderstood. It is insufficient to
say that Comte's innovation was to model the human sciences on the
natural. The eclectics, we have seen, also insisted that philosophy is a
science of observation. The real difference is in the way that the natural
sciences, and hence the human sciences, are conceived. The eclectics held
that certain universal and necessary principles of a metaphysical nature—for example, cause and substance—were an essential part of explanation in the natural sciences. Hence basing philosophy on science
would not eliminate metaphysics, but rather help distinguish between
competing metaphysical theories. Comte, on the other hand, held that
the positive sciences were devoid of metaphysics and that the moral
sciences retained it only because they had not yet reached the positive
stage. The natural sciences had indeed once had such entities, as the
history of such studies as alchemy and astrology proves. Psychology is
to mental physiology what alchemy is to chemistry. The issue between
positivism and eclecticism, therefore, was not simply between science
and philosophy, as Comte well knew. It was between two different
conceptions of science and two of philosophy. This explains why, in
addition to reaffirming the independence of philosophy with respect to
the natural sciences, one of the chief tactics of the later spiritualists
in their effort to resist scientism would be to criticize the positivist
understanding of the natural sciences. Nevertheless, another commonly
cited feature of the battle between positivism and its enemies remains
valid; this was the positivist insistence that the principle of determinism
be applied to moral phenomena. Such a principle could not be admitted
by spiritualist philosophers.

Once metaphysics and psychology are eliminated, however, philosophy as a science of observation still looks very different in Comte than
it does in Cousin. Logic can no longer be an analysis of the forms of
thought, for this would involve self-examination. It cannot even be a
critique of the methods of the positive sciences, for this would assume
a state superior to positivity, and such a state does not exist. Positivist
logic is therefore the study of method, but through the observation of

the positive sciences, not through the abstract analysis of deduction and induction as in the eclectic syllabus.[104] Indeed, one can say that the positive sciences are the only object of philosophy, for in addition to method, positive philosophy consists only of the systematization of the sciences and reflection on the most general results of the sciences.

Also excluded from positive philosophy is ethics. Traditional moral philosophy sought absolute principles of right and wrong, and it shares the fate of all metaphysical modes of thinking. Even restricted to relative principles, ethics as the study of what ought to be done is not strictly speaking a science at all, since sciences do not make normative judgments.[105] Ethics is an art, and it is based on the science of society—sociology. Comtean sociology corresponds most nearly with what in the eclectic scheme were termed the *sciences morales extérieures*—those sciences that studied human behavior from the outside. These were explicitly excluded from philosophy by the eclectics, and in this sense sociology as a science does not conflict with philosophy. But whereas spiritualist philosophers would hold that ethics was in principle independent of the contingencies of human society and history, Comte believed that morality was just the laws of the existence of society at a particular stage in its development. The exterior determines the interior, not vice versa. In this sense, sociology does indeed pose a threat to the study of ethics as conceived by eclectics.

Comte's ultimate aim was in one way fundamentally the same as Cousin's—to regenerate French society after the ravages of the Revolution. Positive philosophy and the positive sciences paved the way for Comte's *System of Positive Politics* (*Système de politique positive*), which detailed the new society based on his principles. Comte explicitly conceived his philosophy and politics as an alternative to, and inevitable replacement of, the parliamentary monarchy of the Restoration and July Monarchy, which he thought merely continued metaphysical modes of thought and patched together bits of incompatible and outdated systems rather than created a truly new and lasting basis for government.[106] It was precisely the eclectic character of the postrevolutionary period that worried Comte most.

Not surprisingly, Comte was unable to find a place within the institutions of academic philosophy, although he tried to solicit one from Cousin's friend Guizot.[107] A product of the Ecole Polytechnique, he remained marginal even to that institution, serving for a time as an examiner, but mostly eking out a living from private lessons. He found few followers among academic philosophers, and his ideas were carried on by Emile Littré, a physician by training who was also excluded from academic recognition for most of his career.

The attack on university philosophy in the name of science was wider than Comte or his direct influence, however, and because his position was so radical, during the Second Empire he was less of a threat than thinkers who found more palatable ways to present their ideas. For this reason, Comte's influence was for a long time negligible beside that of the two writers who epitomized midcentury scientism, Hippolyte Taine and Ernest Renan. These two owed their eminence first of all to their literary skill, which won them begrudging respect even from their enemies. In addition, their object of study was itself literary: Taine wrote literary criticism, Renan biblical criticism and philology. These assets cannot be overestimated in a society in which *culture générale* was overwhelmingly literary. Taine and Renan addressed the heart of French culture in a way that Comte, in his preoccupation with the philosophy of science, did not. When they claimed to apply the methods of science to literature, they were assured of an audience.

Renan's philosophical education began in the seminary where he was studying to become a priest.[108] But it was contact with the Higher Criticism that eroded his faith and made him decide to leave the seminary in 1845. In order to gain academic credentials, he prepared for the *agrégation* in philosophy and placed first. He therefore knew intimately the reigning eclectic philosophy.[109] His own convictions, however, were determined by his contact with text criticism, his study of comparative grammar, his reading of Hegel, and his friendship with the chemist Marcellin Berthelot. These influences combined in Renan to produce a faith in the methods of science together with a worldview that saw the origin and gradual emancipation of mind from matter through history. Renan agreed with Comte in denying that metaphysics was a science, but he did not therefore condemn it to superannuation.[110] On the contrary, Renan held that metaphysics represented the eternal yearning of the human mind for the ideal, and as such it helped to realize that ideal. Metaphysics was poetry rather than science, an expression of the human soul rather than a reflection of external reality. While this position retained a role for metaphysics, it was almost as objectionable to eclectics as Comte's brusque dismissal.

Backhanded compliments also fill Renan's essay on Cousin, written in 1858 as a review of the latter's *Fragments et souvenirs*. Renan did not disagree with the content of spiritualism; he too thought that spirit was the noblest aspect of reality, although he thought that its origin lay in matter. But he did not think that Cousin was a great philosopher, and certainly not a scientist. "M. Cousin belongs more to literature than to science. He is above all a writer, an orator, a critic, who has taken an interest in philosophy."[111] Renan does not even acknowledge Cousin's scientific pretensions. Defending Cousin against those who thought he was not a deep thinker, Renan says that Cousin did not want to create an original philosophy, "but to give an eloquent and in a sense popular form to the great truths of the moral order" (ibid., 68). He thinks that Cousin did not understand science in any case. Cousin's active participation in public life also prevented his being a real philosopher. As a politician and administrator, he had to make too many compromises with popular opinion and religious authority to retain his independence as a thinker.

If Renan's critique is clothed in deference and circumlocution, Taine's is direct and satirical. He too knew his subject well, and his humiliation by the philosophical establishment added venom to his pen. A brilliant student at the Ecole Normale, he failed the *agrégation* of 1851, apparently because he showed too much sympathy for Spinoza.[112] After being sent to teach in the darkest provinces and in a lower class, Taine rebelled and abandoned the Université. He returned to Paris, gave private lessons, and completed a dissertation on the *Fables* of La Fontaine. Eventually he was able to make a living from his writings on literature, which attracted the attention of the most prominent critics of the day. By 1864, he had gained sufficient respect to be offered a chair in aesthetics and art history at the Ecole des Beaux-Arts, achieving a position within the academy but outside philosophy.[113]

His attack on university philosophy did not wait until he had achieved academic security. His *Classical Philosophers of the Nineteenth Century in France (Les philosophes classiques du XIXe siècle en France)* appeared in 1855 and 1856 as a series of articles in the *Revue de l'Instruction Publique* and was published in book form the following year. Many of Taine's criticisms anticipate those of Renan. Taine considered Cousin an orator rather than a philosopher, someone more concerned to persuade than to demonstrate.[114] He analyzed the writings of the eclectics in great detail to show how rhetorical effect and figurative language masked logical holes in the arguments. He thought eclecticism was more literary than scientific, and he accused its followers of not knowing how science really worked. The eclectics substituted abstraction for analysis, generalities for facts (ibid., 293). In addition, he condemned eclectics for subordinating science to morality, of refuting doctrines because they were immoral rather than invalid (ibid., 290–92). These features had led eclecticism to sterility, immobility, and impotence. Taine summed up his characterization of official philosophy in the following terms:

> No one would want to compare it, like the old [philosophies], to a river that waters and overturns; no noise, no movement, no effect. It is a bath, very clean, very quiet, and very tepid, where fathers, as a health precaution, put their children. (ibid., 311)

If Comte attempted to paint eclecticism as an obstacle to social regeneration, Taine made it a laughingstock. Unlike Comte, however, Taine thought that eclecticism reflected all too well the needs and desires of society, to the point that he saw little hope of dethroning it.

Taine differed from Comte in other ways, as well. If he agreed that moral phenomena should be subjected to the methods of the natural sciences, he did not think that the idea of cause should be banished from science or philosophy. It should instead be redefined as the fact or law from which an effect can be logically deduced (ibid., viii–ix). Taine also believed in the possibility of psychology as a science; he defended Condillac against the attacks of the eclectics, and when *Classical Philosophers* appeared, he was already at work on a systematic treatise of scientific psychology, published in 1870 as *On Intelligence (De l'intelligence)*. Above all, he did not think that metaphysics was impossible: ". . . beyond all these inferior analyses called sciences, which reduce facts to a few particular types and laws, there can be a superior analysis called metaphysics which would reduce these laws and types to some universal formula" (ibid., ix). Comte had explicitly excluded the possibility of such a reduction, and for this reason Taine portrayed himself as offering a middle way between spiritualism and positivism.

CONCLUSION

Eclectic spiritualism, dominant in the 1830s and 1840s, remained entrenched in the educational system under the Second Empire, and even emerged once again as a quasi-official philosophy after the trials of the 1850s. Nevertheless, thinkers from both the spiritualist and positivist sides began to pull apart its uneasy synthesis of metaphysics and observationalism. Spiritualists denounced the claim of eclectic philosophy to be a science of phenomena; positivists ridiculed it.

2

Eclectic Buddhist: Théodule Ribot

The *Revue* keeps a place for ... [metaphysics], for it does not profess pure empiricism; but even from metaphysicians it will ask for facts, being persuaded that nowhere can one do without experience and that wherever experience is lacking, there are only logical quibbles, imaginary creations, and mystical effusions.
—Théodule Ribot, Preface, *Revue philosophique*

ONE of the first to establish a foothold in the French universities for the positivistic human sciences was Théodule Ribot (1839–1916), who left a career as a professor of philosophy to found a psychology grounded in empirical research and independent of metaphysics. This obviously threatened eclectic philosophy, for not only were psychology and metaphysics linked in this tradition, but psychology was the foundation of eclectic philosophy. Ribot noisily announced the bankruptcy of eclectic psychology and introduced the French public to traditions he considered more scientific—English associationism, German psychophysics, and the native French tradition of psychiatric psychopathology. He also produced original works that subjected traditional topics in philosophical psychology to a scientific treatment. His efforts drew widespread support from the scientific community and from those who favored a more scientific approach to moral phenomena. Such support eventually enabled him to return to the academy as the first professor of experimental psychology in France, first at the Sorbonne and then at the Collège de France. He inspired a whole generation of young philosophers to become experimental psychologists, and their success in attaining academic positions ensured that the research program of experimental psychology would continue. For all of these reasons, Ribot has generally been considered the founder of scientific psychology in France.

Nevertheless, his work was more of an inspiration to, than a foundation of, scientific psychology in France. Even though Ribot denounced speculative psychology and demanded that psychologists observe for themselves, he himself neither experimented nor observed. His psychological works consisted of observations culled from the works of others—primarily physiologists and psychiatrists—and interpreted systematically from a biological and evolutionary point of view. When the Laboratory for Physiological Psychology was created at the Ecole Pratique des Hautes Études under his patronage in 1889, he conferred its direction upon a doctor, Henri Beaunis, rather than assume it himself. Some of his contemporaries noted the secondhand character of Ribot's information, and successors have for this reason found him a somewhat problematic founding father.[1] Moreover, in his opposition to speculative psychology, Ribot supported and drew from a wide variety of research traditions. As a result, he has been taxed with not being selective or systematic enough and accused of failing to create a coherent paradigm for scientific psychology.[2] Perhaps ironically, Ribot's legacy can be characterized as eclectic.

Ribot's philosophical opponents, however, generally did not find fault with his facts. Indeed, they were very much impressed with his scientific learning, and few of his critics denied that psychology could benefit from a wider range of more objective studies. What they objected to was the way in which he seemed to equate objective with physiological, denying autonomy to the purely phenomenological study of consciousness. Ribot often claimed in his early work that psychology could only be scientific on the assumption that every mental process had a physiological counterpart. He asserted that this was simply a necessary working hypothesis, but his critics charged that he was replacing a spiritualistic with a materialistic metaphysics. Debate was joined not on the question of whether facts should be the basis of psychology, but on the issue of whether and how one should distinguish a scientific from a metaphysical hypothesis.

This debate is relevant because, despite Ribot's concern to keep metaphysics out of science, his scientific outlook was informed by a metaphysical monism that he claimed was consistent with the results of modern science. This predilection can be traced back to his days at the Ecole Normale Supérieure, where his interest in things Eastern got him into trouble. The story is told that one day a classmate, the future historian Gabriel Monod, seized upon the presence of a small Buddha in Ribot's room to write on the walls of the Ecole Normale, "Ribot is a Buddhist!"[3] The next day Ribot was called into the office of the director and informed that "no self-respecting young man allows such a dishonorable epithet to be associated with his name" (ibid.). Among the aspects of oriental philosophy that fascinated him were its deterministic monism, stressing the oneness of all things and the strict fatalism governing all events, and its search for enlightenment by stripping away the illusions of consciousness. Much of Ribot's work was dedicated to revealing the illusions of spiritualist philosophy—the simplicity of consciousness, the freedom of the will, the reality of faculties, and the predominance of the intellect over the emotions.

In terms of both theory and practice, then, the work of this eclectic Buddhist has an indeterminate status between philosophy and science that can be placed in either category—or both—according to the standards of the reader. This peculiar character coincides with the position of the author as a rebellious product of academic philosophy, reproducing the traits of the institution he opposed. It also reflects the complex task he set himself in creating a scientific psychology. In wresting psychology away from philosophy, he justified himself by bringing medical and physiological research to philosophy. His polemical works were written for a philosophical audience, and it was to that audience that he presented the findings of researchers who cared little for academic philosophy. Ribot's psychology in some ways constitutes a compromise between spiritualist and evolutionist psychology, a compromise represented, as Jacqueline Thirard has suggested, by Ribot's founding the *Revue philosophique* as both a forum for the new psychology and a journal of philosophy open to all schools.[4]

But the extent of the divergence between scientific and philosophical psychology, and hence the extent of Ribot's compromise, should not be exaggerated. On the one hand, in the infancy of experimental psychology, the standards of what counted as empirical were much less rigorous than they later became. However much the new psychologists talked about the need to distinguish between fact and interpretation, in practice the line between them was quite indistinct. On the other hand, the eclectic tradition Ribot challenged was one that explicitly claimed to be a science of observation. Hence it could hardly ignore someone who claimed to be bringing new facts to bear on the domain of psychology. Despite the rhetoric of rejection and revolution, therefore, the relationship between eclectic and scientific psychology was never one of complete opposition.

In this chapter I explore the career of Ribot until his appointment to the Collège de France in 1888. It can be characterized by three successive movements—(1) away from the academy, which Ribot left to gain his independence and to establish his intellectual position; (2) when he came to an uneasy truce with academic philosophy; and (3) when Ribot, having established his intellectual position, moved back into the academy on his own terms. At no point, however, was his relationship with academic philosophy completely severed, and this relationship constituted the framework within which these movements took place.

BREAK WITH ACADEMIC PHILOSOPHY

Théodule Armand Ribot was born 18 December 1839, in the small Breton town of Guingamp.[5] After showing promise at the lycée of nearby St. Brieuc, he entered the Administration de l'Enregistrement, the branch of the Ministry of Finances that registered deeds and transfers of real property and collected the taxes imposed on such transactions. This was not his choice of career; he was compelled by his father, who wanted for his son a secure if undramatic position. Ribot remained in this employment for two years, until he manifested the independence of spirit characteristic of him by abandoning his career as soon as he reached legal majority and going to Paris to prepare for the entrance examinations for the École Normale Supérieure. He was accepted at the Ecole Normale in the entering class, or *promotion,* of 1862, where he soon attracted the attention of Elme Caro, an eclectic professor who encouraged him to pursue philosophy. Despite the fundamental differences of opinion that would later separate them, they always maintained a respect for each other that transcended their doctrinal divergence.[6] His other instructors in philosophy included Albert Lemoine, another eclectic who specialized in problems of the relation between mind and body, and Jules Lachelier, who arrived in Ribot's final year. Lachelier's precision impressed Ribot, but his uncompromising idealism dismayed the student.[7] These professors were mentors, critics, and institutional forces with whom Ribot would have to deal for many years.

As important as his professors were the classmates who would make up the academic elite of his cohort. Henri Joly (entered 1860) would later write on comparative psychology from a spiritualist perspective and compete with Ribot (as well as Espinas) for several positions. Ludovic Carrau (entered 1861) would assume Caro's chair at the Sorbonne. Gabriel Monod, author of the Buddhist prank, and Ernest Lavisse (both entered 1862), would become the most prominent historians of their generation, promoting the positive spirit in the study of the human past. Félix Alcan, who also entered the same year as Ribot, would found a prestigious academic publishing company that assumed the publication of the *Revue philosophique* and of Ribot's later works. Several other classmates went into educational administration, and many less influential career philosophers would form the core of those who read and contributed to the *Revue philosophique.* The value of these contacts cannot be overestimated; despite the rivalries that would divide some of them, the *esprit normalien* often overrode differences of opinion. Among themselves they might disagree, but when *normaliens* faced competition from non-*normaliens,* the solidarity of the elite could make strange bedfellows. This was as true of alliances between professors and students as of relations among students.

One of Ribot's closest friends at the Ecole Normale was Alfred Espinas (entered 1864). Upon his departure from school, Ribot began a correspondence with Espinas that constitutes one of the major sources of information about Ribot and about academic politics in Paris during the early Third Republic.[8] In his first letters, from the autumn of 1866, he was already grumbling to Espinas about the life of the professor of philosophy in a provincial lycée. His first assignment was to the lycée at Vesoul, a small town about thirty miles north of Besançon near the border with Switzerland. The only positive thing Ribot could say about it was that at least there was no bishop.[9] This was not an advantage to be ignored. Relations between academic philosophy and the church were tense, as we have already seen (chapter 1), and in the conservative provinces the lycée professor cut a small figure beside the local clergy, who in conjunction with practicing parents watched their schools carefully for the slightest deviations from orthodoxy. The situation was particularly difficult for an ardent young professor who was strongly anticlerical, as Ribot seems to have been already by this point. It became worse when Ribot was transferred in 1868 to Laval, fifty miles east of Rennes. Laval was a much larger town and closer to home, but it boasted a bishop and three seminaries. In Ribot's estimation, it was "rotten with Catholicism."[10] Having already been denounced by the chaplain of the lycée at Vesoul, Ribot anticipated more travails in Brittany, known for its religiosity.

Worse still, Ribot was attracted to positivism. Having passed the *agrégation* in August, he was using his spare time to read Taine, John Stuart Mill, and Herbert Spencer, the latter whose *Principles of Psychology* he decided to translate. He was developing an admiration for English positivism, and the notes he took became the foundation for the study of *English Psychology (La psychologie anglaise contemporaine)* he would publish in 1870. This work was designed to introduce to the French public associationist psychology, the experimental or a posteriori school, "as it is unknown, or very nearly unknown, in France."[11] It contained chapters on each of the chief representatives of that school.[12] He excluded from his account members of the a priori school, such as William Hamilton and William Whewell, presumably because the sort of psychology they represented was what the French already had in the form of eclecticism. And lest there be any doubt about the implications of this school for eclecticism, Ribot prefaced his exposition with a polemical introduction on the nature and future of psychology.

He began by retracing the familiar story of the separation of the individual natural sciences from philosophy, which had originally been the universal science. The condition of this emancipation was that a discipline abandon metaphysical questions—of first causes and principles, of fundamental concepts, et cetera—and restrict itself to the explanation of facts on the basis of whatever axioms seem to work.

Thus then, everywhere and always, particular sciences which have a special object are only constituted by leaving a balance of unsolved questions aside

at their outset. . . . They must start from some postulate, from certain rational or experimental truths; they must not stop at questions of principles,
and they must leave discussions to philosophy. . . . This is the absolute condition of their existence as exact sciences capable of progress. (ibid., 9)

Ribot acknowledged that "to those who judge them as philosophers,
their point of departure is ruinous, ill established, not discussed; but if
philosophy condemns, experience absolves them. . . . Debates on principles prevent arrival at consequences" (ibid., 9–10). This applies to the
moral as well as the physical sciences, and Ribot cited the examples of
linguistics and political economy as moral sciences that had begun to
progress by leaving insoluble questions behind.

What will philosophy be when all possible sciences are detached from
it? *"It will be metaphysics and nothing more"* (ibid., 10). But this means
that philosophy will not be a science, since metaphysics deals with the
kinds of questions that the sciences cannot address. In words reminiscent
of Renan, Ribot maintained that philosophy would be closer to art and
poetry than to science. This does not mean that it is illegitimate or that
it will disappear. Taking issue with Comtean positivists, Ribot thought
that metaphysics would remain a necessary and noble occupation of the
human mind. These definitions of science and philosophy obviously
conflict with the eclectic view. Ribot called the "ordinary sense" of the
word "philosophy"—including psychology, logic, morals, and metaphysics—"a rather incoherent assemblage of four or five sciences" in
which no unity could be found (ibid., 5–6). This incoherence resulted
from the fact that the moral sciences had not yet completely freed themselves from metaphysics.

Such a characterization of the nature and future of the moral sciences
carried obvious implications for psychology. Ribot criticized the eclectic
definition of psychology as the study of the soul through the means of
consciousness on several grounds. First of all, it confuses metaphysics
with science by postulating as its object an entity called the soul. Next,
by making introspection its exclusive method, it effectively restricts itself
to the souls of philosophers, "so that, finally, psychology, instead of
being the science of psychical phenomena, has simply made man, adult,
civilized, and white, its object" (ibid., 17). Ribot agreed that introspection—the "subjective" method—was a necessary condition of scientific
psychology, and he specifically denied the Comtean thesis that the study
of mental phenomena could be reduced to brain physiology. Nevertheless, he argued that the subjective method was not a sufficient condition
for science; the "objective" method—employing comparison and induction—must control and verify the testimony of consciousness. Finally,
eclectic psychology reifies the similarities among different types of mental phenomena into faculties. It assumes that because we can apply the

same name to different phenomena, there must be something corresponding to this abstract name that acts as an underlying cause.

If these are the flaws of eclectic psychology, then scientific psychology must avoid them. It must restrict itself to phenomena, so that, paradoxically, psychology cannot be the science of the soul. It must be objective as well as subjective, meaning that it should study mental phenomena in the "facts which translate them, not in the consciousness which gives them birth" (ibid., 25). And it must be comparative, studying the manifestations of mental life across sex, age, cultures, and species. The English of the a posteriori or experimental school have done the most to advance this conception of the scope and method of psychology.

They have also made substantive contributions to the content of scientific psychology. The guiding ideas of English psychology gleaned from Ribot's expositions may be gathered under three general headings: the reduction of mental processes to the interaction of elements; the evolution of individual and specific mental capacities from simple organic reactions by small increments; and the necessity of relating psychological manifestations to their physiological conditions. Each of these is an implicit rebuttal of eclectic doctrine. Associationism contradicts the eclectic assertion that the fundamental categories of thought cannot be derived from experience. Evolution is inconsistent with the assumption that higher forms cannot be derived from lower. And the search for physiological conditions of mental phenomena undermines the eclectic dualism of mind and body. No one representative of the English school combined all these principles, in Ribot's opinion. The Mills championed associationism, but they did not have the concept of organic evolution. George Lewes had the most direct grasp of physiology, but his psychological writings were not extensive enough to constitute a complete doctrine. Alexander Bain best represented the use of physiology in psychology, but he too did not make full use of the comparative implications of evolutionary theory. On the whole, Spencer received the most praise from Ribot.

This raises a problem, for of all the psychologists surveyed, Spencer is the one Ribot most explicitly calls a philosopher rather than a scientist. Ribot says of Spencer that he epitomizes the philosophical mind, which ". . . is a certain manner of thinking, not acquired, but developed by culture, which has its characteristic traits, just like the poetic or the scientific mind. If there be a definition which expresses its qualities and its defects, which may be accepted by everyone, and agreed to by all the schools, it appears to be the following:—It is the mind which generalizes" (ibid., 125). This is certainly a fair characterization of Spencer, but one might ask how it compares with the definition of philosophy as metaphysics and what it means for the status of scientific generalization. In the passage that follows, Ribot identifies generalization with synthesis, implicitly associating analysis with science. This conveniently ignores the fact that scientific laws are also generalizations. With Spencer, although his powers of analysis are great, synthesis runs beyond analysis. His generalizations surpass the domain of verified fact, and hence become philosophical.

Ribot does not consider Spencer metaphysical for this reason. Comparing Spencer's theory of evolution to Leibniz's idea of the continuum and Hegel's dialectic, he notes that all are hypotheses, in that they go beyond the facts. Leibniz's theory, however, was "a view of the future by a genius, an hypothesis not then verified by facts," and Hegel's was "completely subjective, . . . boldly bending facts to its a priori conceptions," whereas Spencer's theory rose from the study of the sciences and was verified by innumerable facts. As important, however, was that Spencer considered his theory to be *only* a hypothesis, subject to verification or refutation, rather than an a priori construction immune from empirical confirmation. His philosophy was therefore "free from metaphysics, and supported by the experience of nearly two centuries" (ibid., 127).

There would seem to be several different criteria here for distinguishing between metaphysical and scientific hypotheses. One would lie in the origin of the hypothesis, whether it was objective or subjective, suggested by the facts or imagined by the mind. In this sense, Leibniz and Hegel were originally metaphysical. But another criterion seems to be the number of facts actually covered or contradicted. In this sense, Ribot suggests that Leibniz's theory has become scientific because it has turned out that his hunch was correct. A third criterion consists of whether in principle one accepts facts as evidence for or against one's hypothesis. In this sense, Ribot implies not only that the dialectic has not proved to coincide with the facts, but that Hegel would not accept *any* fact as a refutation of his theory.

Aside from the fact that these criteria are not necessarily consistent with each other, the first two allow a given hypothesis to pass from the status of metaphysical to nonmetaphysical and back again. Only the third criterion imposes a difference of kind between metaphysical and scientific hypotheses. It is the basis on which Ribot excluded the a priori school. And the way in which Ribot couches his characterization, the claims of the a priori school are based less on their inherent exemption from empirical falsification than on their stubborn refusal to recognize it. Even here, then, the demarcation between the two kinds of hypothesis is not as sharp as Ribot would have it. Moreover, "scientific" philosophers have the privilege of going beyond the facts without becoming metaphysical. Whereas in his introduction Ribot divided knowledge between science and metaphysics, in his exposition he uses an intermediate category of "philosophical" between science and metaphysics, a category including hypotheses that under certain conditions are scientific and under others metaphysical.

The inconsistency in Ribot's use of "philosophy" is not the only troubling aspect of his presentation. Ribot praised the English psychologists for the wealth of observations upon which their theories were based, above all Bain, whose work Ribot considered "the most complete repertory in existence of exact and positive psychology, placed in the light of recent discoveries" (ibid., 252). Ribot included examples of some of the facts cited in favor of this school, but he provided no definition of a fact, much less a good or scientific fact, nor did he seek to examine the facts on which these theories were based. One must therefore take on faith that these facts are carefully gathered and critically examined. Moreover, Ribot's notion of scientific method, of the connection between fact and theory, is somewhat vague. Although he constantly invokes the experimental method, his definition of it seems to lack rigor. " . . . Collecting an innumerable multitude of facts around some fundamental principles, and submitting principles to verification by facts, is a truly experimental method" (ibid., 211). Ribot does not always distinguish between observation and experimentation, nor between a confirming instance and a demonstration.

But Ribot came to associationist psychology as a dissatisfied academic philosopher, not as a critical scientist. The English met the criteria of natural science he had learned from his academic philosophy: they accumulated facts, they made generalizations, and they verified their generalizations against other facts. They seemed to have more facts and more verifications, and they paid more attention to the natural sciences than did the eclectics. Moreover, these general criteria did not contradict the standards of science prevalent in the natural history of the day. He could reasonably conclude that their theories were more scientific.

Despite the militant tone of the introduction, *English Psychology* was politely received by spiritualist philosophers. Ribot had to admit that he was "well received in Paris, even by Caro, Janet, Lachelier, etc. . . ."[13] The book enjoyed even greater success with those favorable to its subject and was quickly translated into a number of languages including English, in which version it went through a large number of printings. Appearing in the same year as Taine's book *On Intelligence,* it played a large role in making associationist psychology and evolutionary biology important forces in French philosophical discourse.

Ribot did not rest long on his laurels. He soon began to read German philosophers and psychologists, a program that would eventually produce a book on *German Psychology of Today*. But for the immediate future, it was more important for him to complete his dissertations, the topics, methods, and data for which were drawn from his work on the English. For his Latin dissertation he wrote on the eighteenth-century associationist David Hartley, and for his French dissertation he chose the problem of *Psychological Heredity (L'hérédité psychologique)*. One can judge what Ribot learned from the English better through this work than from the earlier study, since his thesis required him to appropriate and defend rather than simply summarize.

Psychological Heredity argued that the biological law of heredity, "in virtue of which all living beings tend to reproduce themselves in their descendents," applies to all mental phenomena.[14] Ribot divided his work into four parts—on the facts, laws, consequences, and causes of heredity. First, drawing on a number of literary and historical sources, and especially on the works of Francis Galton and Alphonse De Candolle, Ribot marshaled a large number of instances in which not only instincts but peculiarities of emotional temperament and intellectual ability were passed from one generation to the next. In other words, not only could general characteristics of the species—in human language, reason, sociability, et cetera—be transmitted, a proposition that was not controversial, but so could individual differences of personality and capacity, a proposition that went against the beliefs of many spiritualists that bodies are governed by heredity but souls differ only by education and circumstance. Moreover, Ribot maintained that acquired characteristics could be inherited, so that the habits of the parents could be transmitted to their offspring. This raised the possibility of an evolutionary change in the species, a proposition Ribot put forth only as a hypothesis, but one that he clearly favored. If this hypothesis proved correct, then heredity played not only a "conservative" role, ensuring the replication of the species according to the same type, but a "creative" role, accumulating changes in the psychological makeup of the species (ibid., 37).

In the second part, he argued that these facts demonstrated that "heredity is the law, nonheredity the exception," in the transmission of psychological traits (ibid., 167–68). The characteristics of the parents would necessarily be reproduced, unless accidental causes such as blending, reversion, or arrested development intervened. Ribot specifically opposed the theory that alongside the law of heredity there is a law of "innateness" that would be an internal source of spontaneous variation.[15]

In the third section, on consequences, Ribot argued that on the hypothesis of evolution, the natural processes of adaptation and heredity could account for the highest and most individual properties of the human mind. He presented Spencer's theory that the categories, by most standards the highest forms of human thought, were the result of the collective experience of the species as a possible solution to the debate between rationalists and empiricists.[16] He also asserted that the self, *le moi*, the core of human individuality and character, was subject to the laws of heredity. The self is simply the sum of intellectual, emotional, and physical characteristics of an organism, and since all of these parts are determined by heredity, their sum is as well. "The hypothesis of a principle of individuation distinct from phenomena is one of those that the new psychology tends to eliminate. When one has considered in the individual his intellectual activity, his affective life, lastly that reverberation of the life of the body that serves as the basis of all the rest, it is difficult to see what there would be beyond this to look for" (ibid., 324). Ribot claimed that most historians and politicians had overlooked the power of psychological heredity in human affairs and had therefore exaggerated the effects of the environment, even though as a Lamarckian (more accurately, a Spencerian) he thought that over generations the environment could alter the organism.[17] Although the slow rate of evolution might dismay some meliorists, the fact of evolution suggests that intelligent control of reproduction could improve the human species.[18]

In the final section, he argued that psychological heredity could only be explained on the assumption that psychological phenomena depend on physiological conditions. It is these physiological conditions that are passed on through the reproductive process, not some mental substance. When the new organism is sufficiently developed, the psychological phenomena once again appear. In other words, mind arises out of matter. The reverse hypothesis, that physiological conditions depend on psychological, runs into two difficulties, one conceptual, the other empirical. Conceptually, Ribot claimed that the idea of one spiritual substance engendering another was unintelligible. Empirically, "experience shows that mental development is, everywhere and always, subject to organic conditions, and . . . there is nothing in experience to demonstrate that the reverse is generally true" (ibid., 397). Ribot thought that the burden of proof should rest with the idealists, since the other camp has the evidence on its side. In a certain sense, though, Ribot considered the entire mind/body problem to be misconceived. Mind and body, originally just two different sets of phenomena, had become hypostatized in philosophical argument into a dualism of antithetical and irreducible substances. "Contemporary science tends toward a unitary doctrine, monism. It considers the manifestations of psychic life to be merely one example, the highest and the most complex, of vital activity" (ibid., 399).

To assert that science leads to monism is to raise again the question of how one distinguishes between science and metaphysics. Ribot is in general extremely careful to distinguish between facts, laws, hypotheses, and causes. Facts are the material of science, but "science begins only with the search for laws" (ibid., 157). Ribot thinks that the law of psychological heredity is a legitimate induction from experience, not hypothetical, even though he admits that it is an empirical generalization rather than a scientific law.[19] Hypotheses are theories that are not currently capable of being confirmed or disconfirmed. Ribot is careful to caution repeatedly that he considers evolution to be only a hypothesis. Causes are also part of science, Ribot maintains, siding with Mill and against Comte. "If science begins with the search for laws, it ends only with the determination of causes" (ibid., 389). Nevertheless, he himself admits that the issue of causes—in this case, whether mind causes body or body causes mind—lands him in the middle of metaphysical hypotheses. Even though he qualifies his discussion of monism by noting that while he is "leaning on experience," he is also "surpassing it a little," he does not refrain from taking that step (ibid., 398).

In part this can be excused by saying that, after all, it was a dissertation in philosophy, and therefore Ribot was allowed, even obliged, to address philosophical issues. But hypotheses pervade *Psychological Heredity,* guiding the selection of facts, the formulation of laws, the drawing of consequences, and the determination of causes. Moreover, Ribot claims that science can decide between metaphysical hypotheses like monism and dualism, if not actually then at least potentially. Just as in his work on English psychology, so in his dissertation the distinction between science and metaphysics is drawn more clearly in theory than in practice.

Such was also the judgment of his dissertation committee. Although Ribot gave his dissertation to his committee in May of 1872, it was over a year before he defended it. Ribot wrote to Espinas that his advisers were "favorable enough" toward it but that they feared that the public presentation of such a positivistic thesis at the Sorbonne would cause a scandal.[20] Despite such trepidation, they eventually allowed him to defend it in June of 1873 and awarded him the doctorate. Their willingness to accept such a militantly antispiritualist tract represented a tremendous change from the days of the July Monarchy and Second Empire, when such temerity on the part of a candidate would never have been granted a public hearing, much less official approval. The more liberal atmosphere of the Third Republic may have taken some of the governmental pressure to enforce spiritual orthodoxy off of the members of the committee. Twenty years of work by eminent friends of positivism such as Vacherot, Taine, and Renan surely softened the ground by making the idea of naturalistic moral sciences more respectable. But this does not explain why they would approve a work with whose ultimate theses they could not agree.

The answers may be discerned in some of the responses of the members of the *jury de doctorat.* From the summary in *Le Temps,* it appears that Caro took the leading part in the discussion.[21] He and Adolphe Franck in fact did *not* approve the ultimate implications of Ribot's work. Caro affirmed the increasing ascendency of spontaneity over heredity as one approached the higher regions of mental life. He denounced Ribot's attempt to explain the self by the organism and warned that "the abyss of materialism" lay in such a position. Other members of the jury, however, seem to have been persuaded that psychological heredity was at least a real phenomenon, whatever they thought of Ribot's interpretation. Charles de Rémusat was reported to have said that heredity is a fact, although one could argue over its causes and consequences. In other words, the committee of philosophers approved a dissertation in philosophy at least in part on the basis of the "scientific" facts rather than on the "philosophical" interpretation. This allowed them to avoid the charge of either materialism or illiberalism.

Caro expanded his criticisms in his presentation of *Psychological Heredity* to the Académie des Sciences Morales et Politiques not long after. Although he praised the "abundance of facts analyzed, the extent of the learning (*science*), and the sometimes specious rigor of the method" in Ribot's work, he did not find Ribot's data conclusive. First of all, he thought that Ribot allowed his preference for the evolutionary hypothesis to guide his arrangement of facts. He also found Ribot's examples more anecdotal than compelling, his examples of illustrious families vitiated by loose definitions and glossed-over exceptions.[22] Caro believed that many of the cases could be as easily explained by environment (milieu) as by heredity. In terms of the law under which Ribot subsumed the facts he presented, Caro found no reason to assume that it must hold with equal rigor for higher as for lower functions, nor to eliminate the auto-transformation of the self by its own efforts. Only Ribot's idée fixe prevented him from acknowledging that these facts could be interpreted otherwise. It is not the case, then, that eclectics simply allowed the appearance of science to blind them to defects of method. They were capable of judging the factual basis of their positivist opponents. They were also able to show that despite claims to the contrary, metaphysical positions did underlie scientific findings. This was not a problem for eclectics, who thought that science and metaphysics were connected anyway, but it did raise a problem for someone like Ribot who tried to make a sharp distinction between them. Despite his opposition to the thesis, Caro approved it because he respected the quality of the work and the talent of the author. In other words, professional competence could be used as a means of certifying a candidate without endorsing his ideas.

The discussion that followed Caro's presentation reiterated and amplified the same themes raised at Ribot's defense. This is not surprising, since some of the same people were involved in this debate. Franck maintained that there is no psychological heredity at all. He denounced Ribot's hypothesis as "this fatalism, this empiricism, this materialism that takes away men's liberty," and he presented as a counterexample "the semitic race, first pastoral, then warlike, then agricultural, then commercial and industrious. . . . This race has passed through all the social and political forms."[23] As a Jew as well as an eclectic, Franck saw in spiritualism the best defense against racism. His position was not shared by the other discussants, Nourrisson, Rémusat, and Marie-Louis Parieu, who seemed convinced that psychological traits could be inherited, although they denied that *all* such traits were, as they denied that heredity negated free will.

With his dissertation in hand, Ribot pressed for an appointment in higher education. But although he had the required doctorate and, it seems, the sincere support of his faculty mentors, the administration considered him too controversial to be rewarded with a position that would in its eyes—and in the eyes of the church—imply approval of positivism.[24] When it became evident that he would not receive a university position in the near future, Ribot requested a leave (*congé*) with no intention of returning until he was offered such a position. His ties with official philosophy were thereby attenuated almost but not quite to the point of breaking.

Uneasy Truce with Philosophy

Even before his defense, he had become disenchanted with his academic career and had taken a year's leave, during which he moved to Paris to get the education in science he had not received at the lycée or at the Ecole Normale. He enrolled in the Faculty of Medicine and took courses at the Faculty of Sciences, although he also studied philosophy at the Sorbonne under Janet, Caro, and Lévêque. He began anew the translation of Spencer's *Principles of Psychology,* with which he persuaded Espinas to help him.[25] As important as these projects, were the contacts with influential individuals that he was able to make as a result of his residence in Paris and his newfound notoriety. After the publication of *English Psychology,* he had entered into correspondence with a number of leading positivists, both English and French, and not long after he began to study the Germans he was in contact with Wundt. When *Psychological Heredity* appeared, he began to get invitations to the soirees of eminent Parisians such as Taine. These contacts were to be extremely valuable for his future career.

The first publication to appear during this period, and also the first fruit of his German studies, was a small volume on Arthur Schopenhauer that appeared in 1874. This choice of subject is not as strange as it seems for an antimetaphysical psychologist, and it makes perfect sense for someone in rebellion against academic spiritualism. For Schopenhauer was the antithesis of eclectic philosophy. To the intellectualism of academic philosophy, Schopenhauer countered with the thesis that representation was illusory and will alone real, a position that also undercut spiritualist dualism. To those spiritualists who would argue that they too found in human will the primary evidence of spiritual reality and freedom, Ribot could point out that will for Schopenhauer was blind and completely determined. To the duty-based optimism of the one school, Ribot could oppose the quietist pessimism of the other. Not the least of Ribot's attraction to Schopenhauer lay in the fact that the German pessimist despised academic philosophy. Ribot did not accept Schopenhauer's philosophy: he did not agree with his pessimism, his idealism, or his metaphysical character. But he did admire his emphasis on the primacy of will and passion, and he thought science could accomplish what Schopenhauer tried to do through metaphysics—dispel the illusions of consciousness and put the human species in its proper place in nature.[26]

Ribot's publications on English psychology and Schopenhauer helped expand the horizons of French philosophical discourse, bringing formerly unknown or excluded points of view into the universe of polite discussion. He ensured that this expansion would continue when he founded the *Revue philosophique de France et de l'étranger,* the first issue of which appeared in January of 1876. He announced to Espinas in April of 1875 that his publisher, Germer Baillière, had agreed to found a journal that would include original articles by French and foreign philosophers as well as critical reviews, and that he would be its editor.[27] His extensive connections inside and outside the university enabled him to assemble a cast of contributors that during the first year included spiritualists such as Paul Janet and Jules Lachelier, positivists such as Taine, and foreigners such as Wilhelm Wundt, Spencer, Mill, and Lewes. In addition to such luminaries, he enlisted younger and lesser-known philosophers, most of them former classmates from the Ecole Normale, to write the reviews. The *Revue* was an immediate and unqualified success.

It was also quite a *coup* for a young and disaffected ex-academic philosopher. The *Revue philosophique,* which has been published continuously ever since, was the first French journal of philosophy that was not the organ of a particular school.[28] Ribot explicitly intended it this way, declaring in the preface to the first number that "the *Revue philosophique* of which we are beginning the publication proposes to be open to all schools."[29] Since it appealed not to ideology but to competence and contribution to the field as criteria for inclusion, the *Revue* can be viewed as the first professional journal of philosophy in France.[30] As such, it quickly became the primary outlet for academic philosophers, who until then did not have their own journal.[31] This made Ribot a powerful figure in academic philosophy.

He did not use this power to discriminate against spiritualists or to impose demands on them. Indeed, he held scrupulously to his declaration of editorial neutrality, and even the most ardent idealists acknowledged his fairness.[32] Most likely he did not think it necessary to work actively against academic spiritualism. By giving positivist philosophy a forum beyond the ranks of the converted, and by forcing spiritualism into juxtaposition with it, he undoubtedly assumed that the philosophy of the future would win on its own merits. Ribot did use his editorial discretion to weigh the *Revue* in favor of the partisans of scientific philosophy. During the first year, of the fifteen authors of original articles whose philosophical inclinations could be readily identified, ten were broadly positivist, four spiritualist, and one neo-Kantian. Nevertheless, if the *Revue* was the forum for the positivist assault on official spiritualism, it was also the arena in which spiritualist rebuttals appeared. The openness of the journal could work both ways, to the point that, as Meletti Bertolini has argued, the new *Revue* became a site of dialogue and integration between scientific and philosophical approaches to psychology as much as a site of confrontation.[33] And the coexistence of different schools gave the *Revue* a decidedly eclectic character that made it more acceptable to academic philosophers, despite Ribot's protestations to the contrary.[34]

Thirard has noted the irony of a philosophical review founded by someone who claimed to emancipate himself from philosophy and to advocate a purely scientific psychology.[35] It is true that the lion's share of the *Revue*'s articles were on psychology, and that the majority of these came from the advocates of scientific psychology.[36] But by placing these articles in a review explicitly dedicated to philosophy, the implication could hardly be avoided that psychology was a part of that discipline. Far from becoming "metaphysics and nothing more," philosophy in the *Revue* virtually reproduced the extensive and heterogeneous character of the lycée syllabus. Ribot divided the subject matter of philosophy into five categories: the *connaissance théorique de l'homme* (including psychology, logic, and aesthetics), morals, the philosophy of science, metaphysics, and the history of philosophy.[37]

The alternative would have been to establish a journal of psychology *tout court,* from which Ribot could have banished metaphysics and eclectic psychology, but it is likely that such a venture would simply have underlined the fragility of scientific psychology. Most of the articles on psychology came either from philosophers like himself, other literary figures such as Taine, or foreigners. There were few contributions from French doctors, physiologists, and psychiatrists, although Ribot clearly hoped to attract them. Nevertheless, in 1876 he lacked the medical and scientific credentials to lure them into such an enterprise, nor did the scientific community yet find such a venture attractive enough to abandon established journals like the *Annales médico-psychologiques* or the *Journal de physiologie*. For the time being, the best he could manage was a compromise grafted onto philosophy rather than medicine, physiology, or biology.

The tactical compromise with academic philosophy represented by the *Revue philosophique* did not prevent Ribot from continuing to criticize metaphysical spiritualism or to promote an independent and scientific psychology. In an article on "Philosophy in France" written for the English journal *Mind,* Ribot dismissed eclecticism as "a doctrine without originality, and standing absolutely aloof from the discoveries of science."[38] He denounced its psychology as superficial, literary and its metaphysics as timid, commonsense deism. Frequently citing Taine's *Classical Philosophers,* Ribot held that the continuing power of eclectic spiritualism depended entirely on its institutional position in the lycée and in the faculties. He did acknowledge that the spiritualists were trying to adapt themselves somewhat to more contemporary doctrines in response to their critics, which he divided into two principal groups: the "mystic spiritualism" and the "Positivists." The former, including Ravaisson, Lachelier, and Alfred Fouillée, were characterized by a greater reliance on the philosophy of Maine de Biran, who stressed the primacy of spiritual activity over abstract concepts in the self-definition of the soul. Ribot characterized them as subordinating the sciences to the moral point of view and of "preferring the methods of art to those of science" (ibid., 372). They lacked method and indulged in obscurantism. For the latter reason, they had achieved little recognition outside the academy, where however they constituted an enemy from within that threatened eclecticism. Ribot divided the positivists into the orthodox followers of Emile Littré and the wider outlook common to men of science, which was inspired more by "contemporary English philosophy" than by Comte (ibid., 374). In this group Ribot included Taine, the chemist Marcellin Berthelot, and the physiologist Claude Bernard. He thought that this philosophy had gained the support of most scientists, and he declared his hope that the *Revue philosophique* could help further this sort of speculation, grounded in "facts, thorough study, and scientific culture" (ibid., 386).

Two years later, Ribot published his *German Psychology,* which like his earlier study of English psychology had grown out of his attempts to educate himself in this tradition. In this work, he presented the psychology of Johann Friedrich Herbart, Gustav Theodor Fechner, E. H. Weber, and Wilhelm Wundt for the first time to the French public.[39] Ribot's introduction was even more scathing in its treatment of philosophical psychology than his preface to *English Psychology* had been. The old psychology is "doomed," "radically false," "old and feeble," and, piling up contradictory epithets, "childish."[40] "In vain its wisest representatives attempt a compromise, and repeat in a loud voice that it is necessary to study facts, to accord a large share to experience. Their concessions amount to nothing" (ibid., 2). Metaphysics and psychology are not only separate, they impede each other: " . . . talent in metaphysics bears an inverse ratio to talent in psychology. . . ." (ibid., 3). No compromise is possible between two such antithetical habits of mind.

The vehemence of Ribot's denunciation may have something to do with the greater contrast he found between eclectic and German psychology than between eclectic and English psychology. The experimental rigor of the Germans caused him to reevaluate English psychology. Whereas in his earlier study he had called associationism explanatory compared with purely descriptive faculty psychology, now he characterized English psychology as descriptive compared with the truly explanatory German studies. English psychology was above all natural history, a systematic description and comparison of mental phenomena. Although the English increasingly made use of physiology, their method was still essentially observational rather than experimental, continuous with the ideological analysis of Locke and Hume. The Germans, on the other hand, "accord little place to description" (ibid., 12–13). Instead of collecting a wide range of observations with a view to classifying them and making inductions from them, psychologists like Wundt took only a narrowly defined problem, such as the least perceptible difference between two degrees of sensory stimulus, and submitted it to rigorous experimental control, manipulating the variables to isolate the causal mechanism. In this way they were able to define the phenomenon under investigation and to identify its causes much more precisely than the British.[41] The Germans had advanced psychology from natural history to natural science.

Although the use of experimentation was the decisive contribution of the Germans, Ribot did not generally call their discipline experimental psychology. Instead, he preferred the term *physiological* psychology, and the difference is significant. Experimentation was made possible only by the assumption that "every psychical state is invariably associated with a nervous state" (ibid., 6). This assumption, which in a general form was an underlying theme of *English Psychology,* became more sharply defined in the study of German psychology. The success of psychophysics also reinforced the monism already present in Ribot's earlier works, since it allowed him to define psychological and physiological manifestations as two faces or aspects of a single phenomenon. This solved for him the problem of how the mind related to the body, for under this hypothesis there was no point at which physical motion suddenly disappears into mental substance, to reappear as suddenly with a new determination impressed upon it by mind. At every point in the chain from sensory stimulus to motor response, physical action leads to physical reaction. The difference between physiology and psychology was simply this: "Nervous process in its simple aspect belongs to physiology; nervous process in its double aspect belongs to psychology" (ibid., 8). Ribot denied that this definition absorbed psychology into physiology; he maintained that psychology could no more be reduced to physiology than physiology to chemistry.

Nevertheless, the postulate of psychophysical parallelism created the possibility of experimentation and redefined the nature of explanation in psychology. Using this postulate, one could fix the fleeting data of consciousness and measure them by their physical counterpart. The method of "concomitant variations" could be fruitfully used to see how psychic phenomena varied with changes in their physical causes. Moreover, to explain in psychology now meant to find the particular physiological structure or event with which a psychological phenomenon was connected. Laws relating psychological phenomena to one another on a purely psychological level—such as the law of association—Ribot no longer considered explanatory, as he had in his earlier work.

This is not to say that either Ribot or the Germans ignored the need for observational or comparative data that was not of a strictly physiological nature. On the contrary, Ribot gave one of the earliest accounts in French of the Völkerpsychologie practiced by German scholars such as Heymann Steinthal, Theodor Waitz, and Wundt himself. He considered the study of the collective products of the human species—such as language, myth, religion, and custom—to be a natural and necessary extension of physiological psychology, in that it provided objective data for the higher mental processes that were at present inaccessible to laboratory technique. He even admitted that collective products often presented characteristics not present in the individual members of the group, although he found the concept of the Volkgeist somewhat "mystical."[42] In general, he preferred the English school of E. B. Tylor, John Lubbock, John McLennan, et cetera, which avoided such flights of speculative fancy.

Academic philosophers reacted to *German Psychology* in much the same way as they had to Ribot's earlier works. Although some were appalled at what they considered its materialism, most agreed in broad outlines with the review by Thomas-Victor Charpentier, a Parisian lycée professor and eclectic whose comments appeared in Ribot's own journal.[43] Charpentier praised the "truly scientific method" Ribot exhibited in his discussion of German psychologists. He even agreed that the physiological method was a legitimate and necessary part of psychology, one that the eclectics had hitherto neglected. But he denied that it was the whole of psychology. He also denied that one could completely separate psychology from metaphysics. He found Ribot's own admission that "it is perhaps a necessity inherent in all psychology, even the experimental, to set out with some metaphysical hypothesis."[44] Moreover, Ribot himself acknowledged that in most cases, he had to select the experimental psychology out of the metaphysical context that had inspired and guided it. Thus the works of Herbart, Fechner, Steinthal, and Hermann Lotze were impregnated with metaphysics in a way that contradicted the claim of Ribot that psychology could and should be independent. Unlike Ribot, who simply condemned the old psychology, Charpentier and other eclectics acknowledged the merits of the new psychology but maintained their own rights as well. They refused to view the issue in the terms in which Ribot described it, as an either/or, all or nothing struggle.[45]

Although Ribot was much impressed by the Germans, he made no attempt whatsoever to conduct the kind of experiments they had created. The series of monographs on which he embarked after the publication of his study of German psychophysics made little use of the data or methods he had summarized so brilliantly. Their principal effect seems to have been to reinforce the connection between psychology and physiology he had already received from the English, and to heighten his appreciation of the rigor of experimental method. At first glance, these monographs, published between 1881 and 1885, do not seem to bear much relation to the English psychologists, either. Rather than concentrating on comparative and normal development, Ribot's next studies used the evidence of mental pathology as natural experiments to expose the mechanisms underlying normal psychological phenomena. In turning to this method, Ribot was undoubtedly influenced by his contact with French physiologists and psychiatrists. The science of physiology was in large part founded on the principle that the pathological and the normal differ by degrees rather than kinds, and that the one can illuminate the other.[46] Ribot had trained with some of the eminent physiologists of his day, and he can hardly have failed to pick up this precept from them. In addition, French psychiatry had a long tradition of observing and classifying psychological dysfunctions.[47] Ribot had read one of the premier French alienists, Moreau de Tours, as early as 1870, and he made extensive use of the reports of such psychiatrists in his studies on psychological diseases. For these reasons, Ribot has often been placed in the "essentially French" tradition of psychopathology.[48]

Nevertheless, Reuchlin is undoubtedly right when he claims that this native tradition had less influence on Ribot that did its English counterpart.[49] Although Ribot found much information in the writings of French psychiatrists, in his opinion their professional preoccupation with identifying and treating disease prevented them from relating their findings to the more general issues of scientific psychology.[50] Moreover, French psychiatrists and physiologists were less likely than their English and German colleagues to be influenced by the theory of evolution. Finally, although the English psychologists did not make extensive use of the pathological method, English neurophysiologists were at the forefront of such research. For these reasons, Ribot leaned more on foreign than on native sources. T. H. Huxley, William K. Clifford, William B. Carpenter, and Henry Maudsley are cited much more often than Claude Bernard, Charle Philippe Robin, Félix Vulpian, Jules-Bernard Luys, and Louis-Pierre Gratiolet. John Hughlings Jackson provided Ribot with the concept of mental "dissolution" that guided all of his studies.[51] In short, Ribot remained faithful to his early English inspiration.

The topics Ribot chose for his studies—memory, will, and personality—were all important parts of the eclectic syllabus. By treating them from the method of the "new psychology," he was sure to trample on eclectic doctrine. Although Ribot claimed that his method combined "teachings of physiology . . . with those of intuitive perception," the priority and explanatory power with which he endowed physiology brought him close to epiphenomenalism.[52] For example, he analyzed memory into three related events: conservation, reproduction, and localization in the past. Only the first two were necessary; they were also biological in nature. Only the last was psychological; it was also the least necessary, the most unstable and fleeting.[53] Ribot analyzed will into two components: "the state of consciousness, the 'I will,' which indicates a situation, but which has in itself no efficacy; and a very complex psycho-physiological mechanism, in which alone resides the power to act or to restrain."[54] In his effort to oppose the spiritualist emphasis on consciousness, he gave it as little role as possible in his psychology. Only in his discussion of personality did he relent and acknowledge a causal role for consciousness.[55]

Ribot also used the "new method" to undermine the metaphysical entities with which he thought spiritualism burdened psychology, and to replace them with concrete elements. He studied the phenomena of partial amnesia to show that memory is not an independent faculty but a complex assemblage of different memories attached to different organs.[56] He emphasized the illusory nature of the conscious will in order to call attention to the complexity of organic will. An act of will is not a simple and absolute determination of the self; it is the resultant of the character—in the last analysis, of the entire structure and history of the organism.[57] Moreover, "the unity of the Me is not, as taught by the spiritualists, the unity of one entity manifested in multiple phenomena, but the coordination of a number of states that are continually arising, and its one basis is the vague sense of our own bodies. . . ."[58] As in his dissertation, Ribot located the principle of individuation in the body, not in a metaphysical soul.

Through the study of the pathological manifestations of these various phenomena, Ribot was able to formulate a general law that applied to all of them—in disease, "the functions acquired last are the first to degenerate."[59] This "great biological law" is the law of dissolution, first formulated by Hughlings Jackson, and based in turn on the evolutionary schemata of Spencer. It assumes that the attributes of an organism developed most recently are doubly unstable, both because they are not yet firmly fixed in the structure of the organism and because they are likely to be the most complex and delicate. Hence they are the most likely to break down under the pressure of injury or toxin. Applied to memory, this implies that recent memories are the first to be forgotten in amnesia, and this principle has since been called "Ribot's law." Applied to will, it means that those determinations based on the higher functions, such as reason, are the most likely to be deranged. Ribot organized his data to show the progressive nature of mental degeneration.

Despite the talent with which Ribot analyzed and organized his data, his method still amounted to "collecting an innumerable multitude of facts around some fundamental principles." Moreover, these facts were not gathered from his own observations, but from the books of others. In this sense, it was comparable to the ethnologies being written in the armchairs of Paris and London, from information provided by people who were on the scene but who often had different motives from their metropolitan synthesizers. Ribot was kept from the field by his lack of medical training, which denied him access to the kinds of phenomena on which he based his generalizations. He often expressed regret about this, and he encouraged his protégés to study medicine, so that they would have opportunities he did not. What he *was* able to do was to bring together several different traditions of research, all of which bore in some way on his conception of scientific psychology. In this sense, he was a scientific eclectic.

The secondhand character of Ribot's monographs did not prevent them from being seen as scientific, even by scientists, nor from enjoying considerable popularity.[60] They went through many editions in several languages. Academic eclectics met these studies with the same general strategy as the earlier works—they acknowledged the "scientific" character of the works, although they objected to many points of both fact and theory, and they lamented the excesses of Ribot's assumptions.[61] Indeed, these monographs did much to establish Ribot's scientific credentials and reputation. He was no longer considered primarily an expositor of the doctrines of others; he had his own theory and a record of accomplishment in the scientific psychology he advocated.

RETURN TO THE ACADEMY

With such credentials, Ribot began to think about establishing an academic position that would reflect the new psychology. Technically he was still on leave rather than separated from the Ministry of Public Instruction. He thought first of the Ecole Pratique des Hautes Etudes, an institution created in 1868 to facilitate research and advanced training in the sciences and letters. Ribot asked his former classmate, the historian Gabriel Monod, who had become director of the Ecole, about the possibility of a position. He conveyed the results of his inquiries to his friend Espinas: "the response was that 'it would be very interesting, but that there is a third room of archaeology to found and three chairs of Coptic, etc., and that there is no space."[62] If Ribot's report is exaggerated—one doubts that Monod intended to create three chairs in Coptic studies—the general impression he got is clear—other subjects had higher priority than experimental psychology. Ribot was put off by the less than enthusiastic reception he received and shelved his plans for a university position.

He was doubly chagrined the next year to learn that a position had been created in the "History of Psychological Doctrines" at the Ecole for Jules Soury.[63] Ribot's friends had not double-crossed him, however; indeed, they were probably as surprised as he was. Soury was an archivist and paleographer who had also studied medicine at the Salpêtrière. His main claim to fame was a book in which he attempted to show that Jesus was mentally ill, suffering from "meningo-encephalitus."[64] Anticlerical Republicans enjoyed this sort of scientific debunking of religion, and it brought him to the attention of Paul Bert, physician, scientist, and minister of education in 1881. Bert had the position created for Soury over the objections of the Ecole.[65] Despite Soury's neurological orientation, his course was placed in the section of historical and philological sciences rather than natural sciences, probably as much because of Soury's equivocal credentials as because of uncertainty as to where psychology should be placed. Ribot took comfort in the fact that Soury's course was not a success, but he was clearly upset by the maneuver. He does not seem to have recognized two lessons he could have learned from this episode. One was that there was support in high places for scientific psychology. The other was that although he could not expect much from members of other disciplines, who had interests of their own to protect, an outsider with clout could impose a new position and get away with it.

Over the next few years, Ribot's friends urged him several times to offer a *cours libre* at the Faculty of Letters, and even though it seems likely that he would have had the support of the director of Higher Education, Ribot refused.[66] Such courses were free in more than one sense: in addition to being only loosely connected with the university curriculum, they were not remunerated. In the summer of 1885, however, a more promising initiative came in the form of a proposal to the Paris Faculty of Letters from Paul Janet.[67] Janet had been a member of

Ribot's dissertation committee, and although he did not approve the deterministic implications of Ribot's thesis, by the mid-1880s he seems to have become sincerely persuaded that an experimental approach to mental phenomena was a necessary part of psychology.[68] His private relationship with Ribot appears to have been contentious and stormy, but he now came to Ribot's aid.[69]

Janet suggested the creation of a *cours complémentaire* in experimental psychology for Ribot.[70] His suggestion came in the context of a discussion of how to raise the status of a spiritualist philosopher, Ludovic Carrau. Carrau had taught a *conférence* for some years at the Sorbonne with success, and his tenured colleagues wanted to recognize his service, even though there was no room at the moment for another chair.[71] They therefore suggested creating the title of "Directeur des conférences de philosophie." Janet seized the opportunity to make his own suggestion. He argued that "the moment has come to widen the framework of philosophical teaching."[72] The faculty should open itself to the two new directions philosophy had taken in the recent past, idealist and neo-Kantian on the one hand and experimental and scientific on the other. This would involve two new *cours complémentaires,* which he proposed to offer to Emile Boutroux and Ribot.[73] In this expanded curriculum, it would make sense to designate a director of studies, and hence M. Carrau could be accommodated.

Janet's proposal was obviously designed to be as diplomatic as possible. He tried to forestall spiritualist objections by offering to increase the number of spiritualists and to elevate the rank of one of them. Nevertheless, his colleagues in philosophy were not impressed. Charles Waddington, who held the chair of the history of ancient philosophy, objected to introducing into the Sorbonne a doctrine that tended toward materialism. He stated that the function of the faculty was to safeguard a philosophical teaching that was ultimately destined for lycée students, since the primary audience of university instruction was future lycée teachers. For this reason, the Sorbonne should not embrace doctrines that would be objectionable, even dangerous, to the nation's youth. He also claimed that positivism was not really a science, but only a doctrine, hence by implication not suitable. Elme Caro, who occupied the chair of philosophy, praised the merits of Boutroux and Ribot, but he thought that the Collége de France and the Ecole Pratique des Hautes Etudes were more appropriate sites for new subjects. The dean, the professor of geography Auguste Himly, pointed out that the Collège de France did not offer *cours complémentaires,* and he also warned that the administration would likely give Ribot a course whether the faculty approved it or not. Caro, however, pressed the point of order that the issue of *cours complémentaires* was not formally before the body and called for the faculty to approve his proposal for Carrau without deciding on Janet's proposition. This the faculty did.

This debate illuminates the sentiment of the faculty philosophers in the mid-1880s toward the sort of positivist psychology represented by Ribot. That sentiment was divided.[74] Janet at least supported the introduction of experimental psychology into the Sorbonne, because it represented a powerful new school of thought. For this reason it deserved a place in the university, whether or not one agreed with its findings. Janet sided with those who saw the university as a center for independent thought, as long as the thinker was on the forefront of his field, even if parts of that thought might be objectionable to others. It is significant that Janet called experimental psychology a new direction *in philosophy;* this indicates that for him this new field was still part of his discipline, not a separate enterprise.

The other professors, however, remained faithful to the traditional view of the university as a guardian of classical education and servant of the system of secondary education. They upheld Cousin's vision of philosophy as ideology and the university as defender of the correct ideology. Waddington declared that the faculty had a responsibility to oppose doctrines it considered mistaken and subversive, no matter how sophisticated they were. Caro too had difficulty admitting that the Sorbonne might harbor intellectual diversity, worrying about "the particularly delicate situation that the divergence, or rather the contradiction, of opinions during examinations in general and above all at dissertation defenses can create with respect to the public. . . ."[75] In their view, the university was not supposed to be on the forefront of new knowledge; on the contrary, it was the depository of knowledge that had stood the test of time.

As the dean hinted, the director of higher education, Louis Liard, created the position anyway. Liard was himself a former academic philosopher and no positivist, but he was committed to university reform, and he thought that keeping up with the latest developments in academic research was crucial to the new role of the university.[76] Only two weeks after the faculty meeting of 29 June, Ribot reported to Espinas that Liard had approached him and offered him a *cours complémentaire* at the Sorbonne.[77] An *arrêté* of 31 July confirmed the appointment officially. Two other *arrêtés* of the same date made Carrau Directeur des conférences de philosophie and created for Boutroux a *cours complémentaire* in modern German philosophy. The official action follows so closely Janet's proposal that one suspects collusion between Janet and Liard, although it is not possible to say whether Liard gave Janet a trial balloon to release before the faculty or whether Janet suggested to Liard a way of placating his colleagues. As in the case of Soury at the Ecole Pratique des Hautes Etudes, so at the Sorbonne the first position in experimental psychology was created not at the insistence of the faculty but by administrative fiat. Ribot actually accepted this gift rather reluctantly, grumbling to Espinas that Soury made more than he did, but in December he began the first instruction in physiological psychology ever at the Faculty of Letters.[78]

In his inaugural lecture, Ribot reiterated his claim that scientific psychology was a part of biology, not philosophy.[79] Recent progress in psychology had come when the biological method replaced the "ideological." He outlined a program of psychology that concentrated on observation and experiment rather than on "ideology," or on the analysis of ideas. In his courses, he proceeded to study subjects such as "the sentiments and the emotions" (the topic of his first course) according to such objective methods. Or more precisely, he used the studies of those who employed such methods. Ribot himself did not, nor did his course include an experimental laboratory, either for demonstration or for student exercises. His task was to explain the results of scientific psychology to students in the Faculty of Letters, not to train future psychologists.

Although Ribot was a success as a lecturer, he never enjoyed teaching under the conditions in which he found himself.[80] Because his course was not required, and because faculty courses were open to the public, he had a superficial and ever-changing audience, largely of foreigners, whom he could teach little. His only real students were future philosophy professors, and they were interested primarily in a psychology that would prepare them to pass the *licence* and *agrégation* examinations and to teach the psychology portion of the lycée curriculum in the philosophy course. Despite some changes, those examinations and courses still relied primarily on eclectic psychology, and this was taught by the professors of philosophy at the Sorbonne. Ribot had competition from his philosophical colleagues and the weight of the educational system against his own brand of psychology. Moreover, he found the task of preparing lectures backbreaking.

In 1887, though, his prospects brightened considerably when Renan, by that time director of the Collège de France, told him that he wanted to create a chair for him at that institution.[81] Such a position would represent the pinnacle of academic achievement and a more complete consecration of the new psychology. Nevertheless, even with Renan's considerable influence, it would be a difficult task. At the Sorbonne, Liard had created a new position; at the Collège de France, Renan would have to transform the existing chair in the Law of Nature and of Peoples (*Droit de la nature et des gens*) being vacated by Franck, who was retiring and who we have seen was opposed to Ribot's doctrines. The remaining two philosophers at the Collège de France were Charles Lévêque and Jean Nourrisson, both eclectics. They would ordinarily have a great influence on the selection of a candidate for any position related to philosophy, as the experts in the field. Moreover, there was an informal candidate for the position, Henri Joly, a spiritualist who had substituted for Franck at the Collège. *Suppléants* often benefited from the assumption that they were the heir apparent chosen by the previous incumbent, and even though Renan was proposing to change the chair, Joly's specialization in criminal psychology would make him as qualified for the new chair as for the old. Finally, any attempt to change the chair away from the philosophy of law or away from Joly might arouse the opposition of Jacques Flack, who held the chair of comparative legislations, and more generally of the legal profession, which had an interest in seeing itself represented in the Collège de France.

When the issue came up at the faculty meeting of 4 December 1887, the first speakers, Lévêque and Flack, argued in favor of keeping the chair in its present form.[82] Others who supported retaining the name of Law of Nature and of Peoples included Guillaume Guizot, professor of Languages and Literatures of Germanic Origin; Paul Leroy-Beaulieu, professor of Political Economy; and Emile Levasseur, professor of statistics and geography. On the other hand, H. D'Arbois de Jubainville, who held the chair of Celtic Languages and Literatures, thought that the vacant chair was essentially the same as Flack's chair of comparative legislation, and hence redundant. Several other members thought that the question of individual merit should outweigh the question of the title of the chair. The Collège should seek out the best person possible, regardless of area of expertise. It was Renan himself who suggested that the chair be changed to experimental psychology and who recommended that Ribot be named to the new chair. The Collège voted by a comfortable majority to change the title, and one battle was won.[83]

The next step was to accept and vote on nominations for the chair. The Collège de France, as did most academic bodies, presented the minister of Public Instruction with a list of two candidates, ranked in order of preference. In principle the minister could choose whichever of the candidates he preferred, but in practice he almost always chose the candidate the Collège preferred, since it would be an affront to the dignity of the Collège for its candidate to be rebuffed. As predicted, Ribot and Joly were the two principal candidates.[84] At the next meeting of the Collège, on 22 January 1888, Lévêque presented the qualifications of Joly, who he maintained had scholarly credentials equivalent to Ribot's but greater seniority.[85] Lévêque also proposed Espinas for the second spot. This may seem surprising, since Espinas was no less a positivist than Ribot, but it clearly indicates his desire to placate those who might favor a positivist while making sure that the only serious candidate—Ribot—was not considered. Of the remaining discussion,

Renan reports only that Edouard-Gérard Balbiani, professor of comparative embryogenesis, criticized Joly's work on comparative psychology as inexact and unscientific, adding that he would be an embarrassment to the Collège. It is clear that the Collège was not impressed with Joly. They nominated Ribot on the first ballot by a wide majority. Joly then became a candidate for the second position, but it took three ballots before he could get a majority. Another battle was won.

This was not the end of Ribot's fight, however. Because the new chair was considered part of the "sciences morales," the Académie des Sciences Morales et Politiques had the right to submit a candidate to the minister as well. This body had a larger contingent of philosophers, nearly all eclectic spiritualists. The other members were fairly conservative and mostly committed to notions of free will and spiritualism that they thought positivism denied. In addition, unlike the Collège de France, there were no natural scientists at the Académie. It is not surprising, then, that when Franck himself, a member of the Académie, presented the credentials of Joly to this body, his candidate received more sympathy than he had at the Collège.[86] Despite the support of the philosophers Janet and Ravaisson, Ribot was soundly defeated, and Joly presented on the first line.[87]

This presented the minister with a dilemma, since two equally august academic bodies had nominated different candidates for the position. According to Ribot, the minister was disposed in favor of Ribot, and Renan and Liard lobbied hard on his behalf, but Joly had a brother-in-law in the Chambre des Députés and tried to bring parliamentary pressure to his aid.[88] Eventually Renan and Liard carried the day for Ribot, and on 9 April 1888, Ribot gave his inaugural lecture in the chair of Experimental and Comparative Psychology.[89] With this success, experimental psychology had finally won a decisive victory over its opposition in the highest forum of the academic system.

This episode makes it clear that the positivist approach to the moral sciences still faced opposition from academic philosophers, even though this opposition was no longer monolithic. In fact, at least one academic philosopher, Paul Janet, went to considerable lengths in support of Ribot, even writing an article in *Revue des deux mondes* defending the action of the Collège de France, of which he was not a member. He argued, first, that the Collège de France had the right to transform chairs in any way it saw fit, and, second, that "objective" psychology of the sort Ribot advocated had originally been created by spiritualists and that it need not conflict with the principles of spiritualism, provided both stayed within their legitimate boundaries.[90] This sort of argument would become a way for eclectics to support a positivist against a spiritualist, given the merits of the particular issue and the particular candidates. Nevertheless, Ribot had to reach outside philosophy for help in attaining the highest positions in academe. At the Collège de France, the philosophers voted against him; at the Académie des Sciences Morales et Politiques, he got more support outside the philosophy section than within it.[91] The academic elite of philosophy were still not ready to embrace the positivistic human sciences. Even Janet's support had limits. When Ribot vacated the *cours complémentaire* at the Sorbonne, Janet did not try to fill it with another experimental psychologist; instead, he moved that the position be changed to a course on logic, stressing new developments in the field and the absence of anyone competent to teach the subject in higher education.[92] For Janet, experimental psychology was only one new development among many "in philosophy." His colleagues acquiesced without so much as a question, and experimental psychology disappeared from the Faculty of Letters.

At the same time that he was creating an academic position for the new psychology, Ribot was also involved in creating its first professional association, the Société de Psychologie Physiologique. Researchers associated with the Salpêtrière had begun to publish in the *Revue philosophique* from about 1880, and as their work in hypnotism expanded, they apparently became increasingly interested in "making a mark in physiological psychology" as opposed to neurology or psychiatry.[93] For his part, Ribot ardently desired to expand the space devoted to physiology in the *Revue,* and he also wanted the prestigious name of Charcot associated with it. By 1885 enough serendipity and momentum had gathered to launch the new organization, with Charcot as president. Ribot, who was named vice president, began to devote a section of the *Revue* to the Société. The Société organized the First International Congress of Physiological Psychology, which was held in Paris in 1889. This was the first international meeting of experimental psychologists, drawing participants from Europe, England, and the United States. Its overwhelming success led to many more and attested to the growing maturity of the new discipline.

Finally, 1889 also saw the creation of the first laboratory of experimental psychology in Paris.[94] Although it was created at the instigation of Ribot, it was principally sponsored by his rival Soury and housed within the Ecole Pratique des Hautes Etudes. Its direction was given to Henri Beaunis, a doctor and physiologist from Nancy.[95] Ribot gave a separate section in the *Revue philosophique* to the papers produced by this laboratory until 1894, when it had gathered enough momentum to found its own journal, the *Année psychologique,* which was the first journal devoted exclusively to experimental psychology in France.

CONCLUSION

By 1889, Ribot had helped to organize most of the institutions associated with a new discipline: an academic position, a journal, a laboratory, and a professional association. Even so, he remained in some ways marginal to his own creation. His position at the Collège de France, because it was not related to any degree program and had no enrolled students, prevented him from effectively forming a generation of successors. Moreover, he soon abandoned the burden of teaching to return to his books and journal. His journal remained tied to philosophy, and a truly professional journal of psychology could only establish itself by breaking away from him. He did not work in the laboratory he thought essential to the progress of the discipline. And he was not president of the organization that united the friends and practitioners of the new science.

Much of this was the result of his continuing relationship to philosophy. Despite his attempts at scientific education, he retained the literary methods of his philosophical training, and he never attained the scientific credentials that would give him direct access to the factual matters relevant to his interests. On the other hand, it was his general, philosophical, conception of the field of psychology that allowed him to rise above compartmentalized scientific disciplines and to unite elements from various sources. And without the support provided by his relationship with philosophers through their contributions to the *Revue philosophique,* he would never have been able to carry out his campaign against them.

3

Incurable Metaphysician: Alfred Espinas

To prove that one should not philosophize, one has to philosophize; and some metaphysics is necessary even to those who try
to prove the impotence of metaphysics. It is therefore best to take
questions as they come, without worrying too much about the
label they wear, except to bring them back as soon as possible . . .
to definite terms, to experimental research.
—Alfred Espinas, "Nature and Immaterialism"

Rɪʙoт's friend and classmate Alfred Espinas (1844–1922) attempted
to do for sociology what Ribot tried to do for psychology—rid it of
spiritualist philosophy and apply to it the methods and assumptions of
the natural sciences. Like Ribot, he violently attacked the methods and
assumptions of institutional spiritualism, and he elaborated an alternative approach based on the relevance of biology to human social phenomena. In his doctoral dissertation, *On Animal Societies (Des sociétés
animales)*, he advocated the comparative study of human and animal
societies, attempting to show that social life was a feature of the natural
world that could be studied like any other biological phenomenon.
Moreover, like Ribot, he adopted the theory of evolution in its Spencerian formulation. Espinas also adopted Spencer's psychology and applied it to the explanation of social phenomena, stressing the habitual
and instinctive component of human action, and he denounced the efforts of spiritualist philosophers and politicians to interfere with the
organic evolution of society on the basis of reforms deduced from an a
priori set of moral principles. Espinas's efforts to elaborate a scientific
sociology made him a natural choice for what has usually been considered the first chair of sociology at the Sorbonne—the chair of social
economy *(économie sociale)* created in 1893. He also inspired and aided
others, like Durkheim, who were interested in pursuing the same vision
of sociology. In many ways, then, he might be considered the sociological counterpart of Ribot as founder of a new discipline.

Yet this comparison fails in crucial respects. Other aspects of Espinas's
intellectual activity unsettled Ribot. Ribot did not object to his friend's

use of the organic analogy, but he did object to Espinas's most original extension of that analogy, the further assertion that societies possess a collective conscience. While Ribot played down the social and political consequences of his science, Espinas emphasized the benefits of social science for social harmony and progress, contrasting these benefits with the deleterious effects of spiritualist morality. Above all, Ribot objected to the efforts Espinas made to refute spiritualist philosophers in their own terms through editions of the classics, histories of philosophy, and the use of metaphysical arguments against spiritualist metaphysics. Ribot called his friend an "incurable metaphysician" and begged him to stop wasting his time in sterile philosophical pursuits.[1]

For these reasons, Espinas has been considered too biological, too political, and too philosophical: too biological, because he attempted to model sociology directly on biology; too political, because his commitment to the Third Republic made him more interested in an effective ideology than a coherent science; and too philosophical, because he seemed more concerned with a philosophy of evolution than a science of society. Both his contemporaries and his historians have found this blend something of a puzzle and have offered a number of explanations. Paul Janet thought that Espinas confused science and philosophy.[2] Georges Davy thought he erroneously equated science with evolutionism.[3] And Roger Lewis Geiger has maintained that his orientation was "essentially philosophical" rather than scientific, and that he was blinded by the political implications he wanted his sociology to have.[4]

Nevertheless, terms such as *confusion, blindness,* and *essentially philosophical* are inadequate characterizations of Espinas's position. For one thing, his position follows logically from his fundamental assumptions about the nature of science, philosophy, and morals. From the beginning to the end of his career, Espinas insisted that science and philosophy are inseparable; for him, the choice was not between philosophy and science, but between scientific and antiscientific philosophy. Indeed, within his definitions of philosophy and science, it is impossible to tell where one stops and the other starts, since philosophy is simply a higher degree of generalization of the results of science, which is itself a series of generalizations derived from facts. This conception may not ultimately be the most fruitful, but it is hardly inconsistent. Nor is it essentially philosophical, because only certain conceptions of philosophy imply this relation to science.

Moreover, if for Espinas the proximate goal of science is disinterested knowledge, its ultimate goal is usefulness. Science that does not serve action is useless in his eyes. In this sense, the whole purpose of sociology was to base decisions about what ought to be done on the nature of society. For this reason, Espinas saw no inconsistency in drawing moral implications from social science. This constitutes what has been termed the *naturalistic fallacy,* but it was the driving force behind most social theory. In short, Espinas's anomalous position does not result from his confusion about the relation of science and philosophy, or between science and morals, but from the fact that his refusal to delineate the boundaries between them put him at odds with others who, for one reason or another, had an interest in distinguishing them more sharply.

Espinas's peculiar practice of philosophical science and scientific philosophy had precedents in nineteenth-century thought, but the distinctive traits of his work—his preoccupation with the history of philosophy, his ontological naturalism, his critique of individualism, and his concept of the collective conscience—cannot simply be interpreted as the logical extension of his intellectual forebears. His original ideas are the product of many influences, and hence the motivation for his particular synthesis must rely in part on something other than the intrinsic characteristics of his intellectual predecessors.

I would like to suggest that what organizes Espinas's thought and accounts for its distinctive features is a particular relation to academic philosophy. Espinas's position, both institutionally and intellectually, was one of opposition from within. Unlike Ribot, Espinas carried out his critique of spiritualist philosophy from within the academic establishment. He enjoyed a successful academic career, abetted by the very philosophers he condemned. He identified himself in opposition to his spiritualist colleagues, and his writings are all organized around an opposition between a caricatured spiritualism and his own naturalism that Geiger aptly termed *Manichean.* Nevertheless, he continued to address academic philosophy in its own terms—for every eclectic doctrine, Espinas constructed a naturalist doctrine. This is very different from the effort of Ribot, who tried to abandon philosophy altogether. Espinas, in his rebellion against one philosophy, constructed a counterphilosophy as much as a social science. In doing so, he remained involved with academic philosophy more than he would ever admit. His positions were in some fundamental ways not that different from eclectic philosophy, despite the disavowals from both sides. Both insisted that science depended upon and demonstrated metaphysical propositions, and both maintained that one's attitude toward the relation between human and natural phenomena had moral consequences. In his quest to make society an object of science and of secular moral obligation, he gave it attributes of being that bordered on the metaphysical.

EARLY EDUCATION AND CAREER

Alfred Espinas began his lycée education at Sens in 1855, but his academic promise encouraged his family to send him to Paris, where he

entered the Lycée Louis-le-Grand in 1861. His early education coincided with the policy of bifurcation in the French secondary system,
when students aspiring to the baccalaureate in science were separated
from those pursuing the degree in letters at an early stage.[5] Espinas
chose the literary option, avoiding the advanced mathematics that constituted the majority of the preparation in science. The literary option
included some science, but Espinas's knowledge of science was shaped
above all by his enthusiasm as an amateur naturalist, an interest that
began early in his childhood.[6] The nature of Espinas's early exposure to
science is significant, since throughout his career his image of science was
that of the naturalist rather than that of the mathematician or physicist. If
his formal training in science was limited, his training in philosophy
was complete. The philosophy class was rehabilitated the year before
Espinas entered the Ecole Normale, and so he was exposed to the entire
eclectic syllabus, taught at the Lycée Louis-le-Grand by Paul Janet,
rather than the truncated *classe de logique*.

When Espinas entered the Ecole Normale Supérieure in 1864, he
originally intended to study literature, but he quickly became attracted
to philosophy instead. He studied under two spiritualists, Albert Lemoine and Jules Lachelier, and was evaluated at the end of each year by
a third, none other than Janet, who in 1864 had become professor of
philosophy at the Sorbonne.[7] Espinas was apparently a competent but
not spectacular student in philosophy.[8] Perhaps more important to Espinas than the teachers he had at the Ecole Normale were his classmates,
many of whom were to become leaders in French education: Henri
Marion (entered 1865), the first professor of the Science of Education
at the Sorbonne; Louis Liard (entered 1866), future director of Higher
Education; Elie Rabier (entered 1866), future director of Secondary
Education; Emile Boutroux (entered 1865), a personal friend whose
philosophy Espinas nevertheless vigorously opposed; Edmond Perrier
(entered 1864), a biologist whose later work closely paralleled that of
Espinas; and above all Théodule Ribot. Although Ribot and Espinas
were only at Normale for one year together, they formed a close intellectual relationship that was to remain throughout their long lives.

Espinas graduated in 1867 and began to work his way slowly up the
hierarchy of French lycées. He was assigned first to Bastia in Corsica,
and then to Chaumont, where the reports of his administrators indicate
that he spent more time in the cafés than was respectable for a young
professor, and that he consorted with people of "advanced" political
opinions.[9] Until 1870, however, there is no indication of his future
positivism, nor that his advanced political opinions were incompatible
with the spiritualism of the academy, which also opposed the Second
Empire. Indeed, an administrative report of March 1870 states that "his
doctrine is spiritualist."[10]

In that year, however, two things happened that affected him profoundly—the Franco-Prussian War broke out, and Espinas discovered Spencer. Espinas was mobilized with the local militia but never saw action. He later maintained, though, that the defeat of France and the Commune turned his thoughts to the study of society.[11] When he did so, he had just begun to read Herbert Spencer, perhaps introduced by Ribot, who congratulated him on his interest.[12] From 1872 to 1874 Espinas collaborated with Ribot in the translation of Spencer's *Principles of Psychology*, and when Ribot began his *Revue philosophique* in 1876, Espinas helped by contributing over forty reviews in the first five years of the journal, primarily of Italian positivists. In Spencer and the Italian positivists, such as Roberto Ardigo, Cesare Lombroso, and Enrico Ferri, Espinas found thinkers who combined a respect for the natural sciences with an evolutionary framework.

The Italian positivists also provided an example of a philosophy of science that combined empiricism with metaphysics. In 1880, Espinas synthesized his reviews in a book on *Experimental Philosophy in Italy (La philosophie expérimentale en Italie)*, parts of which first appeared in the *Revue philosophique*. In his introduction, Espinas acknowledged that

[the Italian naturalist philosophers] will be accused of wanting to restore metaphysics. We are not convinced of the seriousness of such a reproach. The position adopted by the positivists who affect ignorance of metaphysical problems is really not tenable; . . . either these problems have solutions and it is necessary to find them, or they don't have any [solutions] and it is necessary to demonstrate this.[13]

Either way, one is doing metaphysics, even if the point is to prove that one should not do metaphysics.[14] Espinas separated himself from the position of Comte, Littré, and Ribot, who argued that metaphysics is erroneous and irrelevant. Espinas maintained that scientists do in fact deal with metaphysical questions; they just call them by other names. But whatever you call them, they won't go away. They constitute the most serious problems of science—the nature of causality, of knowledge, et cetera—and should be addressed rather than swept under the rug. They also lie behind the desire for unity that drives rational inquiry and provides the view of the whole that the specialized sciences can no longer attain. Again, it is a matter of indifference to him whether these "grand hypotheses" are elaborated by philosophers or scientists; indeed, it is better if the scientists do it, since they are closer to the facts from which they arise. But Espinas does not think the desire for such overviews or their utility will ever disappear. Ribot had berated Espinas for such views as early as 1871 and continued to do so, but with little effect.[15]

Espinas's growing attachment to an evolutionary metaphysics was difficult to reconcile with the requirements of his lycée teaching. His views got him into trouble in the classroom, where even a neutral attitude toward Darwin could upset provincial parents. In 1875, Espinas allowed a class discussion on the mental faculties of animals to wander into the question of the evolutionary hypothesis. Although he avoided endorsing evolution, he did not repudiate it, either. Word of the discussion leaked out, and a conservative local paper denounced him. The rector defended Espinas, although he counseled "moderation," but the inspector general, A. Chassang, thought that Espinas had committed an indiscretion by allowing the discussion at all.[16] Such incidents were fairly common in the careers of young philosophers, and the school officials generally sided with their employees, more to defend the authority of the Université than the views of their professors. But the incident highlights the incompatibility of Espinas's views with the restrictions placed on lycée philosophy at this period.

In general, however, Espinas was a good teacher, even by the standards of eclectic orthodoxy. Despite his lackluster speaking skills—one report notes laconically that "he was not born an orator"—he was apparently an effective teacher, producing students with outstanding records of success in the *concours* and the baccaluareate. It is therefore not surprising that he moved up the ranks in the secondary system. Even before he completed his dissertation, though, he was soliciting a position in higher education, where he would be freer to teach the evolutionary philosophy he had embraced.[17] He had no real chance at such a post until he received his doctorate, but by the mid-1870s he was already at work on the dissertation that would give him access to higher education.

On Animal Societies

If Espinas was interested purely in academic advancement, he could hardly have chosen a worse topic or a worse approach than *Animal Societies,* a study of social life in the animal world. In this work, Espinas attempted to show that moral phenomena were a part of the natural world, that they were in fact simply the biological conditions of existence of society, and that they could therefore be studied by the methods of the natural sciences. His thesis was bound to arouse opposition from his spiritualist readers, since it managed to challenge every tenet of the eclectic syllabus. But in challenging eclectic philosophy, *Animal Societies* also incorporated a number of its features, and a close reading can find in this book a negative image of the eclectic syllabus.

Historical Dualism and Ontological Naturalism

Before broaching the subject of animal societies, Espinas wrote a long introduction in which he surveyed the literature on the nature of society. He focused on human rather than animal societies, ostensibly because works on the latter were all but nonexistent. However, since he did not hide the fact that he intended his study of animal societies as part of a general sociology that would include human societies as well, his introduction also served to show where he stood in relation to spiritualist social theory. And since the preface was written in the form of a history of the origin and triumph of the concept of society as a natural phenomenon, it also served to justify his approach to social phenomena. The history of philosophy was for this reason as central to Espinas's project as it was to the eclectics.

But the history he presents is most uneclectic. Instead of attempting to find truth in every school and to combine the partial truths into a whole, Espinas depicts the history of a war between naturalism and artificialism—between the idea that human society is a natural phenomenon and the idea that it is an artificial creation of individual human beings. There can be no accommodation between these two ideas, so that the triumph of one necessarily means the defeat of the other. Moreover, Espinas sides with the naturalists, who in every case are the opposites of those the spiritualists prefer—the sophists over Socrates, Aristotle over Plato, Spinoza over Leibniz, Hegel over Kant. Born among the Greeks, the naturalist thesis overcame many setbacks and obstacles until, by the beginning of the nineteenth century, it had definitively triumphed. "From this moment on no new contribution of any importance will come along to breathe life into the nominalist thesis of society."[18] As disciplines as diverse as linguistics, history, literary theory, and biology, all converged on the naturalist thesis, the rise of a natural science of society became inevitable, and Espinas concluded his historical survey with the founding achievements of Quetelet, Comte, and Spencer.

Espinas's history of philosophy is also uneclectic in that, whereas eclecticism located the source of each philosophical school in some general tendency of the human mind, Espinas offers no explanation for either the existence of the artificialist, spiritualist thesis or the success of the naturalist thesis. He simply assumes that the truth of the latter is obvious, and he can only depict the former as an inexplicable force that somehow persists for two millennia. Thus, Greek naturalism was simply overtaken by Christian mysticism, and, in the modern period, the naturalist thesis was making great strides until the idea of the absolute arose with Rousseau and derailed social theory until the nineteenth century.

This inexplicability of artificialism returned to haunt Espinas and became a major theme in his later work.

Naturalism and *artificialism* are loaded terms for Espinas, implying a number of other oppositions that define the general traits of his thought. If one believes that society is a natural phenomenon, one will employ the methods of the natural sciences to study it, whereas artificialists, since they hold that society is ultimately an abstraction of which actual society is only an approximation, believe that society should be studied a priori, through an analysis of the ideal principles on which it is based. Hence, ignoring the spiritualists' claim to study the soul through observation, Espinas places them among the a priorists. Since naturalists believe that society is a concrete thing, they also believe its laws are simply the conditions of its existence and the relations between its parts; hence, they are relative to the structure of the society and to the stage of its development. Artificialists believe that the laws of society are absolute and timeless ideals to which society must be brought into conformity. And since artificialists consider society as an abstraction, they are invariably individualists. This is why Espinas calls them nominalists.

As important as the particular oppositions is the status given them. More than once he claims that the naturalist thesis can support a variety of different metaphysical positions. He argues that Leibnizian monadology is not incompatible with it, and he cites de Maistre as an example of a spiritualist who rallied to it (ibid., 38, 66). It would seem, then, that Espinas is making a claim only about the phenomenal world, leaving the question of ultimate reality and ultimate causes beyond the realm of science. But when he claims that the resolution to the conflict between naturalism and artificialism is the thesis that "there is nothing outside nature," he is making more than a claim about appearances (ibid., 137). Espinas calls his position naturalist rather than positivist for a good reason: it is a metaphysical position rather than just a methodological postulate. For Espinas, the choice between naturalism and artificialism precedes and determines the methods employed, and naturalism implies more than the bracketing of transcendent terms. This is perfectly consistent with Espinas's views on the relation between metaphysics and science, but it conflicts with the positivism of Comte, who denied that the question had any meaning, as well as with the positivism of Spencer, who left open the Unknowable as a realm beneath and behind nature. The Unknowable has no place in Espinas's cosmology; his opposition to spiritualism does not allow him the luxury of that concession.

If Espinas is a naturalist, it does not follow that he is a materialist. Far from reducing mind to matter, or psychology to biology, Espinas explicitly reassures spiritualists that he does not deny the reality of psychological phenomena.[19] Indeed, he extends psychological phenomena into the furthest reaches of the biological world, opposing "exterior or mechanical" to "interior or psychic" forces much as the spiritualists. Thus, for example, he opposes Darwin's explanation of the slave-making instinct in ants as the result of the chance mutation of favorable behavioral mechanisms. Espinas thinks that a rudimentary form of intelligence is necessary to account for the adaptive character of such an extraordinary practice (ibid., 195). Since societies in higher animals depend on psychological relations, there is also a difference between "social" and "material" or biological organisms.[20] A society is "an organism of ideas," and Espinas claims that he has no intention of explaining the superior by the inferior, the social by the material (ibid., 530).

On the other hand, he does think that consciousness emerges from biological phenomena, as opposed to being an independent spiritual substance. "One passes imperceptibly from the outside to the inside, from a more or less complicated play of movements (is life anything else?) to a correspondence of representations and of desires, to consciousness" (ibid., 527–28). For a spiritualist, of course, there can be no passage from outside to inside, from movement to idea, from matter to mind. Moreover, although he calls the psychological a new "order" of existence, he also describes it as a new "aspect" of reality, defined by the manner in which we know and define it. Whereas material phenomena are known immediately through the senses, psychological phenomena, at least those of others, "are only known through interpretation and must, so to speak, be translated as a function of consciousness after being collected through their material aspect" (ibid., 528). He uses the language of double-aspect monism to retain the connection between the material and the mental and to contain both within the same set of natural laws. Espinas's naturalism therefore allows room for a distinction between mind and matter, but it also covers his uncertainty about whether that distinction should be viewed in terms of a double-aspect monism or an emergent dualism.

Espinas's Philosophy of Science

The opposition of inductive naturalists to deductive artificialists reveals a sense in which Espinas's understanding of natural science agrees with that of the eclectics, despite his claims to the contrary. Like them, he found the use of hypothesis and deduction contrary to the spirit of the experimental method. He thought that general laws would arise from the comparison of particular facts, and *Animal Societies* is organized as an exercise in comparative method. Espinas emphasizes the similar in comparison and the general in explanation. The book begins with the assertion that "the fact of animal societies is known; it is not explained. It has not been attached to any general law of nature" (ibid., 7). He adds that animal and human societies will not be explained until they have been "subsumed under the same law by the discovery of their common characters" (ibid., 8). Espinas assumes that the two sorts of phenomena can be thought under the same category, and this assumption guides his entire work. He thinks it follows from his assertion that society is a part of nature. Yet at the same time, he distances himself from the casual analogizing that applies human names to animal societies and vice versa. One must find a *rational* principle of comparison, and this entails studying each side of the equation independently before attempting to relate them (ibid.). Nevertheless, he is not questioning the validity of the analogy; he is simply rationalizing it by submitting it to the comparative method. By simultaneously invoking and deferring the comparison of animal with human societies, by invoking the most general level of similarity while deferring the question of particulars, Espinas assures himself that his disinterested examination of animal societies will produce answers couched in terms comparable to human society.

This is not to say that Espinas ignores difference and analogizes without restraint. On the contrary, his evolutionary point of view demands that there be a full range of different stages of society in the animal kingdom. But the very emphasis on incremental change has the effect of eliminating any radical difference of kind between human and animal societies, or between societies and organisms. An evolutionary chain links the most primitive aggregation and human society, the most loosely connected society and the most highly structured organism. Espinas notes that the cell theory has led "illustrious physiologists" to describe the individual organism in social terms (ibid., 214). If this is the case, it is because the same laws are at work in both domains. The distinction between sociology and biology becomes one of convenience rather than a reflection of some absolute difference of kind between organisms and societies (ibid., 220). "From the one to the other group there is no longer anything but a difference of degree; or rather they both form only one single ensemble governed by the same laws, which are those of biological evolution" (ibid., 138–39).

This conviction that the human should be compared to the animal extends to psychology. The first edition of *Animal Societies* carried the subtitle *Study of Comparative Psychology (Etude de psychologie comparée)*, and Espinas claimed it to be a "postulate of all comparative psychology" that animal minds should be understood by analogy with our own (ibid., 312). He also assumed that consciousness is proportional to the development of the nervous system, and that wherever we find a nervous system we may postulate a correspondingly developed consciousness (ibid., 255–56). Whether or not he was justified in these assumptions, he was in any case not alone. Darwin made them both, and others went even further.[21] Alfred Binet, creator of the IQ test, wrote on the psychic life of micro-organisms, and his was only one of several works on the subject.[22] Spiritualists objected to this correlation of mind with matter as leading to monism, to a strict dependence of mind on brain, when taken to such extremes. However, they also recognized the need of a physical support for the mind in this world, and they had long abandoned the Cartesian distinction between thinking humans and animal automatons. For this reason, it was possible for a spiritualist to write a *Comparative Psychology (Psychologie comparée)* that appeared in the same year as *Animal Societies*.[23]

To explain animal minds by analogy with our own might seem to be a clear if debatable methodological principle of psychological science. However, Espinas also uses it to define the distinction between science and metaphysics. "Every being offers two aspects: one is a succession of phenomena following each other according to a law, and the other is an efficacious virtuality from which these phenomena incessantly emanate."[24] The first aspect defines modern science, which Espinas claims is necessarily mechanistic, since it defines things in terms of their material appearance. The second aspect characterizes metaphysics, which is necessarily idealist, because the only example of an efficacious virtuality we have is our own mind, which lies behind our visible acts. We therefore always interpret the real nature of phenomena on the analogy of the relation between mind and act.

This correlation of science with mechanism and of metaphysics with idealism puts Espinas in an extraordinary position. To be scientific is to be mechanistic, and hence to be consistent with itself sociology should presumably limit itself to mechanistic explanations of social phenomena, without engaging in speculation about the internal thought processes of social actors. Indeed, Espinas indicated at the beginning of the book that the existence of regularities among social facts was a sufficient basis for a science of society. Nevertheless, he has made it a "fundamental postulate of comparative psychology" that the minds of animals should be understood by analogy with our own, and he defines higher societies in terms of their psychological character. This would seem to make sociology essentially metaphysical, according to his distinction between science and metaphysics.

Espinas does not indicate that he is aware of the consequences of making this distinction. The most consistent explanation would seem to lie in Espinas's relative indifference to the distinction between these two realms. He is talking about what we call science and what we call metaphysics, but since mind is a part of nature, the distinction is ultimately verbal rather than real. It is nevertheless significant that this distinction was only introduced at the end of the book, in a section of the conclusion devoted primarily to the "nature" of society, in a passage in which he attempted to show how he could avoid the charge of explaining the superior by the inferior, the mental by the material. It is a passage in which he addresses the gravest potential objections of spiritualist philosophers directly, trying to show why his views of science need not be incompatible with spiritualist metaphysics. Rather than dismiss his opponents, he attempts to engage and persuade them by using the language of metaphysics while he situates metaphysics within nature. In the process, he defined his project in such a way that his opponents could denounce it as metaphysical, using his own words to convict him.

The laws governing such a variety of instances can only be of an extremely general and qualitative nature, and in his conclusion Espinas summarizes them. They include the following: all societies depend on cooperation, all societies grow naturally, organized societies effect a division of labor, different functions must be hierarchically organized, there is a direct relation between size and organization, et cetera. These might seem like general principles rather than scientific laws, and Espinas himself anticipates this objection, responding that even the laws of physics began as qualitative statements and become qualitative again once they reach a high degree of generality. Moreover, many of the laws of biology are completely qualitative (ibid., 515–16). Scientific laws do not have to be quantitative.

Espinas's insistence that everything ultimately be explained by general laws of evolution and organic development is most directly inspired by Spencer. But nothing in his comparative, inductive approach contradicts eclectic philosophy of science. If his critics would stress differences rather than similarities in particular cases, they would not condemn his method as fundamentally unscientific, as the reaction to his work, discussed in the following section, would show.

Being and Consciousness: The Social Individual

Social life is a universal fact of nature, and it exhibits regularities. This, Espinas claims, "is enough for science to attach itself" to these phenomena (ibid., 9). But in addition, "more delicate and far-reaching questions will come intermingle with this experimental research and will increase its difficulties while making it doubly worthwhile" (ibid., 10).

We will see . . . that as the constitutive elements of the living body form by their participation in one biological action a single whole that has in the plurality of its parts only one and the same life, likewise the individual animals that constitute a society tend to form, through the exchange of their representations and the reciprocity of their psychic acts, a single consciousness that is more or less concentrated but that is also apparently individual. (ibid.)

This *conscience collective* raises two questions, the answers to which Espinas hopes his classification of societies will provide:

1° What is the relation of individuals with the psychic center to which their activity attaches them . . . and if the whole forms a veritable individual, how, in the animal world, is a collective conscience possible?
2° What sort of being is society? Is it a being properly speaking, something real and concrete, or must we see in it only an abstraction, a conception without object, a word? In short, is society a living being like the individual, as real, and in this case even more real than he, or is it only a unity of collection, a verbal entity whose substance is formed entirely by the individual? (ibid., 10–11)

One might ask of these questions, are they scientific? When he says that they are of a "portée supèrieure" to the questions about the regularities of social life, does he mean to say that they are philosophical rather than scientific? It is possible to make this interpretion, and many of his readers did. Indeed, given the distinction he later draws between metaphysics as psychology and science as mechanism, questions about a collective conscience would be necessarily metaphysical. In any case, the real or nominal existence of society would seem to be an ontological question par excellence. On the other hand, these questions "mix with this experimental research," and the exposition of the facts will "prepare the solution" of these questions. The status of the questions is ambiguous, and this ambiguity informs the organization of the facts.

These two questions are also linked to the concept of individuality, the physical and spiritual unity that is a condition of existence. It is not surprising therefore that Espinas attempts to reformulate the concept of individuality in such a way that it applies to society. In doing so, he undermines the spiritualist understanding of individuality, making it a relative and variable concept. Nevertheless, he uses it in much the same way they did—psychological individuality is the highest form of being and the only ultimate guarantee of being. He employs an evolutionary framework to depict the emergence of a collective psychological individual among higher species. His reformulation of individuality has two parts. First he uses contemporary science to show that the biological individual is composite and variable. Then he argues the converse, that where there is composite unity, as in animal societies, we can ascribe a corresponding degree of individuality.

According to Espinas, the cell theory had revealed the composition of the organic individual and rendered problematic the distinction between individual and society. Moreover, invertebrate zoology had uncovered cases in which it was difficult to decide whether organisms were single individuals or colonial aggregations. As a result, biologists were forced to admit that individuality was relative and variable. The concept of individuality thus undermined was in any event anthropocentric. "Certain morphological considerations . . . play a preponderant role in our conception of the individual. We consider ourselves to be the model of individuality and deny it to any being that deviates from this type" (ibid., 216). "However," he continues, "there is no reason to believe that we are the absolute measure of things" (ibid., 217).

Our hesitation also finds its roots in the feeling of psychological unity. "Serious objections [to the composition of the individual] can only come from psychology. It seems in effect, that the name of individual implies the existence of spiritual atoms, of absolutely simple beings" (ibid., 214–15). We take our sense of self to be something indivisible. But like morphological individuality, psychological individuality admits of all degrees, levels, and kinds. Espinas cites studies of animal behavior under anesthesia and cerebral mutilation to show that consciousness and the self can be diminished. Indeed, if one grants the correspondence of neural and mental organization, consciousness itself can be described as the coordination of lower levels of mental activity into a higher unity through the transmission of impulses among nerve centers. Consciousness therefore varies according to the complexity and integration of the neural structure.

Having thus extended the notion of society into the individual organism and made of consciousness a social phenomenon, Espinas tries in *Animal Societies* to prove the converse as well. If organic individuals are composite wholes constituted by the interaction of living parts, then societies, which also have interacting and conscious parts, may also be individuals. And if the consciousness of individuals is composite, the composite individual that is society may have a consciousness as well. Espinas uses biology to undermine the spiritualist definition of individuality, but only because he wants to argue that "the nation . . . is an individual. This whole book has no other point than to demonstrate this proposition indirectly" (ibid., 223–24). He attempts to do this by showing how the evolutionary development of social life among animals recapitulates that of the organic individual, with the emergence of both a progressively higher degree of integration and a higher level of social bond, culminating in the collective conscience.

After eliminating associations among different species from his definition of society, Espinas classifies normal societies according to the primary natural function that the social link fulfills: nutrition, reproduction, or what he calls relation. In the first category he places the lowest forms of colonial life, characterized not only by association for nutritive purposes, but by a purely physiological link and by asexual reproduction. Among the nutritive species, some unicellular organisms form simple aggregations that may be easily broken, whereas worms, which

Espinas places at the apex of this category because of their segmentation, are so closely integrated as to constitute veritable organisms.

The next stage is constituted by those species in which sexual differentiation requires that male and female unite, forming a new kind of association—the family. Under this category, Espinas considers everything from the "nuclear" families of birds to insect colonies. Aside from the reproductive nature of the social bond, families are also distinguished by the fact that the bond between them is psychological as well as physiological. Familial animals must communicate in order to mate and care for their offspring. Within the category of familial societies, Espinas discerns three major types: conjugal, in which the male and female join only for insemination and leave each other and their offspring once this has been accomplished; maternal, in which the mother remains united to her offspring, protecting, raising, and educating them without the aid of the male; and paternal, in which the male remains with the female and offspring to help with the care of the young. In terms of integration, one might think that the social insects—bees, ants, and termites—would be at the summit of such societies, but in fact Espinas places the paternal families found among birds and mammals at the top of this class. This is not simply a reflection of nineteenth-century patriarchy, although it may be that as well. The superiority of these families has more to do with the higher intelligence of the individuals, enabling a closer psychological union among them.

> . . . an invisible double chain [biological and psychological] attaches parents to their young and . . . this incessant communication, this constant sharing of all intellectual functions, this tight solidarity of emotions, thoughts, and desires makes of the family . . . an individual moral organism, one and the same consciousness. From this point of view it is not a group of beings, but a single being. (ibid., 432)

Just as the worm represents the limiting case where a biological society becomes a biological organism, so the higher familial animals present a double society, biological and psychological, merging into a single organism linked both ways.

The final category of social life is the *vie de relation,* in which the social bond is constituted solely by communication. Here Espinas discusses such phenomena as schools of fish, flocks of birds, and herds of various mammals, culminating with the systems of dominance among primates. Far from seeing this type of social group—which he terms the *peuplade* (horde or tribe)—as an extension of the family, he argues that a well-developed family is antagonistic to a larger group, since young adults tend to leave their parents and form independent families rather than continue to submit to the authority of their parents (ibid.,

473). Instead, he hypothesizes that the origin of the animal horde lies in the play of the young of different families. The sentiment behind this interaction is the pure attraction of like for like, without any link of kinship or biological necessity. The play among young develops this sentiment in higher species to the point where it can overcome the antagonism of competing family groups and permit larger associations (ibid., 474–80). Among the species that attain this kind of society, Espinas traces degrees of integration from the simple mating groups of birds, in which all are more or less equal and no organization is discernible, to the complex systems of dominance and subordination among primates. Of these latter, Espinas writes that

> manifestations of sad or happy sentiments constantly exchanged from one individual of the *peuplade* to another establish between the different members a close community of emotions and thoughts; one can thus say that the social unity, so clearly represented by the chief, is a unity of consciousness of which physiological relationships are only the condition; it is truly an individual. (ibid., 507–8)

Thus within every class of society, relations run from the relatively loose to the point where individuals and societies cannot be distinguished.

As these citations make clear, communication is the medium by which the collective conscience is created, and Espinas devotes a great deal of attention to the examination of animal communication. Through the exchange of signs—representations, as Espinas generally calls them—the consciousnesses of discrete individuals become occupied with the same objects. The more representations shared, and the longer the contact, the more extensive the identity of consciousnesses. More than just representations are shared, however; actual states of mind are shared. He claims that "it is a universal law in the entire domain of intelligent life, that the representation of an emotional state causes the appearance of this same state in the one who witnesses it" (ibid., 359). Representations are not merely symbolic; they evoke the mental state they symbolize. This affects action as well as emotion, since an idea evokes not only the affect but the response associated with it. Thus an identity of mental states is created among the communicating members of society. But this collective mental state is not simply the least common denominator among the associated individuals; to the extent that communication allows the pooling of information and impulses from all members of the community, the collective conscience actually surpasses the consciousness of each member taken in isolation.

The collective conscience is therefore not an afterthought in *Animal Societies*. It is the organizing principle and the culmination of Espinas's work. But it is not simply a result of his research: it also justifies his sociology, since it establishes the reality of society. Indeed, it is as real as the human individual, since like the human it possesses a conscience, and more real than things that are mere inanimate objects.

... To be regarded as consistent with the collective reason and experience of humanity, that is the sign without which phenomena cannot be considered real. ... [But] if a being, instead of only for others, exists for itself, that is to say instead of being known by a consciousness other than its own knows itself and possesses itself in its own consciousness, that being enjoys a better founded reality. (ibid., 538)

The collective conscience is a projection of the Cartesian *cogito* onto the social body, and it serves the same ontological purpose—society thinks, therefore it is. Moreover, the link established between consciousness and being, between psychology and metaphysics, is one of the defining features of eclectic spiritualism. Espinas was attempting to persuade spiritualists by using against them the very terms they employed for their own arguments. But this meant that he posed his sociology in equivocal terms that skirted the edge of metaphysics.

Animal Morality

The reality of society made it not only an object of science but an object of moral obligation as well. The existence of society as a concrete being made it possible to define morality as the biological conditions of existence of the collective individual rather than a set of abstract ideals. Needless to say, this contradicted the spiritualist definition of morality, and Espinas devoted the end of his historical preface to an attempt to characterize and refute the objections of the artificialists, chief among whom are spiritualist philosophers. These latter grudgingly admit many of the propositions of the naturalist thesis, "but there is one domain where such habits of thought ... are rejected absolutely—that is the domain of traditional morality, itself founded on intuitive (a priori) metaphysics" (ibid., 143). To the objection that the naturalist thesis ends in moral relativism, Espinas answers that on the contrary it makes morality a real function of a real entity rather than an optional deference to a Platonic abstraction (ibid., 148). But Espinas argues that making morality a part of nature does not diminish its force or dignity. He points out that even Christians like Joseph de Maistre have been able to reconcile the natural character of moral law with their belief in a supreme Being. Finally, to the objection that, in disclosing the merely natural basis of moral law, science undermines the social fabric by eliminating the fear of divine sanction, Espinas answers that science increases our commitment to morality by showing the necessity and benefits of moral action.

But even if one agreed to define morality as the conditions of existence of society, one might still refuse to dignify animal behavior with the name of morality on the grounds that animals do not understand the principles on which they instinctively act. In the last section of the book, Espinas argues that the social behavior of animals can be construed as moral, since it is disinterested action directed by an altruistic impulse from within (ibid., 558). He claims that the ethical principle behind the act need not be clearly formulated for the action to be moral. Intelligence does not change the character of the act; it only makes it clearer. And since Espinas uses the concept of reasoning from particular to particular to argue that intelligence extends far down the evolutionary ladder, many species of animal act with some degree of awareness.

Espinas did not invent any of these arguments for a positivist view of morality. They are significant in this context because they show the extent to which he was addressing the fears of his spiritualist mentors in terms they would understand, rather than simply refuse to address the moral and metaphysical implications of a scientific approach. His ultimate aim was to found a new morality, not simply a science of morality. But the fact that he (accurately) anticipated that spiritualists would object to a scientific approach on moral grounds shows that the naturalistic fallacy was not limited to positivists. Spiritualists too assumed that one's views on how the world is would affect what one ought to do.

Reception

Espinas conceived his work as a single sustained argument, a work of biological sociology that rested on the metaphysical postulate of naturalism and demonstrated the equally metaphysical thesis of the reality of society as a thinking subject. Most of his critics did not see it this way; they could not agree that the parts added up to a coherent whole. Instead, they tended to take the book apart and to judge its biology and its philosophy separately.

The first to do this were the members of his dissertation committee. Espinas finished his thesis toward the end of 1876, and by January of 1877 his advisors were in a quandary over the work they had been given to judge. Three things bothered them above all about the thesis: the scientific difficulty of the biology, the references in the historical introduction to Comte and Spencer, and the theory of the collective conscience. The philosophers were clearly discomfited by the biological learning of their candidate. Paul Janet, a member of the jury and the intermediary between Espinas and the committee, wrote to Espinas:

My dear friend, . . . your thesis is not a thesis of philosophy, but of zoology. It was made for the Faculty of Sciences. On each page, there are words I do not know, facts of which I cannot judge, names of animals of which I do not even know the spelling. How can I vouch for all that?[25]

Nevertheless, this inability to judge the biology did not work against Espinas. On the contrary, it looked like science to the spiritualist philosophers, and indeed they praised it as a contribution to zoology.[26]

The favorable references to Comte and Spencer caused more difficulties, to the point that the members of the committee demanded that he suppress them. Espinas refused, suppressing the entire historical introduction rather than eliminating the founders of sociology to which the history led. Since this incident became a minor cause célèbre in the history of sociology, it is important to fix its meaning as precisely as possible. It was not simply a question of the spiritualists disagreeing with Espinas's unorthodox views. They did not agree with his theory of the collective conscience either, but they did not force him to suppress it. Nor did they force him to alter his Latin thesis on Plato's theory of the Republic, although in that work Espinas tried to show that Plato vacillated between a naturalist and an artificialist view of society, and that all the good points in Plato came when he adhered to the naturalist method.[27] Janet, at least, seems to have had no difficulty approving a competent work of positivist morality, even though he disagreed with it. This is in itself a major change from the era of Cousin, when dissenting philosophers could expect to be excluded from the academy, as was the case with Taine. In this sense, the fact that *Animal Societies* was not rejected out of hand represents a major change in academic philosophy. Janet seems to have feared the scandal that would be provoked by the Sorbonne approving a thesis that supported notorious positivists.[28] Lalande points to the tradition of political intervention in the affairs of the Sorbonne and the precarious situation of the Republic at this particular period, with a royalist majority still in power.[29] The academy was therefore not yet safe for positivism, although it soon would be with the accession of the radical government of Gambetta and his minister of education Jules Ferry.

Once these incriminating references were removed, *On Animal Societies* was defended in early June 1877. In addition to Janet, two other prominent spiritualists, Caro and Waddington, took part in the discussion, and although they praised the zoology, they objected strenuously to the philosophy. Of his Latin thesis A. Mézières remarks that "the author of this work knows natural history better than the history of philosophy," and concerning *Animal Societies,* after admiring the science, he complains that "the author concludes with philosophical considerations that can be accused of obscurity and equivocation."[30] Despite their objections to his philosophy, however, the jury found that he defended them competently and declared him worthy of the doctorate. It is clear, however, that Espinas's display of scientific learning played a large role in their decision.

Although he approved Espinas's thesis, Janet elaborated on his objections in a review of *Animal Societies* for the respected French paper *Le Temps*. He considered Espinas's theory of the collective conscience to be "essentially philosophical," since it is a theory of consciousness. He chided Espinas for making such an elementary "mistake" as thinking that consciousness could be divided or composed.[31] Such a blunder could be expected of a scientist, who could not be expected to know psychology, but it was inexcusable from someone trained as a philosopher. Janet thought that this mistake was the result of Espinas's failure to distinguish between science and philosophy.

> ... in the book on *Animal Societies*, the physiological and zoological facts overflow and fill almost the entire work: the philosophical thought is drowned in the torrent ... in a word, in the book of M. Espinasse [*sic*], the philosopher is somewhat masked by the naturalist. Now this is a rather grave fault: in order to speak as a naturalist, it is necessary to be one all the way. However informed the author may seem to philosophers, he can only seem insufficiently informed to naturalists. Scientists are rather jealous: they do not like it very much if one sets foot in their domain when he is not one of them. As for philosophy, it must not be the servant of science. It must speak in its own name and by virtue of its own right. (ibid.)

In attempting to do both philosophy and natural science, Espinas had succeeded in doing neither. This, he concluded, was the "unfortunate (*fâcheux*) effect of the positivist influence, which tends more and more to drown philosophy in the sciences." Janet thought this was a growing problem in philosophy, and he felt that if philosophy did not retain its autonomy with respect to the natural sciences, it would become no more than the "doubtful, contested, and more or less chimerical part of the positive sciences" (ibid.). That Janet distinguished between science and philosophy reveals how positivism had begun to make eclectics distance themselves from their own scientistic rhetoric.[32]

Ribot, who reviewed *Animal Societies* for his *Revue philosophique*, thought on the contrary that this "violation" was one of the strengths of the work, because it took a problem away from metaphysics and set it in the domain of science. He used the problem of individuality as a case in point.[33] Ribot thought *Animal Societies* was a solid contribution to science, although he was silent on the question of the collective conscience, and in private he was critical of it.[34]

The independent philosopher Alfred Fouillée gave Espinas a major place in his 1880 book *Contemporary Social Science* (*La science sociale*

contemporaine).[35] He was in general enthusiastic about the organic analogy, though he too demurred at the collective conscience. He pointed
out that, from a purely logical point of view, it simply did not follow
that if individuals are societies, then societies are individuals.[36] Nor did
he appreciate the arguments Espinas marshaled in favor of the collective
conscience. Fouillée noted that if the sign (quite literally) of a unitary
conscience is the use of the pronoun "I," then the collective conscience
is not such an individual, since we always speak of society either as it
or as we, but never in the first-person singular (ibid., 230–32).

Other reviewers agreed in praising the scientific character of *Animal
Societies*. The reviewer for the British journal *Mind* remarked that "M.
Espinas takes possession, in the name of science, of ground that is all but
unoccupied."[37] Wilhelm Wundt, one of the founders of experimental
psychology, considered Espinas's work comparable only to Darwin's
Expression of the Emotions in Animals as an example of the scientific treatment of animal psychology.[38] But even Wundt, who would go on to
write a multivolume Völkerpsychologie (*Folk Psychology*) on the collective
mind of society, judged Espinas's collective conscience a metaphysical '
entity that marred an otherwise scientific work.[39]

If philosophers and psychologists were unanimous in calling *Animal
Societies* a major work of science, it does not appear that professional
biologists agreed. No French scientific journal reviewed the work, apart
from the excerpt Espinas himself sent to the *Revue scientifique*. This does
not necessarily constitute an exception, however, since this journal was
addressed to the educated layperson rather than to specialists. It is true
that most French biologists were unreceptive to evolutionary theory, but
strictly speaking, Espinas's work did not require a belief in evolution, as
he himself pointed out (although he thought that evolution provided a
better explanation of the phenomena); indeed, most of his data came
from the studies of French zoologists working in a preevolutionary
paradigm. Much less does this explain the silence of Espinas's friend
and colleague, the biologist Alfred Giard, who was one of the earliest
champions of evolutionary theory in France. And it does not explain
the silence of Edmond Perrier, who wrote a massive tome on *Animal
Colonies and the Formation of Organisms* (*Les colonies animales et la formation des organismes*) that came to many of the same conclusions as Espinas, but who mentions Espinas only once, in passing, even though he
knew about *Animal Societies* long before his own work was completed.[40]
Part of the explanation for this silence must be found in the intolerance
of French science for amateurs. Since Espinas did not have formal training or a degree in biology, no one within the profession deigned to
grant his work the dignity of a review. Nor was his scientific influence
abroad any greater. George Romanes, the English disciple of Darwin
who devoted himself to comparative psychology, to my knowledge

quotes Espinas only once, and that in German translation.[41] Ironically, philosophers disagreed with the philosophy and praised the science, while scientists passed over in silence.

The philosophical reviews gave Espinas ample opportunity to respond and to clarify his views. He took this opportunity in a long review article on "Sociological Studies in France" ("Les études sociologiques en France") for the *Revue philosophique*. In this article, he reviewed the biological work of Perrier, who ignored him; Fouillée, who disagreed with his concept of the collective conscience; and for the first time, a contemporary spiritualist, Caro, who objected to the idea of an evolutionary ethics. It is clear that Espinas was hurt by his neglect within the scientific community, and he took every occasion to show where he had anticipated Perrier as well as to criticize both factual detail and theoretical interpretation in the naturalist's work. But his ultimate revenge was to point out the philosophical character of Perrier's work and chastise him, not for being philosophical, but for being a bad philosopher.

> What emerges from this immense work is a new theory of individuality founded on sociology. A philosophical conception is its soul. No new facts, but new insights, a new interpretation, a new synthesis. M. Perrier is basically afraid of compromising himself in the eyes of scientists by confessing the borrowings he has made from philosophers. And this from a philosopher himself! It is through the stubborn development of an idea that his book, which is full of the research of others, well surpasses the partial studies where this research has been recorded.[42]

This characterization could easily be applied to *Animal Societies*. When Espinas finds fault with Perrier's theory of the individual, which like *Animal Societies* sees ontogeny as a form of social evolution but which retains the idea of a special human psychological individuality, he attributes it to the failure of the biologist to break completely with spiritualist philosophy. Perrier's problem was not that he engaged in philosophy but that he engaged in the wrong philosophy. One service philosophy can provide science is to help scientists avoid such errors.

Espinas also clarified his views on science. He reiterated his belief in a dichotomy between inductive and deductive sciences, and his opinion that the natural sciences should be the former rather than the latter. He characterized the deductive method as scholastic, and criticized the contemporary moral and political sciences for continuing to deduce consequences from first principles that were no more than the "hasty and conjectural classification of the forgotten experiments of the individual or the species" (ibid., 357). These summary views are for Espinas a form of hypothesis, and while hypotheses are necessary in science, they are to be surpassed as soon as possible by verified facts and by the generalizations that arise out of systematic empirical investigation. All science consists of two parts, "doctrines and hypotheses in the process of formation, on the one hand, definitively established laws on the other" (ibid., 366). But if hypotheses represent possibilities rather than truth, there are good and bad hypotheses. A good hypothesis depends in large part on facts and rises organically from them, whereas a bad hypothesis is "hasty and conjectural." In this sense, Espinas admits that "the liberal doctrine thus conceived [his evolutionary ethics] is hypothetical, like the assimilation of the social organism to the individual organism, but like the latter it rests upon powerful analogies and assumes the high probability that is attached to a scientific theory in the process of demonstration" (ibid., 510).

If analogy lends probability to an argument, identity is demonstrative, and as Espinas had made clear in *Animal Societies,* explanation in science is essentially the subsuming of different phenomena under an identical law. In "Sociological Studies in France," Espinas responds to the charge, made by Wundt and Janet, that in identifying the individual with society he has muddled both concepts. He denies that he has ignored the differences between them, since in addition to subsuming them under the same general law he has assigned each its specific place in the evolutionary chain (ibid., 595–96). It is clear, however, that this differentiation admits only of degrees, not kind, and furthermore, that within this view of explanation ultimately there is only one law, that of evolution, under which everything must ultimately be subsumed. When Espinas insists that "if the individual is the product of an association, it follows logically that any association can be individualized," against Fouillée's specific objection that this does not follow, the reason Espinas gives is that nature follows the same laws everywhere (ibid., 336–37). The uniformity of nature demands that the processes observed in the formation of individuals must also be observed in the formation of societies.

This view of scientific explanation is also why Espinas can only conceive of sociology as either a priori or biological. Society is either "an independent whole, having no link of origin and nature with the cosmic milieu," or it "is only the continuation and flowering of biology" (ibid., 565, 566). Just as he can only conceive of society as being either sui generis or biological, he can only conceive of it as either idealist, deductive, and a priori or organic, inductive and empirical.

> If society is a set of abstract relationships, a group of immobile ideas, the science that concerns them belongs to deductive logic and must be constructed a priori; if it is a natural whole, a living body, strict affinities will bring sociology closer to the life sciences. It will be the last of the natural sciences. (ibid., 339)

The nature of society determines the kind of study appropriate to it. But by asking what society is, Espinas claims not to be trying to "discover a new metaphysical essence," but simply to "determine in what manner of existence society should be included, and how and with what beings it should be classified" (ibid).

With respect to the collective conscience, Espinas continued to maintain its reality, but he specified the kinds of representations that entered into it. He distinguished between *conscience* and *connaissance,* between self-consciousness, which Espinas maintained is simply the knowledge of one's own body, and knowledge about the external world. If the former is incommunicable, the latter is eminently communicable, because capable of public reference, and it makes up the collective conscience (ibid., 343–45).

With this goal in mind, Espinas took on one of his dissertation committee members, Caro, who argued that spiritualist morality is incompatible with evolutionary morality. Espinas agreed, but he thought that the difference was in favor of evolutionary ethics. Espinas characterized spiritualist morality as the juxtaposition of absolute individualities, each with a priori inalienable rights. This conception of morality makes society impossible, since it makes any alienation of rights and any subordination of one individual to another illegitimate. If on the other hand, individuality is relative, morality is obligation to the collective individual that is the group, and rights are derived from association with that group, then society is no longer an intolerable imposition but a necessary condition of freedom. Spiritualist individualism leads straight to socialism, since it makes any inequality of property or condition illegitimate. Hence evolutionary ethics alone are suited to the modern nation, which demands both differentiation and participation.

University Career

If Espinas was emboldened to attack his judges, he could do so because he had finally attained the security of a university chair. Despite the scandal caused by Espinas's dissertation, not only did he receive his doctorate, but in short order he received the university appointment for which the doctorate was necessary. Moreover, he was supported in his bid for a university position by some of the very people who had opposed his ideas at his dissertation defense.[43] A number of considerations help explain this seeming anomaly. First of all, one must distinguish between personal disagreements and professional competence. Although Espinas held views that were in many ways diametrically opposed to eclectic orthodoxy, he upheld them with skill and became an acknowledged expert in the "scientific philosophy." In addition, he had become a very effective teacher, as even his ideological opponents had to concede. The heterogeneous nature of eclecticism also plays some role in the official acceptance of Espinas—eclecticism was by definition a house with many mansions.

But more than professional courtesy may have been at stake. If Espinas's dissertation had been controversial at a period when the Third Republic was fragile and contested, under the propositivist cabinet of Gambetta and Ferry, it became an asset. Although the minister of education would not in general override the spiritualist philosophers who controlled university appointments, a number of initiatives by the Ministry to reinforce the scientific character of secondary education—some of them aimed at the philosophy class—made it clear that the scientific philosophy was in favor.[44] Among the accommodations to this atmosphere, the elevation of the most notorious scientific philosopher within the educational establishment was one of the most obvious and least offensive, given that he was in many ways one of their own.

For these reasons, Espinas was appointed *maître de conférences* in philosophy at the University of Douai in 1878, and in 1880 he replaced his classmate Louis Liard at the University of Bordeaux. He was well liked at Bordeaux by his superiors, colleagues, and students. In addition to his lectures in philosophy and "experimental psychology," he created a course in pedagogy that was attended by over a hundred primary and secondary schoolteachers every year. He became dean of the faculty of letters in 1887, and honorary dean in 1890. In short, he had achieved a good deal of success within the philosophical establishment.

This did not mean that he did not continue his vociferous opposition to spiritualist philosophy. On the contrary, perhaps his most impassioned attacks came during this period. Two statements from 1884 illustrate his continued hostility toward spiritualist philosophy, which the security of a university chair and a sympathetic government allowed him to vent. In an article on the reform of the *agrégation* in philosophy, Espinas opposed his former classmate Emile Boutroux, who had noted the recent rise of scientific philosophy only to conclude that it should not be included in the philosophy class and therefore not in the *agrégation* either.[45] Espinas replied by taking up Boutroux's distinction between traditional and scientific philosophy, but reversing the valuation of each. Boutroux had argued that whereas traditional philosophy stressed the internal history of the spirit, scientific philosophy viewed human thought from the outside and eliminated the past as a history of error in favor of a scientism based on the latest findings of science. Espinas responded that this may have been true of positivism in its original form, but it was not true of the new scientific philosophy, which was imbued with the spirit of evolution. In fact, it was traditional, or autonomous, philosophy that stressed the immutable character of the human soul. "We still have our ears full of those beautiful words immutability, eternity, indefectibility, that were incessantly on the lips of our professors."[46] In fact, autonomous philosophy was extremely poor in ideas, despite its lofty object; a few phrases could sum up everything it had to say. That is why, according to Espinas, form was more important than content in this philosophy. The best philosophers of this school were *littérateurs,* and one's performance on the *agrégation* was based more on style and taste than on knowledge. Hence a partisan of the scientific philosophy, who cared more about facts than about style, would have a hard time impressing the jury. The trained scientist, who would not have the literary training needed to perform for the philosophers, would have no chance at all.

But of course, it was not only style but substance that separated the autonomous from the scientific philosophy. The latter attempts to connect human phenomena with the rest of nature, to study human phenomena by the methods of the natural sciences. A scientific philosopher would naturally favor those philosophers who did likewise and therefore "glorify Hobbes and Hume, refute Leibnitz by Spinoza, show the merits of the empirical psychologists of the eighteenth century. . . ." (ibid., 596). But this would create a scandal among spiritualists, despite their tolerant intentions. In this situation, the advocate of scientific philosophy has only three choices before the jury of the *agrégation*—to lie, to abstain from the teaching profession, or to resign himself to teaching in the lesser *collèges* (ibid., 598). Such a statement is extremely interesting from an autobiographical perspective, since it is precisely the situation in which Espinas found himself. While it reflects Espinas's anxieties about his position within the academy, it also shows the exaggeration in his claims, since he had in fact been forced into none of the three choices he posed.

If the institutional reign of the spiritualists was still quite firm, the appearance of more and more scientific philosophers on the margins of the academy would eventually expose the "violent character" of this reign (ibid., 599). Moreover, Espinas insisted that official philosophy had stagnated—what French philosophers of the nineteenth century were known abroad? He predicted the eventual triumph of the scientific philosophy, both because of its intrinsic superiority and because it corresponded better to the needs of modern society. Ultimately, Espinas saw spiritualist philosophy as a political instrument created by Cousin as a rallying point for moderate conservatives and liberals after the Restoration. It promised and delivered a moral pabulum that placated both Catholics and lay progressives. It fulfilled this political need brilliantly, but by committing itself to a dogmatism of the *juste milieu,* it not only compromised its intellectual integrity, it prepared its own political obsolescence by linking itself to a historically specific social structure. Social conditions had changed since the Restoration, however; what was needed now were men of facts, not philosophical generalities. Only the scientific philosophy could educate these, the leaders of tomorrow. But beyond this, the triumph of scientific philosophy would herald the return of independent research, unfettered by political fears. It is interesting that Espinas did not oppose autonomous to scientific philosophy as conservative to liberal, but as political to apolitical. He concluded by opposing two systems of culture, humanism and naturalism.

This article amounted to a condemnation of spiritualist philosophy as a corrupt and outmoded institution. In an address to the incoming students at the University of Bordeaux that same year, Espinas went even further in his condemnation of philosophy. He referred to the current crisis in philosophy, characterized by three competing schools of thought—spiritualist, neo-Kantian, and evolutionist. He condemned the stifling effects of a highly institutionalized philosophical profession. He opposed the metaphysical morality taught in the philosophy class, claiming that with the collapse of religion, science and not metaphysics should provide the foundations of morality. He claimed that science, as an intellectual focal point, dispensed governments from having to create and maintain a public morality. "No more official philosophy; free consciences in a neutral state; no other teaching than that of the sciences. . . ."[47] Such would be the eventual character of public education. In the meantime, while they waited for the social sciences to be sufficiently constituted to replace the philosophy class, Espinas recommended that the future teachers of philosophy restrict themselves to the history of their discipline and to avoid dogmatic philosophy in their classrooms. Essentially, Espinas was telling philosophy students that their profession was doomed, and that they should cause as little trouble as possible in the transition from metaphysics to social science.

This speech represents a radicalization of the position he had taken on the relation of metaphysics and science in his writings around *Animal Societies*. In his earlier writings, he already had developed the opposition between scientific and spiritualist philosophies. What has been added is an explicit political perspective that reveals his desire to portray scientific philosophy and social science as apolitical, a rallying point for all, regardless of political viewpoint. He does not think that philosophy needs to be taught, even his own, because it rises naturally out of the natural and social sciences. This Comtean desire to end politics by the coordination of consciences through science has been joined with Spencer's liberalism. Moreover, Espinas has identified spiritualist philosophy with the metaphysical stage in the Comtean historical scheme of things.

In addition to spiritualism, the 1884 address evoked the specter of neo-Kantianism, which Espinas had come to fear as much as spiritualism. Unlike spiritualism, neo-Kantianism was a new phenomenon, and Espinas describes it as a new religious cult, designed like spiritualism to found a rational faith without disturbing received Christian dogma.[48]

His opposition was based on the distinction between science and philosophy, between nature and morals, maintained in neo-Kantian thought. This meant, Espinas noted, that the principle of determinism did not hold in human affairs, and hence that morals could not be treated according to the methods of the natural sciences. Worse still, in its Renouvierist variant, liberty was a feature not only of the moral but of the natural world, which undermined the possibility of natural science. Espinas was not impressed with the epistemological subtleties of critical philosophy, nor with its claim to critique dogmatic metaphysics. Indeed, he saw this merely as a cover for religious ideology that only served to seduce intelligent people. And it was seducing large number of philosophers. From being an isolated crusade of an outsider, neo-Kantianism had, by the blessing of Lachelier, become respectable within the academy. The Franco-Prussian War, creating a moral crisis, had led many to seek an alternative faith, and in neo-Kantianism they found it. Espinas would never lose his suspicion of Neo-Kantian philosophy, although he eventually resigned himself to peaceful coexistence with it.

Despite these vehement denunciations, Espinas used his connections with philosophy to advance his own views and his own career. His friend Ribot worked in Paris to make Espinas better known among the people outside philosophy who would have some say in his eventual appointment to a position in Paris. Ribot even used his own candidacy for the Collège de France to promote Espinas, urging him, when it was still unclear whether the chair would be changed to psychology, to put himself forward as a candidate for the previous chair in philosophy of law and coaxing him, after the chair was in fact changed, to present himself as a candidate for the second choice.[49] Espinas did this, and although he got very few votes, Ribot assured him that he had made himself better known and would do better in the future.[50] Ribot clearly thought Espinas would eventually be a good candidate for any future chair of sociology at the Collège, as well as for one of the chairs of philosophy, and he even made a rough count of the support Espinas had as of 1888: at least half the scientists, several orientalists, a couple of historians, and of course Renan.[51] He could not count on the philosophers, of course, since both were eclectic spiritualists, but since there were only two chairs in philosophy, this opposition could be overcome, as it had been in Ribot's case. Ribot kept Espinas informed of changes in the Parisian academic world, at both the Collège and the Sorbonne, to enable his friend to act quickly should an opportunity arise.[52]

Espinas's opportunity came not at the Collège de France but at the Sorbonne, and he had plenty of support from his colleagues in philosophy. By the fall of 1893, Paul Janet and Emile Boutroux were calling for the creation of a "scientific course" in philosophy.[53] Both felt it was necessary to keep up with new developments "in philosophy," and to do this the Faculty of Letters needed a course either in experimental psychology or sociology. This time Janet was not a lone voice, and indeed the only opposition came from Waddington, who supported the creation of a course in psychology, provided it was sufficiently philosophical, but who opposed sociology on the grounds that it was not yet a constituted science.[54] After some discussion the faculty voted to request formally the creation of *cours complémentaires* both in experimental psychology and in sociology, in that order of preference. There was, then, considerable support for scientific psychology and sociology, although it is significant that these were considered parts of philosophy. However, it is necessary to note the limits of that support, because at the same time that the faculty voted for these new courses, they also placed them at the bottom of their list of priorities. Augustin Cartault, the professor of Latin, asked that before any other new courses were approved his own request for a second course in Latin be granted. This request was backed by Janet and approved by the faculty. The human sciences had to compete not only with other courses in philosophy but with the other disciplines within the Faculty of Letters.

The human sciences also had to compete with the needs of other faculties within the University of Paris—Protestant theology, law, sciences, and pharmacy. Once the Faculty of Letters had made its request, these had to be negotiated with the other faculties at the General Council of Faculties. Accordingly, the dean of the Faculty of Letters, Benjamin Martha, presented his faculty's requests at the council meeting of 23 December 1893. These were sent to an interfaculty committee charged with recommending an order of preference. Martha was a part of this committee, and when the committee met on 19 February 1894, he came armed with reports by Janet presenting the reasoning of the Faculty of Letters.

Janet's report is interesting because it reveals the extent to which he still conceived of sociology as a part of philosophy. He calls it an "annex of moral philosophy and of history."[55] Sociology is necessary because it is the general science synthesizing the results of the particular sciences of jurisprudence, history, political economy, et cetera, which only studied specific aspects of society. But precisely because it is so general and synthetic, it is philosophical in character. Indeed, sociology is only the "objective" side of social philosophy—objective, because it seeks general principles through the examination of facts rather than through the "subjective" method of pure reasoning and the examination of one's own conscience. Hence, while legitimating empirical sociology, Janet tied it to philosophy. As an eclectic, he saw no contradiction between philosophy and fact-gathering.

As it turned out, the discussion at the interfaculty committee did not focus primarily on the need for sociology in the university, which was generally acknowledged, but on which faculty should receive it. The Faculty of Law also wanted a course in sociology and nearly obstructed the Faculty of Letters's efforts. In the end, the Faculty of Law relented, primarily because of a fortuitous development. Between the time that the Faculty of Letters made its request and the time that it arrived for consideration before the General Council, the Comte de Chambrun had donated money for the creation of a course in "social economy" at the Faculty of Letters. This meant that no new money would have to be allocated, always a powerful argument. Hence, the council decided that this course would for the time being satisfy the need for sociology at the Sorbonne.

It is not clear whom the philosophers had in mind to fill the *cours complémentaire* in sociology when they first proposed it. Espinas was the senior specialist in positivist social philosophy among university professors, but he had not himself undertaken any empirical studies in sociology. Moreover, since he was already a full professor and dean at the University of Bordeaux, a *cours complémentaire,* even at the Sorbonne, was not necessarily a step up in his career. For these reasons, the position could equally well have gone to Emile Durkheim, who was already a professor of *science sociale* at Bordeaux but who had only just defended his dissertation at the beginning of 1893. Much depended on how the position was defined, whether as empirical sociology or as sociological theory. Durkheim would be the better candidate for the former, Espinas—because of his seniority—for the latter. Despite its lowly and provisional status—the Comte de Chambrun had only provided funds for a limited number of years—Espinas expressed great interest in the position. Nevertheless, he resisted attempts by the historian Ernest Lavisse to have the chair defined as social and economic history.[56] Espinas lamented that he was not a historian but "a fifty-year-old philosopher," and in the end, the position was defined as the "history of social economy."[57] When the faculty voted on Espinas's candidacy for the position, Janet and Boutroux approved heartily; in fact, the latter presented Espinas's credentials to the council. Whereas at his dissertation defense it had been his claim to science that had made him acceptable to philosophers, in this case it was his credentials as a philosopher rather than a scientist that carried the day.

The courses that Espinas gave at Paris were indeed more philosophical than scientific, by his own admission. Over the years they constituted a survey of the history of social theory. Espinas also had to admit that he was warmly welcomed by the philosophers he had so thoroughly denounced.[58] Indeed, Espinas seems to have gotten along well with his colleagues at the Sorbonne, and he devoted a great deal of time and energy supporting the philosophy program, sitting on dissertation and *agrégation* committees, grading *licence* exams, and coaching students.

Since there were no degrees or careers in sociology, he had little choice but to encourage students to become philosophers with a positivist inclination. He continued to advocate a positivist approach to social phenomena, but he did so primarily from what even he would call a philosophical point of view.

FROM THE SCIENCE OF THE ORGANISM TO THE ART OF ORGANIC ACTION

If his position in the Faculty of Letters obliged him to remain connected with philosophy, Espinas was constrained in other ways as well. He complained repeatedly that his course preparation took up too much of his time, and he felt that the effort was not worth it, because he had only a few students.[59] Moreover, despite the fact that he had written a book on the history of economic theory that had been one of his principal qualifications for his position, the history of social theory was in fact tangential to his real research interest, the history of the theory of action, or "technology," as Espinas referred to it.

Despite his attacks on the spiritualist establishment and despite his demonstration of the individuality of society, "artificialist" philosophies did not go away; if anything, they continued to grow, as the spread of neo-Kantianism indicated. This may be one reason why, after 1884, Espinas had increasingly concentrated on the relation between science and art, or between theory and practice. He began a series of studies on the practical arts—among which he placed medicine, pedagogy, politics, and morals—that occupied him until his retirement from university teaching in 1907. Taken together, they were to constitute a "general technology," and he entitled one of this series a "Sociological Study." These studies therefore represent Espinas's efforts to contribute to sociology in the best way he knew how. Nevertheless, like *Animal Societies,* the studies of technology produced consternation on the part of both philosophers and social scientists, since they were almost exclusively concerned with the history of the philosophical theories of these various practices. Indeed, Espinas believed that the philosophy of practice was entirely neglected by spiritualist philosophers, who thought only about the most abstract problems. Here again, Espinas developed a counter-philosophy rather than an empirical practice.

In developing his own philosophy of action, Espinas also tried to figure out why his scientific theory had so little impact on practical action and to discover conditions under which it would. In the course of this investigation, his ideas about the nature of action changed considerably, as Davy has pointed out.[60] Nevertheless, they represent the development rather than the repudiation of his organic and evolutionary schemata, the application of this schema to his own ideas.

His early ideas were indeed optimistic. He seems to have thought not only that science could provide rules for conduct, but that people would willingly submit to the rules thus scientifically grounded. On the subject of international cooperation, for example, he argued that

> . . . since art and science are correlative, once science finds that the condition of existence of a nation is the subordination of its interests to the interests of a broader group, that is once it is demonstrated that an organic solidarity ties it to other nations, thanks to the multiplicity and the complexity of the commercial and sympathetic relationships, then it will be necessary to admit that, in order to exist, it must at the same time want the existence of the higher organism thus constituted.[61]

He also thought that one could change the will through the action of ideas, and he explicitly differed from Ribot on this point.[62] Although this faith in the ability of science to direct a new morality is typical of the positivist tradition, it is also striking how much similarity it bears to the Socratic dictum that to know the good is to do it. Such moral intellectualism engendered great hopes for the unification of society, but it also made the resistance to scientific morality inexplicable except in terms of ignorance or error.

If Espinas thought that science could aid morality, he did not conflate the two, although he thought that the distinction evolved only gradually. Originally, the two were one, as the early enterprises of alchemy and astrology attest. Since people did not know how nature worked, they formed hypotheses that reflected their desires more than their observations. Science was deductive, because it had the prescriptive form of art. As science developed, it gradually separated itself from art, which needs rules for action even where there is no certain knowledge. But even if science has become differentiated from art, the practical interests in which science originated have not disappeared. "All science, we know, has as its goal the adaptation of our action to the conditions of the environment."[63] Its mastery of the laws of nature makes it an invaluable tool in constructing rules for action.

More striking than the relation of science to art in Espinas is the conflation of means and ends implicit in the assimilation of morals to a practical art. In theory, art is couched in terms of a hypothetical imperative—*if* you want to do this, you must do that. Hence the ends seem to be left open. But Espinas maintains that in reality, the prescriptions of art are effectively categorical, because the desirability of the goal is more or less assumed. There is no art of the bad or undesirable. More fundamentally, Espinas insists that ends are embodied in practice because if they were not, they would have to have the form of abstract and absolute principles, as the spiritualists insisted. By embedding ends in the practices used to realize them, and then studying those practices, Espinas hopes to show that ends are determined by the collective practices of society rather than by eternal principles. Morals depends on politics, not politics on morals.[64]

At this early stage in his career, Espinas also thought that philosophy had an important role to play in the practical arts. In a discussion of pedagogy, a subject in which he was actively engaged, he maintained that philosophers were the proper teachers of pedagogy, since they alone understood the human mind well enough to instruct others how to influence it.[65] Moreover, philosophy also established the most general ends of the arts, as it discerned the most general laws of the sciences.[66] Hence, philosophy had a unifying and directing function in art as well as in science.

Such was the framework within which Espinas began to study the practical arts, a new discipline he entitled "general technology" (*technologie générale*) or "praxiology" (*praxéologie*).[67] The results of this study appeared in several books, on *The History of Economic Ideas (Histoire des doctrines économiques)*, *The Origins of Technology (Les origines de la technologie)*, and *The Social Philosophy of the XVIIIth Century and the Revolution (La philosophie sociale du XVIIIe siècle et la Révolution)*, as well as a number of articles. These works are all essays in the history of ideas rather than the history of technology or practical arts as these would be defined today—economic theory rather than economic practice, political thought rather than political action. Although Espinas acknowledged that *technologie* should begin with a descriptive analysis and classification of the arts, in which the sociologist would procede "like the botanist or zoologist," he declared that he was instead undertaking the history of theories of action. In essence, these works are to *praxéologie* what the historical introduction to *Animal Societies* was to the body of that work. He justified this approach by noting that "the history of the philosophy of knowledge has had its historians; it is perhaps not inappropriate to attempt the history of the philosophy of action."[68]

Nevertheless, these studies were not unrelated to the history of actual practice. Espinas noted that one cannot reflect on practice unless it exists, and that the character of the reflection depends greatly on the state of the corresponding practice. Hence he undertook to describe generally the state of the arts at the different periods he discussed, and to show how the philosophical theory reflects it. In *The Origins of Technology*, Espinas traced the philosophy of action in ancient Greece from primitive times to Socrates. He found in this period three successive theories of action—physico-theological, artificialist, and naturalist. In the physico-theological theory of action, practical arts were seen as the creation of the gods, not of man, as the myth of Prometheus illustrates. They were rudimentary and traditional, with moral and religious sanctions to prevent their change or appropriation. The corresponding political practice was the hereditary and divine monarchy or oligarchy. As Greek commerce developed and practices began to change in the seventh and sixth centuries B.C., people began to conceive the idea that such practices were created by men rather than by gods. This led to a utilitarian view of practice, and an artificialist theory of action became possible. In politics, tyranny replaced traditional forms of government. Finally, in the fourth century, a naturalist theory developed in the thought of Aristotle that combined the best of both earlier theories—the idea that collective practices have their own logic independent of individual initiative and that the laws of practical action are not transcendent gifts of the gods but rational principles derived from experience and experimentation.

In all of these works, Espinas inserted the dichotomies made familiar in *Animal Societies*—absolute/relative, artificialist/naturalist, mechanical/organic, et cetera. Thus the history of economic theory is depicted as the struggle between a utilitarian theory of the acquisition of wealth and a moral/theological objection to the indefinite accumulation of riches. The theory of action involved a prolonged debate between the artificialist and the physico-theological schools, and Espinas championed the sophists as defenders of the artificialist thesis against what he called the reactionary theologism of Socrates. Espinas continued to develop the themes of his earlier work throughout his career.

He also continued to develop the explanatory model by which he tried to explain social phenomena—the organic analogy. He compared the practical arts to the instincts of the individual organism, as inherited adaptive behaviors of the social individual. He traced the development of the practical arts as an evolutionary process of increasing differentiation and coordination of the collective organism to its environment. Moreover, the arts, as adaptations to the environment, preexisted clear and disinterested scientific knowledge of the world. Since societies have to act, collective organisms develop impulses that force them to respond allowing them to survive long before they have acquired scientific understanding of the world. This is fortunate, since if we waited to act until we had such knowledge, we would surely perish.[69] Although Espinas recognized the increasingly rational character of practices, he held that they took the form of rationalizing and improving inherited forms of action rather than displacing the fundamental goals of the practical arts. Interestingly, he never invokes the social industry of animals, although he had featured this in *Animal Societies*. The middle term of *animal societies* drops out of Espinas's later works, having served the purpose of making the analogy between human society and individual organism credible.

As Espinas developed the implications of this view of the practical arts, he realized that they undermined his earlier optimism in the ability of science to modify conduct. He expressed his reconsidered position in his inaugural address at the Sorbonne. The motivating ideals of society are developed from preconscious adaptive impulses rather than from the results of science and rational deliberation. And since these impulses are designed to override the hesitations of reason, reason has only limited influence on them.[70] Moreover, the relation between the various practical arts, and between the goals of action that arise from them, is organic rather than logical, the result of slow adaptation and adjustment rather than deduction from first principles. This means that action is ultimately irrational, if by rational one means logically deduced from a set of self-consistent principles. Such irrationality threatens not only the possibility of scientific influence on social action, but the very possibility of social science itself, since it renders action unpredictable.

> Science is truth or it is nothing; it has value only if it is objective, impersonal, universal, and reproduces as in an indifferent mirror the order of completed things. Every doctrine of action, on the contrary, seizes the future for the good of a given individual or collective being; not only does it predict this future, it creates it in conformity with an ideal that the needs of social life suggest to it. (ibid., 330)

Espinas remarks that it was only gradually and unwillingly that he came to this view, which "shocks our intellectualism" (ibid., 332). But it was implicit in his Spencerian organicism. Spencer was a firm believer in the principle that reason is and ought to be the slave of the passions, since the passions represented the inherited adaptive responses of the species. Moreover, Espinas wanted to use social science to restrain the revolutionary impulses of radical reformers by showing the limitations of deductive reasoning in social matters.

How, with such a view of action, are social science and rational influence on action possible? Social science remains possible, because the collective will, if not rational, is nevertheless not undetermined. Espinas resists the idea that the will, individual or collective, is free. It is determined precisely by the past impulses and adaptations of society, and these can be studied and predicted. It remains true that this view of action creates a gulf between the natural and the social sciences, since inert matter merely follows the laws of nature without interposing the determinations of a will. But social science itself must be organic rather than geometrical, observational rather than deductive. Espinas's new view of social action reinforces the opposition between organism and syllogism that informs so much of his work.

If science cannot pretend to dictate ends to action, it can nevertheless have a more modest influence on action. It can show that a certain course of action will not produce the desired effect. It can excite countervailing emotions to offset the dominant social impulse. And it can, given time, inculcate alternative channels along which fundamental impulses can flow. This is how education can change social action (ibid., 334).

One overriding question remains, however. If the highest social ideals are formed gradually by the evolutionary process of organic development and adaptation, how is it possible that competing and even diametrically opposed ideals can coexist in the same society? In his inaugural address, Espinas tries to explain this phenomenon, which he simply stated as an inexplicable fact in *Animal Societies*. One explanation is provided by the Comtean notion of a transitional stage of history. When societies change, there is often a period when old ideas linger while new ones struggle for ascendancy. Thus, egalitarianism arose in opposition to the tyranny of absolute monarchy, which it eventually conquered, and by the end of the nineteenth century it still reigned, even though its extreme individualism was now beginning to be harmful rather than helpful.

But the opposition between individualism and organicism, as Espinas had repeatedly shown, was a permanent feature of human thought, even if the ascendency passed from one to the other. It was this permanence that threatened the complete victory of organicism. To the solution of this problem Espinas predictably applied the perspective of organicism. Society is not a monolithic, logical system. Action is the resultant, not of one ideal translated deductively into commands, but of "the harmonious conflict of tendencies regulated by the social milieu" (ibid., 330). One should therefore expect opposites to coexist in any society. This provided Espinas a way of reconciling himself to the continued existence of spiritualism, neo-Kantianism, and even Catholicism, as long as they accepted the basic goal of national survival.[71]

Reaction

Espinas's studies in *technologie,* with their mix of philosophy, intellectual history, and sociology, proved even more baffling for critics than did *Animal Societies*. Ribot wrote Espinas that the consensus among his friends was that he should "leave all this erudition behind and return to the path where he had walked so brilliantly."[72] Durkheim thought that "Espinas's book is very embarrassing. It is a work of sociology and then again it isn't; it's just as much a somewhat eccentric (*fantaisiste*) history of philosophy."[73] Paul Fauconnet, a protégé of Durkheim who assessed *The Origins of Technology* for Ribot's *Revue philosophique,* also considered it more a work in the history of philosophy than a work of sociology, although he admitted that it was conceived in the spirit of contemporary sociology.[74] He found the book a mixture of philosophy, psychology, and sociology and concluded that *technologie* as Espinas defined it was not a distinct science. Even the spiritualist *Revue de métaphysique et de morale* criticized Espinas for concentrating on the history of ideas rather than on the sociology of the practical arts.[75] Whatever Espinas was doing, it was not empirical sociology as readers by the 1890s had come to expect it.

To some extent Espinas seems to have concurred in this assessment. He defended himself against the charge that he had not put enough facts into these studies by reiterating that such had not been his purpose.[76] But if Espinas's ambiguous status between philosophy and science served him well in his career, since it allowed him to be a scientist or a philosopher as circumstances required, it ultimately turned against him, since even his friends were increasingly dissatisfied with his approach. Nor did the Parisian students and public find that he responded to their expectations, either. His first lectures met with much dissatisfaction, and he managed to retain only a small number of students during the years of his teaching in Paris.[77] His research into the philosophy of action fell on deaf ears, and Durkheim's rising star soon eclipsed Espinas's setting sun.

4

Becoming Philosophy: The Transformation of the Academic Tradition

> ... philosophy shuns the search for purely scientific facts, even those with the name of psychic or psycho-physiological to recommend them. We must state in advance: here it [philosophy] will contribute not facts, but ideas.
> —[Alphonse Darlu], Preface, *Revue de métaphysique et de morale*

ESPINAS and Ribot were not alone in abandoning academic spiritualism for a naturalistic approach to moral phenomena. An increasing number of young academic philosophers, often inspired by these two, became dissatisfied with their discipline and tried to establish an alternative. They joined the chorus of those who criticized philosophy as an institution and as an intellectual method during the first years of the Third Republic, posing a serious challenge that academic philosophers could ignore only at their peril.

As we have seen in chapters 2 and 3, academic philosophers did not ignore the challenge, nor did they continue simply to dismiss and exclude those who advocated the naturalistic approach. They redefined academic philosophy to include a variety of different schools of thought and changed the criteria for access to university positions from philosophical orthodoxy to professional competence. In this way, they were able to make room for the positivists. However, they also made room for other points of view, and academic philosophy did not succumb to the attacks of positivism. Positivists never became a majority of French academic philosophers; despite defections, nonpositivist philosophy expanded even as psychology and sociology established themselves on the margins of philosophy. Philosophers continued to use old arguments against the naturalistic human sciences. They also developed new ones. Perhaps more important, they reexamined the notion of science itself, challenging the preconceptions that underlay the thought of people like Espinas and Ribot. These arguments played a role in the eventual eclipse

134

of the evolutionary and biological approach Espinas and Ribot had championed.

In this chapter I examine the challenges to academic philosophy during the first years of the Third Republic (roughly 1870–90) and the response of academic philosophers to these threats. The relationship between academic philosophy and the positivistic human sciences can only be understood in the context of this more general "crisis" of philosophy, since the challenge to philosophy was often posed in terms of the replacement of philosophy by a scientific approach to moral phenomena.[1] Another part of this crisis involved the end of eclectic hegemony at the level of higher education and the proliferation of different philosophical orientations and systems. The strategy of accommodation toward positivism was in fact only part of this larger change in academic philosophy. Since each major philosophical school—eclecticism, neocriticism, and neospiritualism—had a different response to the challenge of the human sciences, it is necessary to look at each one separately.

THE CHALLENGE OF THE REPUBLIC

In chapter 1, I detailed some of the early opposition to eclectic orthodoxy. That opposition did not end with the collapse of the Second Empire. On the contrary, with the advent of a Republican government committed to an expansion and renovation of the educational system, dissident ideas could find a ready ear among reforming administrators. Republicans considered education to be the foundation of modern society, providing citizens with the basic skills needed to survive in a commercial society, enabling them to participate in a democratic polity, and inculcating the values needed to sustain that polity.[2] Many French sincerely believed that the cause of their defeat in the Franco-Prussian War was the superior education of the Prussians.[3] But if France was to remain a Republic, not just any education would do, however advanced its technical competence. That education would have to be controlled by the state rather than by the Catholic Church, which was perceived to be hostile to the Republic. Hence the Republic embarked upon a massive campaign to reinforce public education, of which the most famous fruit was the Loi Ferry of 1882, which made primary education obligatory, free, and laical.

The Republican emphasis on education also affected secondary and higher education. Money was poured into these systems, despite the fact that they were already fairly well developed, firmly entrenched throughout the country, and fulfilling the social demand for higher education. This demand was in any event effectively limited by the cost of higher education, which Republicans did little to alleviate. The reason for this investment, as Victor Karady suggests, had more to do with the fear of Catholic competition than with an increase in the demand for higher education.[4] The state had to improve its schools in order to attract the bourgeois clients who might be tempted to place their children in Catholic institutions.

Whatever the motivations, the results were dramatic. To begin with the secondary system, the increase in the number of professors in the Third Republic, increased their visibility and presence; and the higher levels of remuneration raised their status from marginal *déclassés* to respectable bourgeois. Moreover, the expansion of the universities led to increased career possibilities for lycée professors. With these advantages came a demand for greater qualifications, and by the turn of the century, the majority of professors held the *agrégation*. These improvements affected professors of philosophy, since philosophy remained the culmination of the secondary system and the pathway to higher education. Indeed, this period came to be seen as a golden age of the professor of philosophy, whose august presence and noble thought indelibly impressed the minds of the nation's elite.[5]

But if philosophy maintained its preeminence in the lycée, it was not without challenges, provoked in part by increased expectations of a better education. The disagreement came over what one meant by "better." Falcucci identified three axes of issues along with the demands for educational reform at this period: (1) Ancients versus Moderns, (2) humanities versus sciences, and (3) disinterested versus utilitarian education.[6] One might add another—(4) object versus method—which separated those who thought that different subjects should be taught from those who thought that the same subjects should be taught, but that teaching methods should be modernized.

The Ancients versus Moderns dispute questioned the primacy of Greek and Latin within the lycée curriculum and suggested that the study of French authors, especially those of the seventeenth century, could teach the same values without encumbering the student with unnecessary linguistic burdens. This idea was put forth even by some classical scholars and linguists, such as Michel Bréal, who thought that ancient languages would be more profitably taught in the universities.[7] Advocates of such reforms did not necessarily disagree with the goals of classical education—to expose the mind to the best in human experience, to exercise the intellect and elevate the conscience, and to develop an interest in knowledge for knowledge's sake. They simply thought that French authors could do the job as well as the ancients. This issue came up in regard to two concrete questions: the number of hours devoted to classical as opposed to modern languages in the lycées, and the status of French language studies in the schools of secondary special education originally established as practical alternatives to the lycée for students destined for agricultural, technical, and industrial careers. The former question resulted in a tug-of-war over hours in the school day that was eventually resolved in favor of the classics. The latter also initially met with resounding defeat for those who proposed that special education be made comparable. One such effort involved a proposal that the title "special secondary" be changed to "classical French." Typical of the resistance to this suggestion was the report of the Conseil, drafted by the lycée professor Elie Rabier. His committee maintained that "there is only one truly classical education, the education of which the base is the study of ancient languages."[8] The proposal was defeated. The situation was only partially remedied with the reform of secondary education in 1902, when special education was given the status of the lycée and modern languages became an optional specialization on a nominally equal footing with the classical option.

Many academic philosophers upheld the primacy of classical education, for a number of reasons. Greek and Latin thinkers made up a large portion of the history of philosophy, and professors wanted students to have a familiarity with the original language of these authors, even if the majority of readings were done in French. Furthermore, the philosophy class was linked institutionally with classical education, and any attempt to circumvent the classics was likely to circumvent philosophy as well. To include more "moderns" might also be taken to imply including contemporary philosophers, and this would open the philosophy class to criticism from both Catholics and positivists. For this reason, the syllabus of 1874 ends with the eighteenth century, and even the syllabus of 1884, set after the Republic was firmly established, includes only Cousin among nineteenth-century philosophers (Appendix B).

Most important, though, the whole ideology of *philosophia perennis* was part of the humanistic notion that the ancients had lessons to teach the moderns. This was not inconsistent with Republicanism, despite the fact that conservatives also defended classical education.[9] Greece and Rome were republics, after all, and their values could be brought to bear on modern France. For this reason, many liberals also defended the classics, and an academic philosopher such as Alfred Fouillée could argue that the French Republic needed an elite with a background in the classics, and that the moral values of classical education should be extended to all citizens.[10] Espinas and Durkheim both attacked the ahistorical theory of human nature inherent in classical education, but many people—liberal and conservative alike—still adhered to it.

Instead of the argument that the Moderns are as good as the Ancients, opponents of classical education sometimes claimed that modern studies are more relevant and useful than classical. At the root of this claim was a conception of education as preparation for the world that was radically opposed to the idea of education as a Platonic contemplation of the true, the beautiful, and the good. Education must respond to the needs of the modern world, and the modern world needs scientists and businessmen more than writers and philosophers. This argument never gained much of a following among philosophers, who were generally opposed to viewing education in utilitarian terms. Not that they opposed utilitarian education altogether. On the contrary, they thought it necessary and desirable for those who aspired to nothing other than practical careers.[11] But for the elite, they thought that, like Plato's philosopher-king, students could only be truly effective as leaders if they first pursued truth for its own sake.

If philosophers defended the disinterested character of education, and the traditional range of subjects taught, they were more receptive to charges that the methods used in the lycée could be vastly improved. Many who were sympathetic to the goals of classical education thought that the lycée emphasized rote memorization over intelligence and rhetoric over facts.[12] In this sense, the early Third Republic witnessed a genuine revolution in pedagogical technique, as educators tried to encourage their students to attend to facts and to think for themselves. In philosophy, this translated into a certain distancing from the syllabus and from the goal of philosophy as catechism.[13] Thus in 1880 the council put into the syllabus the note that "the order adopted in this syllabus must not restrict the freedom of the professor, provided that all the questions indicated are covered."[14] One consequence, as Fabiani has so well documented, was the development of a new style of philosophy, in which the professor's class became a personal "oeuvre" in which the process of thinking through problems became more important than the position reached.

This new style created problems of its own, however, since it required an unprecedented sophistication on the part of the student. In one famous incident, Hippolyte Taine pulled his son out of a philosophy class after the first day when the professor began with a lesson on the nature of the infinite. But such a style also alienated many professors and administrators, who saw no need for such obscurantism in an introductory lycée class. Another confrontation saw Rabier, as inspector, upbraid the legendary Jules Lagneau for the obscurity of a lesson he audited in 1889.[15] Lagneau was known for his unorthodox method of addressing all parts of the syllabus simultaneously. Rabier, who had written a successful manual of philosophy, thought that Lagneau would lose all but the brightest of his students with such an approach. For reasons both of pedagogy and of science, Rabier preferred clear and distinct ideas. In addition to the question of style, however, there was a disagreement over the nature of philosophy. Lagneau's method put metaphysical reflection at the heart of philosophy at a time when many philosophers were still trying to stress the similarities between philosophy and the natural sciences.

This disagreement over the nature of philosophy came at a time when the role of the humanities and the sciences in the lycée generated more controversy than ever. It has been noted that the natural sciences were not a significant part of the lycée curriculum, and that those preparing for the scientific *grandes écoles* followed a separate track in which they were exempted from the philosophy class. Positivists maintained that the natural sciences provided a better training for the mind than the traditional literary education and proposed to increase the representation of the natural sciences in the classical curriculum. Since this would necessarily involve a decrease in the time devoted to humanistic studies, the issue became another tug-of-war in which the humanists gave some ground, but not much. Defenders of the humanities argued that the sciences did not teach the moral values that the classics did, or even worse, that the sciences could only teach the values of materialism. This concern with moral education was a major reason for the retention of the humanities.[16]

The issue of the humanities versus the sciences had important consequences for the philosophy class, which claimed to be a science at the same time that it culminated the study of the humanities. Recall that many of the criticisms of eclectic philosophy dismissed it as a literary game and ignored its claims to scientific status. For this reason, the revision of the syllabus in 1880 is significant, because it involved an attempt to reinforce the scientific character of philosophy (Appendix B). Metaphysics was placed at the end of the syllabus, "to better mark the separation between problems allowing positive research and scientific solutions, and ultimate questions of an entirely different character."[17] This was done at the suggestion of Paul Janet, but it met with unanimous approval. Henri Marion, reporting on the deliberations of the council, insisted that this change would especially benefit psychology, which could now establish itself as a science of facts and laws.[18] Moreover, the syllabus incorporated questions that referred directly to Ribot, such as somnambulism, madness, and comparative psychology. In logic, the syllabus of 1880 placed induction before deduction, stressing the experimental method over the syllogistic. This echoes the instructions that accompanied the syllabus, stressing that facts should come before theory. This reaffirmation of the scientific character of philosophy, then, came as a direct response to the emphasis on science in the administration of Jules Ferry. It made use of the eclectic claim that philosophy is a science of observation. At the same time, academic philosophers could only do this by distinguishing between science and metaphysics in a way that compromised the eclectic connection between them.

Another area of concern in the reform of public education was the

teaching of ethics. French education at this period made no attempt to
be value neutral. It was assumed that a major function of education was
to instill moral principles. With the laicization of the schools and the
conflict between Catholics and anticlerics, however, the issue of what
kind of moral principles the schools should teach regained force. As
always, the Catholic Church objected to the idea of lay persons teaching
morals, and even more to the idea that morality could have a basis
independent of religion. Republicans were convinced that a lay morality
was possible, although they disagreed over its content and foundations.
Ferry thought that the morals taught by the school would be the same
as that taught by the church; nevertheless, because morality was a social
fact and not a transcendent ideal, it could be detached from its religious
roots.[19] Others held that the content of a Republican morality should
teach values different from Catholic morality: independence rather than
obediance, reason rather than authority, et cetera.[20] But there were only
a few liberals, like Espinas, who wanted to abolish the teaching of morals
altogether as an imposition of the state.

These debates impinged upon the philosophy class, since one of its
major functions, ever since Cousin, had been to provide moral principles
to the elite of tomorrow. Moreover, Cousin had tried to give a rational
foundation to morals, though one that acknowledged the legitimacy and
necessity of religion. Far from dissimulating the ideological intent of
ethics, many philosophers during the early Third Republic wanted it
increased and made more explicit. Marion claimed that most of the
philosophers consulted during the revision of the syllabus "expressed a
desire that a greater place be made for ethics, above all for social ethics,
and more generally for questions of practical interest."[21] The section on
the state and civic duties was therefore enlarged, and some questions
on political economy were included. Moreover, placing metaphysics
after ethics implied that the latter could be studied independently of the
former. When Catholics controlled the syllabus (most recently in 1874),
metaphysics, culminating in proofs of the existence of God, came before
morals, to emphasize that there could be no morals without a divine
sanction.[22]

But even after all these particular issues were resolved, there remained
the fundamental question of whether philosophy belonged in secondary
education at all. An increasing number of critics felt that the proper
place of philosophy was in higher rather than secondary education.[23]
The constraints of a syllabus and the didacticism of the lycée classroom
were incompatible with the nature of philosophical research, which demanded total liberty of thought. Moreover, many critics insisted that
the subject matter of philosophy was too difficult for lycée students.
This issue was raised most directly in 1894, in an article by the journalist
Fernand Vandérem entitled, "Philosophy: A Class to Abolish" ("La phil-

osophie: une classe à supprimer"), but it had already begun to appear in the early 1870s. The article provoked on the part of university professors a number of defensive responses, which Vandérem collected in book form.[24]

It is no coincidence that suggestions to reserve philosophy for higher education should begin to appear at this period. A new role was emerging for higher education that increasingly distinguished it from secondary education. Until the 1870s, higher education was devoted largely to the service of secondary education. The duties of the faculties consisted largely in providing juries for the baccalaureate, and courses were designed primarily to instruct future professors in the material they would present in lycée classes.[25] In this context, it makes little sense to speak of differentiating the content of secondary from higher education, and even when such a distinction was made, the force of the institution pulled the two back together.[26] Indeed, we have seen professors oppose the introduction of "new" doctrines into the university on the grounds that the role of the university was to preserve and transmit tried and true knowledge suitable for the nation's youth.[27] However, at the end of the Second Empire a reform movement began to develop that soon changed the nature of higher education.[28] These reformers advocated the ideal of the university as a center for original research and often compared the French faculty system invidiously with the German university. With the defeat by Germany, this argument gained a large and influential audience, and soon the establishment of research universities became part of the Republican educational agenda.

The development of universities had a number of consequences for academic philosophy. In the simplest terms, it expanded the field considerably; indeed, according to Karady, university positions increased twice as fast as lycée positions during the first thirty years of the Republic.[29] The provincial faculties did not expand as rapidly, since they were already overextended as a result of the establishment of one faculty for each academic district to provide juries for the baccalaureate. Nevertheless, with the Third Republic effort to turn these regional faculties into real centers of research, provincial universities became a respectable alternative to a position in Paris. As an indication of this, among the notable philosophers born between 1851 and 1870 (who would have achieved notability between 1891 and 1910), we see for the first time appear some whose final career position was in a provincial faculty (table 6). This widened the apex of the educational hierarchy, permitting more junior members to aspire to higher positions, and as a result the status of the profession as a whole was enhanced.[30]

Moreover, expansion reduced the ability of any one member of the elite to control access to the top. Philosophical orthodoxy became increasingly difficult to enforce, with the result that of the generation of elite academic philosophers born between 1851 and 1870—the first educated during the Third Republic—less than half were spiritualist, much less eclectic (table 4). In any event, ideological control was against the ideal of the research university, which valued independence and freedom of thought. The result was a diversity of philosophical thought within the university of which some approved and others disapproved, but which no one could deny. Even positivism came to have representatives in elite academic philosophy, who came grudgingly to admit positivism as a legitimate philosophical position. Espinas had shown that it was possible for one to have a career on the margins of academic philosophy and yet remain an outspoken positivist. In 1907, Lucien Lévy-Bruhl would assume the chair in modern philosophy at the Sorbonne. As one who admired Comte and generally (though not always) approved the Durkheimian project, Lévy-Bruhl represented the institutionalization of positivism within the elite circles of academic philosophy.

The expansion of philosophy was perhaps one reason why academic philosophers did not oppose the creation of university chairs in the human sciences, nor discourage academic philosophers from taking those positions. There were only two professors of philosophy at the Sorbonne in 1870; by 1890 there were five, and by 1910 seven.[31] If philosophers are getting what they want, they are less likely to begrudge their colleagues a chair in the human sciences than if that chair were created by suppressing a corresponding position in philosophy. Furthermore, many of the new positions in psychology and sociology were being filled by individuals trained as academic philosophers, so that the human sciences became another career opportunity for students trained in philosophy. Of the generation born between 1851 and 1870, nearly a third of academic philosophers who achieved notability became social scientists (table 7). Academic philosophers actively assisted those philosophers who sought positions in psychology and sociology, especially when the competitor was a candidate from some other discipline.[32] Whatever their disagreements within philosophy, in new disciplines the prestige of philosophy was at stake in the appointment of individuals with a philosophical pedigree.

Finally, many academic philosophers, because of the traditional association of psychology and ethics with philosophy, continued to view the new approaches to these subjects as somehow part of philosophy, even if the new practitioners themselves vehemently denied the connection. We have already seen that Janet and Boutroux considered empirical sociology a new direction in philosophy.[33] Seen in this light, the creation of the human sciences was not a challenge to, but an expansion and renewal of, philosophy, however problematic. Moreover, in the absence of specialized degrees in their new disciplines, social scientists often found themselves obliged to recruit their students from philosophy and to push aspiring social scientists through the philosophy degree program. Hence they were more or less obliged to cooperate with philosophers. Indeed, this dependence may have been the best strategy for limiting the challenge of the human sciences, even if, as I think, it was largely unconscious. If experimental psychology and empirical sociology were merely adjuncts to philosophy, there was no justification for a large number of them. One or two would suffice. Whatever the underlying intentions, it remains true that for a long time positivists remained a small presence within academic philosophy.

The distinction between secondary and higher education should not be exaggerated, however. Fabiani is substantially correct in asserting that academic philosophy remained centered on the lycée classroom. Moreover, most elite academic philosophers still began their careers as lycée professors (table 2). Although it remained difficult to obtain an elite post without an academic background, the auxiliary positions (*maître de conférences, professeur adjoint*) needed to permit philosophers to go directly into careers in higher education were only just beginning to appear, and competition for them was such that most aspirants still had to bide their time in the lycées before obtaining even a junior university position. This in itself served as a limitation on the number of positions in psychology and sociology, since even if they were allowed to be considered part of philosophy, they constituted only a small portion of the syllabus.

RESPONSES OF ACADEMIC PHILOSOPHY

These concessions at the secondary and higher levels should indicate that the relation between academic philosophy and the new directions in the human sciences was complicated and ambivalent. The increasing diversity of academic philosophy during the early years of the Third Republic makes the situation even more complicated. For this reason, I will consider the responses of each of the major currents in French philosophy separately. In each case, the response included a different definition of philosophy and of science, as well as detailed responses to the theories of social scientists.

Eclecticism

The nature of the changes in the syllabus of 1880 should be sufficient to show that eclectic spiritualism did not die with the Second Empire. Aside from the fact that several prominent eclectics retained high university positions until the end of the century, academic philosophy necessarily became eclectic as an institution as more and more different schools of thought were represented within it. In Espinas's terms, academic philosophy became an organism of competing tendencies that balanced each other and that reflected the range of tendencies in society as a whole. From representing a single school of thought that claimed to extract the truth from every system, academic philosophy became a field within which those systems coexisted. Indeed, academic philosophy became more reflective of the trends in French philosophy as a whole, as a comparison of tables 4 and 8 makes clear. This disturbed some eclectics, who remained inflexibly hostile to new developments to the very end of their long careers. Others, however, realized that they were in a unique position to act as arbiters among their more dogmatic counterparts. Eclecticism was, after all, by definition forced to accommodate all competing schools of thought. Two examples of this eclectic accommodationism may be found in the later work of Paul Janet and in the philosophy manual of Elie Rabier; the latter was a professor of philosophy at the lycée Charlemagne in Paris and director of secondary education.

We have seen that Janet, who had been a student of Cousin, gave considerable support to the human sciences, supporting the creation of positions and advocating the individuals who sought them.[34] We can get a better idea of why he gave this support by looking at some of his later works. I have already mentioned Janet's *Elementary Treatise of Philosophy* (1879) as an example of the continuity of eclectic philosophy.[35] It also embodied new features that responded to the challenges posed by positivism in general and by the positivistic human sciences in particular. This book, together with his *Principles of Metaphysics and Psychology (Principes de métaphysique et de psychologie,* 1897), which he called his "philosophical testament," represented Janet's effort to incorporate the insights of positivism into the elastic framework of eclecticism. In the preface to the *Treatise,* Janet notes that philosophy has changed, and he has changed his manual to reflect these changes. For example, he begins his psychology with a section on "l'homme physique," in which he gives a rapid survey of the human body, concentrating on the sense organs and the nervous system. The reason, he claims, is because "any philosophy must begin with what really exists: and what exists in fact, is the *entire person,* soul and body."[36] Philosophy has had a tendency to consider only the soul. For similar reasons, Janet concludes his psychology with a section on "l'homme social." "Conventional psychology has perhaps been a bit too inclined to cut everything up, and if on one hand it made too great a separation between mind and body, it has also made too great a separation between the individual and society" (ibid., vii). Janet also places practical before theoretical ethics, because one should study moral facts before moral laws. Finally, as the council did at his suggestion, Janet eliminated metaphysical propositions from his psychology, logic, and morals, and moved such higher questions to the end of his work, in order that the other parts could be studied as sciences of fact. In this way he distinguished between science and metaphysics, even though he defended the legitimacy of the latter.

If these concessions to the positivist spirit of the age were sincere, they did not involve radical changes to the central tenets of eclecticism, which had always advocated the method of observation in philosophy. Furthermore, if for example Janet admitted the need in psychology for the kind of observation and experimentation practiced by physiological psychologists, he held that it could supplement but not replace introspective psychology. Janet called comparative and experimental psychology objective, since it treated psychological phenomena as objects of the external senses, and introspective psychology subjective, since it treated those phenomena that could be perceived only by the thinking subject. In the *Principles,* he explicitly attacked the position of Ribot that subjective psychology required the control of objective psychology to be a science.[37] Janet characterized Ribot as having made two principal objections to the autonomy of subjective psychology: that it is descriptive rather than explanatory, and that it is observational rather than experimental. To the first objection, Janet replied that since subjective phenomena follow laws, such as association, there are indeed explanatory statements in psychology that do not require reference to physiological states. Indeed, turning the tables, Janet claimed that physiological explanations were usually no more than the projection of subjective laws onto some physiological model.[38] To the second objection, Janet replied that most of the natural sciences are observational rather than experimental. Moreover, in most sciences, including subjective psychology, one observes phenomena directly. Objective psychology is the only science in which one observes not phenomena but their signs, since on the one hand one cannot see into the head of the object of one's study, and on the other, physiological phenomena are usually inferred from the concomitant subjective states (ibid., 1:165–66).

Sociology does not appear in Janet's *Treatise.* Espinas's work was still too recent to have made Comte's conception of the science of society a force with which to reckon. If Janet begins his section on ethics with a survey of moral facts, he nevertheless defines it as a deductive science. The moral sciences other than philosophy consist of philology, history, and the *sciences sociales.* Under this last rubric, Janet places political science, jurisprudence, and political economy. All of these sciences, with the exception of jurisprudence, are inductive.

Although he welcomed empirical studies as a useful partner in philosophical investigation, Janet remained confident that they could not threaten the spirituality and liberty of the soul or the universal character of moral law. His gestures in the direction of empirical psychology and sociology therefore have the character of intercalations rather than a radical rethinking, and they met with scorn from Ribot.[39] His fidelity to eclecticism is even more clear in his defense of philosophy than in his discussion of the moral sciences. Janet devotes the greater part of his *Principles* to the defense of the conception of philosophy as a science against the objections of positivists and neo-Kantians alike. He specifically objected to positivist attempts to characterize philosophy as the science of the unknown, as the study of the highest generalities of the natural sciences, or as an art rather than a science. Janet claimed that all these definitions "agree in removing all positive content from philosophy and reducing it either to an empty form or to a pure activity of the mind."[40] Indeed, Janet defended the legitimacy of metaphysics as a science. Nevertheless, the fact that he distinguished sharply between metaphysics and the other parts of philosophy indicates that he acknowledged the independence of the inductive sciences from metaphysics (ibid., 1:236).

Elie Rabier tried even harder than Janet to accommodate both positivism and the new spiritualism that took its cue from Ravaisson.[41] His *Philosophy Lessons (Leçons de philosophie,* 1884–86) is interesting both because it was read by philosophers and psychologists in addition to lycée students, and because as director of secondary education Rabier had a very powerful role in the definition of academic philosophy. Rabier's debt to eclecticism is apparent in his attempt to reconcile opposites. But this effort led him further away from Cousin than Janet was led.

Rabier called his philosophy an "intelligent empiricism."[42] By "empiricism" he meant that all our knowledge is ultimately derived from the senses. Rabier opposed the theory of innate ideas, in either its classic rationalist or its Kantian form. The mind is not born with categories; it constructs them out of experience. Nevertheless, Rabier held that the process by which the mind elaborates the categories cannot be explained in mechanical terms, since no amount of association will give the products of experience the necessity and universality of concepts. Concepts are relations, and Rabier held that relations are not simply given in the way that sense impressions are. Relations are discerned through judgment, a creative activity of the mind irreducible to association (ibid., 1:249). Hence this empiricism is intelligent, and Rabier invoked the dictum of Leibniz, "There is nothing in the intellect that was not first in the senses, except the intellect" (ibid., 1:392). Rabier's intelligent empiricism abandons the argument for spirituality of the soul based on innate ideas that was a prominent feature of Cousin's philosophy, and it also shuns the arguments of the neocriticists. Nevertheless, it is consistent with the strand of French spiritualism that placed the active spirit above static ideas, and that also found inspiration in Leibniz rather than in Descartes. But like the neo-Kantians, Rabier does not want to accept the further spiritualist position that the external world is a creation of mind.

This intelligent empiricism also had consequences for Rabier's philosophy of science. In striking contrast to Cousin, Rabier held that the status of all scientific laws was hypothetical. For Rabier, every idea is a hypothesis, because every idea surpasses the sensory fact from which it arises.[43] Ideas can be verified by experimental procedures, "but precisely because an idea needs demonstration and proof, it follows that it was not from the first the expression and translation of the observed facts" (ibid.). Rabier was not the first spiritualist to stress the role of hypothesis in science; Elme Caro had stressed the irreducibility of science to induction in his *Materialism and Science (Le matérialisme et la science,* 1867), and spiritualist philosophers had begun to stress the creative activity of the mind in producing scientific knowledge.[44] Rabier removed any transcendental vindication of this hypothetical nature of science, relying solely on experimental verification. This applies even to the principle of the uniformity of nature, which is merely a hypothesis elaborated by reason and confirmed by experience.[45] One reviewer complained that Rabier had conceded too much to the empiricists.[46] Rabier himself, however, stressed the fact that his account thwarted the attempt by empiricists to derive general principles mechanically from experience.

In the human sciences, Rabier tried to use his intelligent empiricism to bring eclectic psychology into accord with the most recent findings of the new experimental and physiological school, notably Ribot. Rabier maintained with Janet that psychology was the science of consciousness, and that psychological phenomena were irreducible to physiological. Nevertheless, Rabier embraced the extreme consequences of equating psychology with consciousness. He conceded that any mental phenomena outside present consciousness must be physiological. The idea of an unconscious representation makes no more sense than that of an unconscious pain.[47] Memory, therefore, was a physiological phenomenon in which the brain stored not psychological ideas but neural modifications that could reproduce ideas in consciousness when stimulated by a new sensation. Association too is a function of the brain, a process by which one nerve, stimulated by sensation, transmits its energy to a connected nerve, which is thereby stimulated and reproduces the corresponding idea.

Rabier refused, however, to follow Ribot in reducing all mental processes to association. Rabier held that judgment was inexplicable in these terms. Physiological processes provide the material for thought, but they do not explain thought itself. Rabier's distinction between mechanical and active mental processes elicited praise from a number of psychologists who were also dissatisfied with the associationist model.[48] Pierre Janet was among them.[49]

Rabier, writing just a few years after Janet, devotes considerably more attention to sociology, although it is still not called by this name. Within the *sciences politiques*, Rabier distinguishes between practical and theoretical disciplines. Within the theoretical disciplines, he distinguishes between *la politique théorique idéale*, which deduces the proper ends of social and political action, and *la politique théorique réelle*, which studies the laws of actual social and political action. This latter corresponds roughly to sociology in the Comtean sense, as becomes apparent when Rabier divides it into *la physique sociale* and *la dynamique sociale*, which study the laws of coexistence and succession of the social body respectively. Rabier acknowledges that there are indeed laws of society—he cites the work of Comte, Spencer, Albert Schaeffle, Henry Thomas Buckle, and Espinas—and hence a social science in the strict sense is possible.

Nevertheless, two important qualifications separate Rabier from positivists. Unlike Comte before him and Durkheim after him, he thinks that the determinism of social phenomena is limited by human free will. It is only to the extent that impersonal factors intervene in society that social science is possible. And unlike Comte and Durkheim, Rabier denies that normative statements can be derived from statements of fact. *La science politique idéale* cannot simply be abstracted from *la science politique réelle*.

In its updated form, eclecticism found ways of accommodating the experimental human sciences. Open-minded thinkers like Rabier and Janet thought such studies were a necessary part of any complete science of the human species. As with Cousin, however, both positivists and spiritualists were upset by such compromises. Ribot dismissed Janet's efforts to update lycée philosophy,[50] and other philosophers thought the eclectics conceded too much to the positivists.

The willingness of eclectics to accommodate new developments in philosophy marked a decided change from the attitude of Cousin, who felt that since he had chosen what was true from each system of thought, he did not have to tolerate those who held stubbornly to what was false in them. Cousin gave eclecticism a dogmatic character bitterly denounced by those excluded from the academy, and the successors of Cousin were not able to and did not care to maintain the strict orthodoxy upon which he had insisted. External pressures from an administration favoring other forms of philosophy and structural changes in the ability of the elite to control access undoubtedly played a role in this change of attitude, but the rhetoric of eclecticism itself was also an important factor. Eclectics became increasingly sensitive to appeals to freedom of thought, especially since it was a necessary part of any philosophical doctrine that claimed the right to examine and choose from different systems.[51] Of course, the traumatic experience of wholesale repression during the Second Empire also tempered the dogmatism of Cousin's successors. As a result of this changed attitude, philosophical tendencies that had been excluded before the Third Republic were able to gain places within the university, and academic philosophy became more reflective of French philosophy as a whole.

Neocriticism

Among those who benefited from this change were the various proponents of neo-Kantianism who began to appear during the Second Empire. Cousin, it will be recalled, rejected Kant as a skeptic, and both Janet and Rabier continued to oppose critical philosophy, because it denied the possibility of scientific metaphysics and insisted on a doctrine of a priori categories that they found unnecessary. Nevertheless, interest in Kant revived, primarily from two sources. One, ironically, was from the teaching of the eclectics themselves, who in their role as historians of philosophy presented Kant to generations unconvinced by eclectic objections.[52] The other was from the work of Charles Renouvier (1815–1903), who styled himself a neocriticist and who produced a series of *Essays of General Critique (Essais de critique générale)* from 1854 to 1864 that put forth his own highly original form of critical philosophy.[53] Neo-Kantianism in France went from insignificance in 1850 to parity with spiritualism and positivism by the end of the century. Its progress can be documented statistically through the penetration of its adherents into elite academic positions (tables 4 and 8) and impressionistically through the comments of anti-Kantian observers, who were struck by the rapidity of its rise, the dominance it achieved within the university, and the almost religious fervor with which its converts embraced it.[54] To see the implications for the human sciences of French neo-Kantianism, one can look at the work of Renouvier, who never held an academic position, and of Louis Liard, who began his career as a professor of philosophy and who then became director of higher education, supervising the reform of French universities.

Renouvier was not a part of academic philosophy. He was educated at the Ecole Polytechnique and trained in math by Comte. His only direct contact with eclecticism came when he took the philosophy class at the Collège Rollin in Paris. He was already under the influence of the Saint-Simonians, however, and claimed to have learned nothing from his professor.[55] From his association with utopian socialists he developed a concern for the poor that separated him from the staunchly individualistic liberalism of Cousin, for whom he had nothing but contempt.[56] From Comte he gained an interest in mathematics that explains why a doctrine such as the "law of number" should be at the heart of his philosophy. During the Revolution of 1848, Renouvier wrote a *Republican Manual (Manuel républicain)* which, appearing after the June Days, was sufficiently radical to cause the fall of the education minister who had authorized its publication. After the coup d'état, Renouvier retired from public life. An independent income allowed him to devote himself to the private pursuit of philosophy. It was during this retreat that he discovered Kant and wrote his *Essays of General Critique.*

Renouvier used critical philosophy for much the same reason as Kant: to check the pretensions of both empiricists and rationalists by defining the precise limits of human knowledge. By restricting positive knowledge to phenomena, Renouvier acknowledged the power of science and denied the claims of metaphysics, but by asserting the reality of human free will, he left open a domain inaccessible to determinism. Renouvier took issue with Kant on a number of specific issues, however, so that his critical philosophy looked quite different from that of his predecessor.

In the first *Essay,* for example, he began by eliminating Kant's thing-in-itself as an incoherent and superfluous relic of scholastic substantialism, undercutting the distinction between phenomena and noumena that was central to Kant's philosophy. Renouvier preferred the term *representation* to *phenomenon,* and although he divided representations analytically into the "represented" and the "representative" (*représenté* and *représentatif*), this division was within the phenomenal field and was not held to indicate that there was anything beyond phenomena.[57] The "represented" was simply the phenomenon considered as an object of perception, while the "representative" was the same phenomenon considered as a subject independent of perception.

Moreover, although Renouvier thought that there was an a priori element to all phenomena, he held that this a priori element or category was the most general law of those phenomena. Categories are therefore given in phenomena and not imposed by the mind. They are a priori in the sense that they condition the appearance of phenomena, and that the content of representations cannot specify the forms under which they appear (ibid., 2:370). The categories thus have a very different status in Renouvier than they do in Kant. Moreover, Renouvier thought that a transcendental deduction of the categories was impossible for an intelligence implicated in them; any list of categories would necessarily be incomplete. Nevertheless, he elaborated a set of categories that went from abstract and impersonal to concrete and personal: the first was that of "relation," the most general category under which any representation could be considered, and the last "personality," the most synthetic and particular way of thinking about phenomena (ibid., 2:370–71). Unlike Kant's, Renouvier's categories did not apply equally to all phenomena; for example, personality applied above all to humans.

Finally, Renouvier held to what he called the "law of number," which held that any real number must be finite, because it must be countable. This theory, developed among others by the mathematician Cauchy, who taught at the Ecole Polytechnique, purported to demonstrate that an actual infinite number is impossible. This led Renouvier to resolve Kant's antinomies of reason in favor of the finite: the world had a beginning, since it could not be infinitely old; the world has finite extension; there is a smallest particle, since matter cannot be infinitely divisible; et cetera (ibid., 2:374–78). The problems associated with pure reason, then, come not because it leads us to antinomies, but because it cannot encompass the totality of the world and hence is doomed to incompleteness.

If there is nothing beyond phenomena, then there obviously cannot be a science of being that would reveal something underlying phenomena. Hence, metaphysics is impossible, and philosophy is reduced to critique. On the other hand, Renouvier's definition of science as the knowledge of the relations among phenomena is very close to Comte (ibid., 1:107). Moreover, Renouvier was extremely skeptical of hypothesis and analogy, especially when they cannot be verified. His anxiety was raised not only by traditional arguments but by what he considered the recent abuse of these procedures in evolutionary theory (ibid., 1:111). He differed from Comte by denying the possibility of a system of sciences: the sciences are never complete, they do not have the same methods, and there is no hierarchy among them.[58] Renouvier divided the sciences into three groups—the physical sciences, the logical sciences, and *la critique*. Critique included the analysis of the a priori element of the sciences as well as *sciences desiderata,* those studies that had not yet been given a positive foundation.[59] Such were the moral sciences, which Renouvier thought could indeed become positive, on condition that the specificity of their object be respected. By this he meant that they must be explained in terms of the higher categories, finality and personality, rather than reduced to the lower categories of relation, number, et cetera.

Hence it is possible to constitute a science of psychology, to which Renouvier turned in the *Second Essay.* He held that a "rational psychology" must occupy itself exclusively with "psychic phenomena and the investigation of their laws."[60] He had no use for the notion of a substantial soul, which was in any event excluded by his phenomenalism. His definition of psychology is strikingly positive, and Renouvier noted that he could have called his psychology empirical, if that name had not already acquired connotations he wished to avoid. Renouvier's psychology may be characterized by three basic positions. First, psychological phenomena must be explained in purely phenomenal terms on their own level. This excluded any attempt to explain them in physiological terms, and it also meant that they should make use of the highest categories, finality and personality. Second, any phenomenological explanation must take into account the a priori element in representation. This led Renouvier to postulate a mental "function" (he avoided the word "faculty" because of its substantialist connotations) corresponding to each category. It also ruled out the possibility that the categories could be derived from experience. Finally, any account of the mind must acknowledge the liberty of human volition. Otherwise, one of its fundamental features would be neglected in order to fit it into a deterministic scheme.

These positions shaped Renouvier's attitude toward other forms of scientific psychology. English associationism had the virtue of constructing purely phenomenological laws, except when it tried to correlate mental with neural phenomena; unfortunately, associationists also attempted to derive mental categories from experience.[61] On the other hand, Renouvier admired the phenomenalism of Hume's psychology; he and his longtime collaborator François Pillon made the first French translation of Hume's *Treatise of Human Nature* (1878). They described their own psychology as combining the phenomenalism of Hume with the categories of Kant.[62] They were much more critical of German psycho-physics, introduced into France in part by Ribot's *German Psychology,* which advocated the study and explanation of psychological phenomena through their physiological concomitants.[63] Ribot's monographs suffered from the same flaw; instead of explaining mental phenomena by other mental phenomena, they tried to reduce mental to neural phenomena.[64] For the same reasons, Pillon objected to the attempt to make association and memory physiological phenomena, specifically criticizing the efforts of Rabier to accommodate physiology.[65] In refuting Rabier, Pillon contested the assumption underlying Rabier's argument that psychology is coextensive with consciousness: "the domain of distinct consciousness is not coextensive with the domain of psychological facts."[66] Ideas remain in their psychological form even when they are not present to consciousness; hence, memory and association can be explained in purely psychological terms. As a result of their opposition to physiological reductionism, Renouvier and Pillon were led to posit a domain of unconscious ideas.

A deeper reason for Renouvier's opposition to the assumptions of positivist psychology lies in his analysis of the will, which he defines as the capacity of the mind to produce or suspend representations.[67] This capacity implies the freedom of the will, which, since Renouvier did not believe in a noumenal realm, must be manifested in the phenomenal world.[68] In this he differs from Kant, who held that the will is phenomenally determined but noumenally free. This freedom cannot be proved, but it cannot be ruled out, and it makes more sense of moral reasoning. Hence, any strictly deterministic psychology would be inadequate.

The introduction of indeterminism into the phenomenal world also has implications for history, evolution, and social science. Renouvier argued in his fourth essay, an *Introduction to the Analytical Philosophy of History (Introduction à la philosophie analytique de l'histoire)*, that the existence of free will implied that there could be no deterministic laws of history.[69] He attacked both historicism and evolutionism, Hegel and Spencer, with vigor, but especially the latter, since Spencer found increasing support among French positivists, notably Ribot and Espinas.[70] Renouvier especially objected to the idea that there existed a law of progress in history. If in some particular circumstances, such as modern European history, something like progress could be demonstrated, which Renouvier did not deny, an examination of human history overall revealed decline as well as progress, fluctuations, and endless diversity, so that any unilinear and necessary improvement in civilization could not be legitimately derived from historical phenomena.

Renouvier did not simply object to particular theses of Spencer, however; he thought that Spencer's thought was flawed at its roots, and he missed no opportunity to refute the English philosopher of evolution. In an article in the *Critique philosophique*, Renouvier (or perhaps Pillon) criticized Spencer's attempt to reduce the syllogism to a four-term analogy. This replaced the principle of identity with the principle of similitude and opened the way to the abuse of analogy.[71] The resulting sloppiness of thought accounted for many of Spencer's particular theories, including the organic analogy of society, for which Renouvier had special contempt.[72] The social organism was only one example of the misguided effort to explain one level of phenomena—social and moral—by another—biological. Social phenomena, like psychological, must be explained in their own terms.

Renouvier's own explanation came in his *Analytical Philosophy of History* and above all in his *Science of Morals (Science de la morale)*. In his philosophy of history he tried to vindicate the idea that there is a realm of morals that transcends vicissitudes of history—morality, in other words, is not merely a function of society.[73] Although Renouvier claimed that his science of morals was a study of the laws governing phenomena like any other science, in practice the phenomena boiled down to two—human beings are rational, and they believe themselves to have free will—from which he deduced the rest of moral theory.[74] Despite the quantity of historical information cited in his philosophy of history, social conditions played a role only in his applied, not his pure, science of morals, which therefore bore little resemblance to what Espinas and Durkheim had in mind.

Renouvier's rigor, his understanding of science, his opposition to dogma of any kind, and his defense of autonomous morals made him attractive to many of the generation of academic philosophers who came of age during the first years of the Third Republic. The journal he edited with Pillon, *La critique philosophique,* helped to spread his ideas and to show how they could be applied to the social and moral questions of the day. It was successful, and this success is the source of the anxiety Espinas betrayed in his opposition to neo-Kantianism. One academic philosopher who owed much to Renouvier's philosophy was Louis Liard (1846–1917), a classmate of Rabier at the Ecole Normale. In his doctoral thesis, Liard argued that geometrical definitions, being formal, generative, and a priori, differ essentially from empirical definitions, which are material, compositional, and a posteriori.[75] This constructivist view of mathematics was largely Kantian, although it also drew upon the ideas of his spiritualist mentor, Lachelier (discussed in the next section). The thesis gained him a chair in philosophy at Bordeaux in 1874, which he left in 1880 for an administrative career that would soon make him director of higher education.[76]

Before he left, however, he would publish several works of philosophy, including *Positive Science and Metaphysics (La science positive et la métaphysique)*, which was also inspired by Renouvier and Lachelier. In this book, Liard hoped to establish that science and metaphysics are entirely distinct.[77] They have essentially different objects: science investigates the laws of phenomena, whereas metaphysics inquires into ideas that cannot be the object of science (ibid., 54–55). Among these ideas, however, are the fundamental categories of scientific investigation— time, space, causality, et cetera. Therefore positivism cannot ground itself; only critical philosophy can (ibid., p. 58). Liard criticized the attempt of associationism to derive fundamental categories from experience, and he also opposed the theory of Spencer by which the categories would be the inherited associations of the species, a theory Liard thought only pushed the question back one step further.[78]

For Liard, the categories not only reveal the limits of positivism; they intimate the possibility of the absolute, "that which exists in itself and by itself," which is the proper object of metaphysics (ibid., 299). But if the categories cannot be an object of science, then neither can the absolute. The only way to consider the absolute is through the only example of self-created being we have, our own voluntary activity. This activity is essentially moral, since it seeks to actualize teleological regulative principles and does not simply obey deterministic laws. Metaphysics is therefore essentially moral, and if it cannot be demonstrated by the methods of science, neither can science provide a foundation for morals (ibid., 483–84). Metaphysics is justified by its moral role, which is to provide society with a unifying ideal (ibid., 485).

Liard's attempt to give science and metaphysics radically distinct domains and fundamentally different truth claims typifies the neo-Kantian solution to the problem of determinism and free will.[79] Like other neo-

Kantians, he tried to give science its due while preserving an autonomous domain for free will and moral responsibility. Through the efforts of inspired professors of philosophy, Kantian ethics became the quasi-official morality of the Third Republic, taught in the philosophy class and theorized in the university.[80] Of course, this meant displacing eclectic spiritualism, since a doctrine that held that the domains of morals and metaphysics were fundamentally unscientific could hardly be reconciled with one that found the basis of metaphysics in observation. On the other hand, since critical philosophy found a place for the partial truths of positivism and spiritualism, it was in many ways a fit successor to eclecticism as a philosophy of the *juste milieu*.

Liard carried his philosophical perspective into his administrative activity presiding over the creation of true universities in France. He supported wholeheartedly the fundamental reasons behind the university—to make higher education a center of scientific production and to bring the different branches of knowledge into closer contact with each other. For some, like Renan and Berthelot, these goals were inspired by positivist ideals: science should be the foundation of education; science should be organized to counteract the tendency to fragmentation and overspecialization; and science should provide the basis for the unification of minds and the organization of society in the positivist age.[81]

Liard too thought that higher education should include participation in original research, that the universities could further science by encouraging more interdisciplinary studies, and even that universities could help in forming a Republican national spirit.[82] But he gave these goals a significantly different interpretation. First of all, his definition of science stressed the difference between science and mere empiricism—science implied the "alliance of theory and practice, of idea and fact."[83] Science contains a priori elements not reducible to sense perception, even if at the same time thought cannot attain truth without the data of perception. Secondly, Liard thought that the university represented the relation between the mind and the world—"one like the mind, multiple like phenomena." Only the mind can impose unity on phenomena. Finally, Liard continued to deny that science could provide answers to moral problems, warning that one must avoid the "illusions" that science could either put unity into the diversity of minds or provide a social and national ideal.[84] Indeed, "it is precisely for that reason that we must raise . . . [our youth] in the full clarity of science, because only science can redress the mirages it causes."[85]

Thus the more our public spirit appears inclined to absorb ideas that appear scientific, the more we are inspired in the things of politics and society by analogies or metaphors drawn from the things of science, the more is it important, at the age when youth forms its convictions, that it live in milieux where all the currents of science circulate freely.[86]

A better acquaintance with science is the best antidote to scientism. Despite Liard's support for science in general and the human sciences in particular, he continued to insist that morals were beyond their reach.[87]

Neospiritualism

As the example of Liard indicates, academic philosophers often came by their interest in Renouvier by way of their study of Kant in the Ecole Normale. But the Kant taught at the Ecole Normale reflected the perspective of the spiritualists who taught him, and this perspective differed from that of Renouvier. It also increasingly differed from the eclectic interpretation of Kant. In 1873 Paul Janet wrote in the *Revue des deux mondes* that he detected "a new phase of spiritualist philosophy" being developed by "young talents" within the walls of the Ecole Normale. If he could not agree with all of the positions of these new philosophers, he nevertheless found it encouraging that they were boldly embracing the great questions of metaphysics.[88] The individuals to whom he was referring—Félix Ravaisson, Jules Lachelier, and Alfred Fouillée—were indeed part of a revival in spiritualist philosophy that recaptured the interest of many young philosophers at the same time that it largely replaced eclecticism as the spiritualism of the academy. Table 4 makes this revival clear. Of those elite academic philosophers born between 1831 and 1850, half could be identified as eclectic. Of the next generation, only a third could be so designated. In both cases, the remainder belonged to the neospiritualist movement.[89] This neospiritualism brought a new rigor and a new audacity to metaphysical speculation, coupled with a disdain for eclecticism that bordered on contempt. Not content to draw a line beyond which natural science could not go, these thinkers questioned the assumptions of positive science and redefined the nature of the world positive science investigated.

The work of Ravaisson has already been discussed in chapter 1. Jules Lachelier (1832–1918) brought Ravaisson's "spiritualist positivism" further into the academy, more by his teaching at the Ecole Normale (1864–75), his authority as inspector general (1876–1900), and his presence as president of the *jury de l'agrégation* than by his writing, of which he produced little. What he did produce, however, set a new standard for academic philosophy with its precision, difficulty, and austerity.[90] In his thesis on *The Foundations of Induction (Les fondements de l'induction)*, first published in 1871, Lachelier attempted to prove that the principle of induction in the natural sciences depends on the notion of final as well as efficient causes. This is because induction assumes the preservation of order in the universe from one moment to the next, and mechanical causes alone can neither create nor sustain order.[91] Moreover, although the mind must think of phenomena as determined by efficient causes, since they are presented in spatio-temporal form, it must also think of them as organized wholes, since the ability to recognize objects requires the capacity to see many phenomena as harmoniously related to each other. But the necessity of mechanistic efficient causes does not extend to final causes, since one cannot predict with certainty the particular way in which order will arise in any case.

The dependence of nature on two distinct principles allows Lachelier to grant a form of the thesis of universal determinism and yet maintain that the world is governed by contingency and design: "At one and the same time nature is a scientist, unwearied in deducing effects from causes, and an artist ever trying out new inventions" (ibid., p. 40). In terms of the moral sciences, this means that Lachelier can admit that human action can be measured and predicted: "These kinds of calculations in which human wills are treated something like physical agents, seem to be humiliating to our nature; nevertheless, they are not only necessary in private transactions, but especially in our time they have become, in terms of statistics, one of the principle elements in the science of government" (ibid., p. 37). He even sees no contradiction between physiological determinism and free will: "In nature where everything is at once necessity and finality, movement and tendency, physiological necessity does not exclude life, and liberty can be reconciled with the determinism of human actions" (ibid., p. 50). It is a question of which perspective—mechanism or finalism—one cares to take. Lachelier argues, of course, that the teleological perspective is superior, since final causes capture the concrete reality of nature, while efficient causes capture only its abstract outlines (ibid., p. 46). Moreover, only the perspective of final causes allows one to appreciate the fecundity and inventiveness of nature, which is mirrored in the creativity of humanity. Thus, echoing Ravaisson, Lachelier concludes that genuine philosophy is spiritualist realism, not materialist idealism.[92]

Lachelier carried out the implications of this philosophy for the science of psychology in his article "Psychology and Metaphysics" ("Psychologie et métaphysique"), which appeared in the *Revue philosophique* in 1885.[93] In this article, Lachelier criticized the attempt to make observation and induction the sole method for the study of the mind. He agreed with empiricists that if one did this, one could never find metaphysical principles beyond phenomena, as Cousin thought and the eclectics taught. But he tried to refute the empiricists by demonstrating the dependence of the phenomenal world on the mind through the a priori concepts of extension and intension. He then argued that the eclectic distinction between sensitive and intellectual operations, which eclectics used to define two parts of psychology, was actually the distinction between psychology and metaphysics. "Sensitive consciousness" constitutes the immediate data of the phenomenal field; "intellectual consciousness" is the reflexive thought that simultaneously gives objectivity to phenomena and constitutes a distinct subject. The latter can study the former through the methods of observation, analysis, and induction. But the latter cannot study itself by the same means, because its truth is not a given fact or idea but the spontaneity of its own activity.[94] ". . . Pure thought is an idea which is produced of itself and whose real nature we are unable to grasp except by *reproducing* it by a process of a priori construction or synthesis. This movement from analysis to synthesis is at the same time a movement from psychology to metaphysics" (ibid., 88). By following thought in its development we can see how it spontaneously moves from abstract to concrete to living and free being. The use of this synthetic or dialectic method permits Lachelier to claim that "we have supplied . . . a rationale for spiritualism but in terms and by processes which are not quite the same as Cousin's" (ibid., 95). As in his previous work, he concludes that psychology and metaphysics are complementary rather than mutually exclusive.

Lachelier's thought and teaching made a lasting impression on Emile Boutroux (1845–1921), who taught at the Ecole Normale from 1877 to 1885 and then moved to the chair in the history of philosophy at the Sorbonne.[95] In his thesis, *On the Contingency of the Laws of Nature* (*De la contingence des lois de la nature,* 1874), and his Sorbonne lectures, *On the Idea of Natural Law in Contemporary Science and Philosophy* (*De l'idée de loi naturelle dans la science et la philosophie contemporaines,* 1893), he argued that universal and necessary laws do not apply absolutely to the real world. The world "manifests a certain degree of truly irreducible contingency," and hence deterministic natural laws and the reason that constructs them do not completely explain reality.[96] He tried to prove this thesis by analyzing the fundamental notions of the natural sciences, beginning with necessity and working through such concepts as being, matter, and living things to human beings. As he ascended the hierarchy of the sciences from simple to complex, he showed that the fundamental concept of each new level cannot be derived analytically from the concepts governing the lower level. Hence, each new level is contingent with respect to the lower, if by necessary one means logically entailed, and indeed, the degree of contingency increases as one passes from lower to higher levels. Moreover, within each level, the fundamental concept or law represents an abstraction from reality, simplified and modified to allow deduction and calculation. Quality is reduced to quantity to facilitate the application of mathematics.[97] The concepts of science are therefore inherently approximative with respect to reality.

Boutroux did not think that contingency invalidated the positive sciences. First of all, laws do hold in an Aristotelian sense—always or for the most part. The world displays a certain stability that Boutroux had no intention of denying. "It is not therefore scientific research, [but] only the pretension of being able to dispense with experience that is condemned by the doctrine of contingent variations. . . ."[98] Since reality contains a degree of unpredictability, science must always rely on observation and experiment. Boutroux thought that the contingency of nature meant that historical sciences could never be reduced to the abstract sciences. It also meant that the science of one level cannot be reduced to that of the next lower level. Boutroux argued that the success of modern science depended on abandoning the hope of a universal mathesis and finding instead a positive concept appropriate to each domain.[99]

Although Boutroux clearly developed themes present in Lachelier, his thesis is much more radical. Lachelier adopted the Kantian position that phenomena must obey the deterministic laws of thought, even if they also express final causes. Boutroux denied that the laws of thought regulate absolutely the appearance of phenomena. By showing the abstract and approximate character of natural laws, as well as the ultimate arbitrariness of their first principles, he played an important role in the development of conventionalism.[100]

Boutroux's notion of contingency obviously had implications for the moral sciences. He denied that consciousness could be derived from physiological phenomena, although they could be a condition for its appearance. Nor did he think that one could measure psychological phenomena by their purported physiological concomitants. As for the determination of psychological phenomena by their psychological antecedents, Boutroux did not deny that statistical regularities can be found in behavior. Like Lachelier, he maintained that it would be inconsistent with the notion of character and personality to say that an individual's actions could not be predicted with some accuracy from a knowledge of his past actions.[101] But this kind of predictability does not exclude contingency, and statistics are designed precisely to obscure the variability of their object by constructing an abstract unity. Beneath this superficial uniformity lie innumerable individual acts determined by will as well as by antecedent conditions. What holds true of psychology holds a fortiori of sociology. Boutroux argued that social phenomena could not be explained solely by historical or physical laws without reference to the intelligence and free will of human actors.[102]

With powerful thinkers like Ravaisson, Lachelier, and Boutroux, neo-spiritualism won over an increasing number of young French philosophers, both inside and outside the academy (tables 4 and 8). A symbol of this development, and of the revival of nonpositivist philosophies in general, may be found in the inauguration in 1893 of a new journal of philosophy, the *Revue de métaphysique et de morale*. It was founded by former students of the lycée professor Alphonse Darlu, an inspiring neospiritualist philosopher who wrote the prefatory manifesto to the first number.[103] While recognizing the "hospitable eclecticism" of the *Revue philosophique,* which brought philosophers and scientists together, the *Revue de métaphysique et de morale* wanted to concentrate exclusively on "doctrines of philosophy properly speaking."[104] This meant the theory of knowledge, existence, and action—broad enough to admit most of the topics covered by the eclectic syllabus and the *Revue philosophique*. But differing from both in its method, the *Revue de métaphysique et de morale* desired "not facts, but ideas" (ibid., 3). Philosophy was thus defined as the analysis of ideas rather than as an observational science. The *Revue de métaphysique et de morale* quickly became a rival to the *Revue philosophique* for the loyalty of academic philosophers, to the dismay of Ribot.[105]

Nevertheless, as Dominique Merllié has pointed out, the new journal did not become the exclusive domain of nonpositivist philosophers, nor did these philosophers abandon Ribot's journal, leaving him with only the positivists. Just as Ribot realized that he had to appeal to nonpositivist philosophers to make his review a viable venture, so the young editors of the *Revue de métaphysique et de morale,* Xavier Léon and Elie Halévy, recognized that a journal excluding positivists would preach only to the converted.[106] For this reason, they invited philosophers of all persuasions to contribute, and the result was that, in terms of contributors, content, and readership, the new review overlapped considerably with the old. Rather than polarizing the philosophical community, the new journal created another forum in which advocates and opponents of the positivistic human sciences could discuss their disagreements. The real criterion for inclusion in the *Revue de métaphysique et de morale* was professional competence rather than doctrinal orthodoxy, reflecting the changes in the philosophical field during the early Third Republic.

Conclusion

The challenge to eclectic spiritualism posed by Ribot and Espinas found a number of supporters among academic philosophers of the next generation. For this reason, the statement of Harry W. Paul that philosophical positivism "had little real strength in the French philosophical communities of the nineteenth century" does not tell the entire story.[107] The strength of positivism is underrepresented because philosophers who were attracted to it tended to end up in the new social scientific disciplines, and hence they are not always counted as part of academic philosophy. It is true, however, that the institutional and intellectual pressures brought to bear on academic philosophy during the early Third Republic forced a response to critics who questioned the validity of philosophy.

The survival and even revival of academic philosophy attests to the fact that it met these challenges, although the diversity of responses obscures to some extent their success. But all of the responses involved an attempt to defend the intellectual prestige and autonomy of philosophy. For eclectics, this meant reasserting the claim of philosophy to be a science of observation with its own object, while incorporating the findings of the positivist human sciences stripped of their deterministic implications. Such a strategy tried to co-opt the prestige of the natural sciences for philosophy, but it was open to the charge that it provided no rationale for rejecting the assumptions of positivism.

As a result, this strategy found less and less support among academic philosophers, who increasingly began to question the assumptions of the natural sciences and to define philosophy as something essentially different from science. Neocriticists emphasized the a priori element in science and defined philosophy as the analysis of this element. They also distinguished sharply between nature and ethics, reserving freedom for the latter domain, whether they placed it in or beyond the realm of phenomena. Neospiritualists went further in their critique of science, claiming that it necessarily missed the creativity and spontaneity of nature. This aspect of reality could only be captured through reflection, something essentially distinct from the method of the natural sciences. Thus the development of French academic philosophy in the nineteenth century contains something of an irony. Until roughly 1870, it claimed to be a science, although it actually had a fairly uncritical idea of science, and it had little contact with the natural sciences. After 1870, it claimed it was not science, although it developed a more sophisticated understanding of science and made closer contacts with practicing scientists.

In taking themselves outside science, the new currents in academic philosophy also claimed to take essential aspects of the human world away from the natural sciences. The inner sanctum of human uniqueness—pure thought and moral conscience—were made inaccessible to experimental methods. To do this, philosophers criticized the biologism, associationism, and evolutionism represented by Ribot and Espinas. The objections of academic philosophers posed a challenge to future advocates of the positivist human sciences, a challenge to which the next generation of social scientists would in turn respond.

One final consequence of the diversification of philosophy deserves note. With more schools of thought represented in the academy, with less pressure to conform, and more pressure to distinguish oneself, individual philosophers increasingly took positions that their peers found difficult to categorize easily as belonging to one school or another. Contemporary and subsequent historians of philosophy disagree more about the classification of the generation of philosophers born between 1851 and 1870 than preceding generations (tables 4 and 8). In some cases this is a question of which of neighboring orientations to place a given philosopher—positivist or neocriticist? Neocriticist or neospiritualist? In other cases individuals developed ideas that could be classified at either extreme of the philosophical spectrum.[108] It is significant that this was the first generation educated under the more open and tolerant auspices of the early Third Republic. This was also the generation of the social scientists who tried to extend the human sciences while responding to philosophical critiques. If academic philosophy posed a challenge to would-be social scientists, it also suggested the possibility of a nuanced response, one that might incorporate aspects of different schools of thought rather than simply reinforcing dogmatic intransigence. The following two chapters trace the nuanced responses of two such social scientists—Pierre Janet and Emile Durkheim.

5

The Synthesis of Philosophical and Medical Psychology: Pierre Janet

> Yet it seems that by borrowing from the different authors their most correct conceptions, in completing them, the one by the other, instead of contradicting them, we have been able to indicate how paralyses and contractures approximate other conditions.
> —Pierre Janet, *The Mental State of Hystericals*

It was in the context of philosophical reactions to the biologism and associationism of Ribot that Pierre Janet (1859–1947) undertook his first studies in pathological psychology. Ribot's efforts at publicizing scientific psychology had succeeded in persuading many philosophers that the "objective" method needed more serious development than the eclectic tradition had allowed. Janet was committed to the psychology of facts rather than of speculation, and from the earliest days of his career he began to experiment with hypnotism and to observe the mentally ill, in accordance with Ribot's belief that mental illness constituted a natural experiment that revealed the mechanism of the normal as well as the pathological mind. Janet's direct observation contrasted with the secondhand character of Ribot's studies, and in this sense he surpassed his predecessor as a practitioner of empirical psychology.

But if Janet agreed with Ribot's insistence on a psychology based on the observation of others—on an "objective" psychology—he differed from Ribot in the fundamental concepts with which he interpreted his observations. Janet resisted Ribot's tendency to explain psychological phenomena in biological terms, and he considerably expanded the domain of psychological phenomena to include subconscious acts that Ribot and most others had characterized as purely physiological. Further, Janet rejected associationist mechanism, insisting on an absolute distinction between association and judgment that had no place in Ribot's psychology. Above all, Janet was conciliatory toward the tradition of philosophical psychology, and rather than reject its formulations because its method was faulty, he proposed to use its concepts to interpret clinical and empirical research. At the same time, the contact with concrete clinical experience would enrich the concepts taken from philosophy, and the result of this interaction would be a psychology that was well founded in both its empirical and its conceptual aspects.

What accounts for Janet's divergence from the direction in which Ribot wanted to point scientific psychology? Some of the answers may be found within the tradition of clinical psychiatry in which Janet gained his empirical training. The research of Jean-Martin Charcot and his assistants at the Salpêtrière into hypnotism and hysteria, for all its preoccupation with objective, somatic symptoms, nevertheless included an increasing conviction that neuroses depended on ideas as much as on physiological conditions.[1] Moreover, the distinction between mechanical and creative mental processes has a history in French psychiatry.[2] Janet distilled these and other features of the psychiatric tradition into his own psychological system.

Nevertheless, I argue that Janet's unique ability to incorporate elements of that tradition came from his involvement with philosophical arguments that used similar concepts and reached similar conclusions. The objections of philosophers sensitized him to problems in the developing field of scientific psychology, and the solutions they offered enabled him to pick out those aspects of the psychiatric tradition that offered a way out of the difficulties posed by the too close affiliation of biology and psychology. Moreover, the categories and distinctions philosophers developed helped Janet interpret the empirical phenomena he encountered in his psychiatric practice.[3] Janet himself, and those who have studied him, have all pointed out his debt to philosophy.[4] But in locating his primary inspiration in Maine de Biran, commentators have looked too far afield. Pierre Janet's philosophical roots lie much closer to home, in his uncle Paul Janet and in the various trends in academic philosophy in the 1870s and 1880s that constituted the substance of the preceeding chapter.[5] If Janet discovered a kindred soul in Maine de Biran, he discovered Maine de Biran through the academic philosophy of his own day. In many ways, Janet's project of reconciling philosophy and science owes its inclusiveness to the liberal eclecticism of his uncle.

This chapter examines the career and thought of Janet until around 1904, when all of the institutional supports of his professional position were in place, and when the last of his major early works, *Obsessions and Psychasthenia (Les obsessions et la psychasthénie)*, had been published. Janet's thought continued to develop over the remaining forty years of his life, and his later work presents a number of new themes without which a complete understanding of Janet is impossible. In particular, he began to redefine his psychology in terms of conduct or action rather than ideas or consciousness, to the point that it has been seen as a parallel to behaviorism.[6] But Janet published no major book from 1904 until after World War I, and since the focus of this book is on the prewar period, his later work is referred to only when it helps elucidate a problem in his earlier life and thought. After examining Janet's long and intimate relationship with academic philosophy, I attempt to distill those elements of his psychology that differ from Ribot, and demonstrate how these differences were shaped by the philosophical tradition.

CAREER

Janet was born and raised in Paris. His family had for several generations been Parisian and upper middle class, his father a legal editor and his uncle Paul Janet, the eclectic philosopher.[7] Pierre himself credited his uncle with stimulating his interest in philosophy and even encouraging him to pursue scientific psychology. This passage of his autobiographical essay is worth quoting at length:

> My interest in philosophical studies was quickened by the example of my uncle, Paul Janet, my father's brother. Paul Janet, to whom I owe much, was an excellent man, industrious and intelligent, and today it seems to me that justice was not done him. He was not only a spiritual metaphysician, the last representative of the eclectic school of Cousin, but he was a great spirit who was interested also in politics and the sciences, and who, with great liberalism, sought to reunite these studies. He understood the importance of medical and anatomical studies to the moral intelligence of man. It was he who, at the beginning of my philosophical studies, presented me to Dastre, Professor of Physiology at the Sorbonne, and started me in his laboratory. It was he who had me, after normal school, enroll at the Medical School in Paris and continually urged me to combine medical and philosophical studies.[8]

Although Janet wrote this passage when he was over seventy years old, it is consistent with the evidence of his early readings and with Paul Janet's own ideas about scientific psychology. As discussed in chapter 4, the elder Janet not only acknowledged the need for "objective" methods—observation of and experimentation with others—to control introspection, but he also recognized the importance of physiology to psychology. After all, he introduced the discussion of the physiology of the nervous system and the sensory organs in his textbook of philosophy, which was written during the 1870s, when Pierre was pursuing his lycée studies, and published in 1879, when the younger Janet entered the Ecole Normale Supérieure. Thus the same innovation that Ribot dismissed as typical eclectic compromise impressed Pierre Janet as a genuine and viable openness. This difference in attitude toward Paul

Janet is one of the best indications of the difference between Pierre Janet and Théodule Ribot.

His uncle's eclectic attempt to reconcile philosophy and science also provided a way of negotiating his own crisis of faith, which seems to have occurred between the ages of fifteen and eighteen—that is, just at the time when he would have first been seriously exposed to philosophy in the lycée. Whatever the particular sources of Pierre's crisis—adolescent crises of faith were quite common for nineteenth-century intellectuals—the outcome was not a blanket repudiation of religious sentiment, although Janet seems to have become an agnostic. Instead, "it was a quest of conciliating scientific tastes and religious sentiments, which was not an easy task. The conciliation could have been effected by means of a perfected philosophy satisfying both reason and faith. I have not found this miracle, but I have remained a philosopher" (ibid., 1:123).

In addition to whatever personal influence the uncle may have had on his nephew's thinking, it is reasonably certain that the philosophy Pierre studied at the lycée would also have reflected Paul's outlook, not only because the latter was himself so clearly identified with the philosophy of the syllabus, but because Pierre's professor of philosophy, at least during his year of preparation for the Ecole Normale Supérieure, was probably Thomas-Victor Charpentier, an eclectic and a former student of Paul Janet.[9] By the time Pierre entered higher education, he was particularly well steeped in the eclectic tradition.

One other adolescent influence deserves note. From childhood, Pierre Janet was an avid collector of plants, a hobby he pursued throughout his long life.[10] Like Espinas, Janet was drawn to the qualitative, classificatory model of science exhibited by natural history, and although he later claimed that observation and classification were merely necessary antecedents to quantification, he remained skeptical of quantitative methods in psychology.

Pierre Janet entered the Ecole Normale Supérieure with the *promotion* of 1879, which also included Emile Durkheim and the logician Edmond Goblot. Henri Bergson and Jean Jaurès had entered the year before. It does not appear that Janet made any close friends among these classmates, who would become influential colleagues. The professors of philosophy at this time were Emile Boutroux, the Catholic philosopher Léon Ollé-Laprune, and, for one year during the suspension of Ollé-Laprune, Elie Rabier. Of these teachers, Janet would later cite Rabier more frequently than either Boutroux or Ollé-Laprune.[11] Rabier was the most similar in outlook to Paul Janet, and in important ways Pierre Janet's psychological theory would reflect the concerns of Rabier's philosophy (see the next section) rather than either Boutroux or Ollé-Laprune. Although he was admitted as a student in letters rather than in science, Janet also studied zoology and physiology under Dastre, who also taught at the Ecole Normale. As a result of his scientific studies, Janet was able to receive a *baccalauréat restreint* in science.[12]

After he passed the *agrégation* in 1882 (the second out of eight candidates), Janet began his lycée career, first at Châteauroux, and then, from February of 1883 at Le Havre.[13] To judge from the philosophy manual he later published, his teaching was straightforward, clear, and thoroughly orthodox. He does not seem to have been affected by the neo-spiritualist obscurantism that increasingly characterized the philosophy class. He published a couple of speeches on the right of property and on teaching philosophy, and he prepared an edition of Malebranche's book *On the Search for Truth* (*De la recherche de la vérité*) with notes and introduction for lycée classes.

Janet did not have many students at Le Havre, however, and in his spare time he began the research in psychological phenomena that would lead to his dissertation. At first, he apparently wanted to study hallucinations as a means of elucidating the mechanism of perception, and he applied to a local physician, Dr. Joseph Gibert, for patients who might be suitable for such research.[14] Instead, Gibert persuaded him to investigate the phenomena of hypnotism with several suggestible patients and former patients. In particular, Gibert enabled Janet to experiment on Léonie, a former patient who had been hypnotized in the past and who could still be easily put into a trance. In this way, Janet was drawn into the studies of hypnotism in a medical setting that would become characteristic of his work.

Hypnotism had been a subject of much curiosity during the nineteenth century, but its respectability had suffered from the extraordinary, even supernatural capacities attributed to hypnotized subjects as well as from the difficulty of reproducing it under suitably controlled conditions.[15] The French Academy of Sciences had in 1840 officially declared that "animal magnetism" was not a real phenomenon and that it would no longer consider research devoted to the topic. As a result, hypnotism was banished from respectable science, although it continued to provide a living for stage performers and to attract the attention of provincial doctors. Gibert was one such doctor with an abiding interest in the subject, independent of the resurgence of official interest during the early 1880s. At that period, doctors and physiologists in Paris began once more to treat hypnotism as a legitimate research topic, and hypnotism may be said to have been officially reinstated when in 1882 Charcot made it the subject of an address to the same academy that had dismissed it forty years before. Janet found himself, therefore, with a dissertation topic that was at the center of psychological research.

His first paper based on research with Léonie was presented at one of the first meetings of the Société de Psychologie Physiologique, on 30 November 1885.[16] The paper was read by his uncle, a member of the society, to an audience that included Charcot, Ribot, the physiologist Charles Richet, and a number of researchers from the Salpêtrière. This was Janet's psychological debut, and it is significant that, here again, his uncle facilitated his introduction.[17] Curiously enough, the topic of the paper was hypnotism at a distance. Janet and Gibert had performed a series of experiments in which Léonie could apparently be hypnotized by Janet's simply thinking the command to fall asleep, even if he was not in the same room. Janet considered and rejected the possibility that subtle cues from one of the investigators could have suggested to Léonie when to enter into a trance. He explicitly abstained from drawing any conclusions about the nature of the phenomena described, but he obviously thought the results significant, since he concluded that "collecting these apparently mysterious phenomena without prejudice of any sort would perhaps be the best means of clarifying the problem and of working on the progress of the psychological sciences."[18]

Although the topic seems curious for a scientific psychologist, in fact it makes sense in the context of the time. The history of paranormal phenomena associated with hypnotism led many researchers to consider the possibility that there was some element of truth to these claims. Richet for one firmly believed in the possibility of telepathy and had written several articles for the *Revue philosophique* in which he defended the scientificity of such research and developed criteria for evaluating such phenomena, including one of the earliest attempts to apply probability theory to the problem.[19] Alfred Binet and Charles Féré had begun to study the effect of magnets and metals on mental phenomena, as well as the possibility of medication at a distance.[20] On a local level, provincial doctors like Gibert had long been convinced that there might be something to the extravagant claims of magnetizers. As a result of all these remarkable phenomena, the line between normal and paranormal had become indistinct. Lastly, the scientific study of parapsychological phenomena also fit well into Janet's personal project of reconciling faith and reason.

The response of the scientific community to Janet's paper also indicates that his research was not considered to be beyond the realm of science. Richet and Henri Beaunis, a doctor associated with the hypnotists at Nancy, both reported instances from their own research that seemed to confirm the fact of mental suggestion, as telepathy was known.[21] Richet and others from Paris, as well as members of the British Society for Psychical Research, descended on Le Havre to reproduce and verify the remarkable facts reported.[22] Despite a reprise of the experiment that was judged successful by all concerned, however, Janet himself apparently developed misgivings.[23] He was overwhelmed and made cautious by the credence given to a report of strange phenomena made by an unknown and inexperienced investigator, and he began to doubt whether his experiments had been conducted under sufficiently rigorous controls to be considered demonstrative. The spectacular nature of his findings had, however, succeeded in making him known to all the prominent psychologists of the day.

Rather than continue to refine his experiments with mental suggestion, Janet turned to a less dramatic but equally topical line of research. His next paper, published in Richet's *Revue scientifique* in May 1886, attempted to verify the existence of the well-defined phases of hypnotism asserted by the followers of Charcot and denied by the followers of Hippolyte Bernheim. The controversy between the school of the Salpêtrière—as the followers of Charcot were known, after the name of the hospital in which he worked—and the school of Nancy—where Bernheim and his supporters worked—had erupted after Bernheim, a member of the faculty of medicine at Nancy, had investigated another provincial doctor named Ambroise Liébault, who used hypnotism in his private practice.[24] Bernheim became a convert, but the picture of hypnotism that emerged from his experience contrasted sharply with that proposed by Charcot. Charcot considered hypnotism to be an abnormal phenomenon, symptomatic of a weakened constitution and closely linked to hysteria. He also found that hypnotic sleep had three different phases—lethargy, catalepsy, and somnambulism—each marked by definite physiological signs. Bernheim, on the other hand, considered hypnotism to be an essentially normal phenomenon. He held that almost anyone could be hypnotized, and that hypnotism differed little from ordinary sleep. Bernheim claimed that no one outside the Salpêtrière had observed the phases described by Charcot, and he attributed them to subconscious suggestion among the doctors and patients at the Parisian hospital.

In his paper, Janet attempted to reproduce the phases of hypnotism described by Charcot on a patient who had never been to the Salpêtrière and who he claimed had never heard of Charcot. This patient was the same Léonie of the experiments with mental suggestion. He managed to induce in her all three states, but he also found a number of intermediate phases, so that in all he was able to produce nine and possibly ten different and stable sets of symptoms. He was led by this experience to a compromise between the positions of the Nancy and the Salpêtrière schools: the phases of hypnotism are indeed phases of sleep, as the Nancy school insisted, but they are also real, stable, and natural phases, as the Salpêtrière school claimed, not the result of suggestion.[25] This attempt at conciliation between two opposed interpretations is typical of Janet's approach to such disagreements, and if he came eventually to side with the Salpêtrière on this particular issue, his general conviction that psychological phenomena should be interpreted in purely psychological terms echoed a tenet of the Nancy school.

If Janet's first two papers seem tentative and unconnected with his later work, his third paper, published at the end of 1886, contained most of the themes that would preoccupy him for the rest of his career. He described his observations with experiments on Lucie, another patient at Le Havre who was not only suggestible but hysterical. More important, Lucie displayed a singular behavior that attracted Janet's attention—she exhibited signs of split personality. In analyzing this phenomenon, Janet suddenly began to use the vocabulary that would guide his entire later development—the "doubling of the self," the insistence that unconscious thoughts were not physiological, the distinction between association or automatism and judgment or synthesis, and the "narrowing of the field of consciousness."[26] From this point on, Janet's work develops with a unity and coherence that is truly remarkable. He had found his theme.

That theme and the influences that informed it are discussed in the next section. His work with hypnotism developed into a dissertation in philosophy on *Psychological Automatism (L'automatisme psychologique)* that was defended in June of 1889. The jury that evaluated it consisted of Boutroux, his former professor; Henri Marion, the philosopher and pedagogue; Waddington, the old eclectic; Gabriel Séailles, a neo-Kantian who was a *suppléant* and collaborator of Paul Janet; and Paul Janet himself.[27] This was an extraordinarily comfortable committee, and it is unlikely that Pierre Janet encountered any of the anguish and uncertainty that Ribot and Espinas had undergone. He did meet with objections—even from his uncle, who questioned the notion of the split ego—but he was enthusiastically approved on the basis of his philosophical knowledge and scientific learning.[28]

The success of *Psychological Automatism,* which was reviewed in France and abroad, and the doctorate he now held entitled him to a better post than a provincial lycée, and in the fall of 1889 he moved to the Lycée Louis-le-Grand in Paris. Even before his defense, the philosophical community was speculating on the movements that would bring Janet to Paris.[29] His translation to Paris enabled him to improve his scientific credentials in two related ways. First of all, he entered medical school to get the training, degree, and access to subjects he needed to establish fully his scientific credentials with the psychiatrists and physiologists who wrote on psychological matters. At the same time, he directed the laboratory of experimental psychology at the Salpêtrière, which Charcot had established for him. Janet managed to shoulder this staggering workload and to write a second dissertation in medicine, which he defended in July of 1893. It was one of the last dissertations over which Charcot presided, for the eminent neurologist died a few weeks after Janet's defense. Janet continued to work at the laboratory of the Salpêtrière for a number of years after the death of Charcot, and he began a private psychological practice that attracted a wealthy and distinguished clientele, including certainly Raymond Roussel and possibly Marcel Proust.[30] He presented papers at all the international congresses of psychology that were held during this period and was president of the congress of 1900, which was held in Paris.

But until 1898 he still taught philosophy at a Paris lycée. This was not because he lacked support among his colleagues in philosophy. His uncle worked hard to create a place for him at the Sorbonne. In 1888, the faculty had let the *cours complémentaire* in experimental psychology lapse when Ribot vacated it, the elder Janet pushing instead for a course in logic. But in the fall of 1893 he asked that the course in psychology be reinstated.[31] Why this sudden change of heart? It may be in part because the philosophers felt that the need for a logician had been sufficiently met in one of their previous appointments to another position.[32] However, one may legitimately suspect a dose of nepotism, because at the same time that he asked for the course he also named the person he considered most qualified to fill it—Pierre Janet. He pointed out that his nephew had both philosophical and scientific credentials, which made him ideal for a position that would involve the preparation of students for examinations in philosophy. This made him more suitable than Alfred Binet, for example, who "has no university title." He also noted that the younger Janet, although scientifically oriented, did not reduce psychology to biology. On the contrary, "the great merit of the candidate [Pierre Janet] . . . is precisely to have freed psychology from mental physiology, to have taken revenge on it [mental physiology] in some sense, and certainly to have enlarged its [psychology's] domain through the conquest of a new domain, that of pathological psychology."[33] Even though Paul Janet, after having made a strong case for his nephew, then said that he was really only interested in the position, not the particular candidate, it is clear that in his mind the two went together.

This was also the response of his colleagues in philosophy. Waddington, so opposed to Ribot, stated that he supported Janet's argument, especially when he argued for "the fundamentally psychological, that is to say truly philosophical, character of the position," and that he also supported Janet's candidate.[34] The other philosophers agreed. They were willing to support a course in scientific psychology, provided it were sufficiently "philosophical"—that is, antireductionist. They also wanted a candidate who would behave as a member of the philosophy program. Janet's request was therefore approved and sent to the General Council of Faculties as a request of the faculty—behind a new course in Latin but ahead of the new course in sociology, which was also approved at the same meeting.

Anticipating objections at the General Council, Janet prepared a report on the new course that was read into the minutes and used to justify the faculty's request. Between the time that he raised the issue at the Faculty of Letters and the time he wrote his report, dated 13 January 1894, he had apparently had second thoughts about the title "experimental" psychology. In his report he argued that the name "objective" psychology would be better.[35] This was not merely a question of semantics. In the context established by Ribot, Charcot, and the Société de Psychologie Physiologique, "experimental" implied "physiological," and Janet took pains to show that the proposed position was not primarily devoted to physiological psychology. It included physiological psychology, but it also included "ethnographic psychology"— the study of mental life in different cultures—criminal psychology, the psychology of language, of literature, of children, of the sexes, et cetera. In short, the new position would encompass all efforts to study human mental faculties "objectively"—that is, without relying primarily on introspection. These other studies were properly conducted in the Faculty of Letters, which had as its purpose the study of the products of the human mind. Hence the new position would be also well placed in that faculty.

In addition to distancing the position from biology, the use of the term *objective* also served to keep it in close proximity to philosophy, where the distinction between the objective and the subjective method had been a commonplace since Cousin. Janet argued in fact that the objective method had always been a part of philosophy.

> But in our time this objective psychology has made such progress and has so extended itself that it has asked to constitute a distinct science, as has happened to all sciences that detach themselves from the mother science, as they developed a richer subject matter and new methods. But even while distinguishing itself from the common stock, it has not been able to separate itself completely from it. It has remained attached to it by its common objective, which is the science of man.[36]

Even as it becomes an independent science, objective psychology remains part of philosophy, incomplete without the insights of the subjective method that together make up "general psychology." In this way Janet tried to give the new science its autonomy while retaining a special relationship with philosophy.

The elder Janet's arguments were quite prescient, because when the issue came up for discussion at the General Council of Faculties, first in committee and then at the plenary session, several scientists raised precisely the question he had foreseen—namely, why should a course in scientific psychology be taught at the Faculty of Letters rather than at the Faculty of Sciences or Medicine?[37] The dean of the Faculty of Sciences, Jean-Gaston Darboux, and the representative of the École de

Pharmacie, the physiologist Alphonse Milne-Edwards, also questioned the appropriateness of the term *objective* instead of *experimental*. For them, only the latter was truly scientific. The representatives of the Faculty of Letters—the historian Ernest Lavisse, the professor of Latin eloquence Martha, and the dean Auguste Himly—used Janet's report to stress the connection of psychology with the other studies of the human mind, which were housed in their faculty. They also emphasized that the course would not be primarily experimental or biological, so that it would not need its own laboratory. The professor would use the insights of clinical and laboratory research—perhaps his own, perhaps those of others—but the course would not include experimentation as part of the instruction.[38] In other words, they accentuated its philosophical character in order to justify its place in the Faculty of Letters.

The same arguments were also used to counter another objection—that there were already enough resources devoted to psychology, with a chair at the Collège de France and several psychological laboratories in Paris, including one at the Sorbonne. The representatives of the Faculty of Letters had to argue that their course would not duplicate the research and instruction being undertaken at other institutions. Other objections were raised, but in the end the council agreed to retain the request of the Faculty of Letters for a *cours complémentaire* in "objective" psychology.[39] Nevertheless, it was made last on a list of eight requests for new positions from the different faculties. The position was not a victim of opposition—only one person voted against it—but of priorities within and outside the Faculty of Letters over which philosophers had no control. Despite Paul Janet's valiant efforts, his nephew would have to wait.

In the fall of 1897, after several years in which the request languished at the bottom of the university's priorities, Janet and Boutroux asked the Faculty of Letters to move the course in experimental psychology to the top of its list.[40] Shortly thereafter, the administration finally approved the position, and Pierre Janet left his lycée to begin teaching at the Sorbonne. But Janet stayed at the Sorbonne only a short time. His destiny was at the Collège de France. He had been a *suppléant* for Ribot since 1895, and in 1901 Ribot announced that he was leaving his chair to devote his full time to research and editing. Times had changed since Ribot's battle for his chair in 1887–88. When Ribot vacated his chair, the philosophers supported the retention of the chair as experimental and comparative psychology.[41] In fact, the neospiritualist Henri Bergson, who had been elected to the chair of philosophy in 1900, wrote the report recommending that the chair in experimental and comparative psychology be maintained.[42] He also presented Janet's credentials to the professors of the Collège when they met to vote on Ribot's successor.[43] In the subsequent election Janet, supported by the philosophers, defeated Alfred Binet, who was presented by the physiologist Marey. The Académie des Sciences Morales et Politiques put up no obstacles, either. It also nominated Janet first and Binet second.[44]

Another indication of the changes that had occurred in the attitudes of philosophers is that the Faculty of Letters retained Janet's course and found someone else to offer it. After Ribot left the same position in 1888, it was allowed to lapse, and the philosophers asked for a logician instead.[45] This time the faculty voted unanimously to retain the course and offered it to Georges Dumas, another philosopher cum physician who defeated poor Binet for the same reason Janet had four years before—Dumas had more philosophical credentials.[46]

Janet retained his chair at the Collège de France for over thirty years, retiring only in 1935. It gave him the security and freedom he needed to develop his psychological ideas, but, true to the nature of the Collège, it was not without inconveniences. Not being part of the university system, it did not confer any degrees, nor did it attract many students. Janet lectured mostly to foreigners.[47] Moreover, he eventually lost contact with the medical world. Charcot's model of hysteria was eventually abandoned after his death, even by his closest followers, and although Janet had been quite critical of several aspects of Charcot's theory, in characteristic fashion he was balanced and conciliatory enough to find merit in parts of it. This was enough to condemn him in the eyes of those who wanted to put as much distance between themselves and Charcot's psychological work as possible.[48] In 1910 the laboratory of psychology at the Salpêtrière was eliminated when a doctor hostile to Janet's work assumed the position once held by Charcot.[49] The laboratory at the Sorbonne, which had been created for Ribot when he was appointed to the Collège de France, was in theory attached to Janet's chair. However, it was directed by Binet, who had no intention of relinquishing his only university position to the person who had defeated him in the contest for the Collège. When Binet died in 1911, instead of reverting to Janet, the laboratory was entrusted to Henri Piéron.[50] Piéron, who taught at the Sorbonne until 1920 when he joined Janet at the Collège de France (with the latter's blessing), trained many more psychologists of the next generation.[51] The rivalry between Binet and Janet may also explain why Janet never published in Binet's *Année psychologique,* founded in 1894, and why instead he established his own journal, the *Journal de psychologie normale et pathologique,* in 1904.

Between 1889 and 1903 Janet published the books that reported his observations and elaborated his early theory—*Psychological Automatism* (1889), *The Mental State of Hystericals (L'état mental des hystériques,* 1893–94), *Neuroses and Fixed Ideas (Névroses et idées fixes,* 1898), and *Obsessions and Psychasthenia (Les obsessions et la psychasthénie,* 1903). A hiatus of over fifteen years separates this last book from his next major work, *Psychological Healing (Les médications psychologiques)*. In the 1920s, Janet's psychology took a different turn, as he began to develop his psychology of conduct, which he wanted to make completely externalist, objective, and behavior-oriented. His early work, however, develops with a great continuity, and an examination of the psychological theory contained in these works is needed to assess the ways in which Janet borrowed from his philosophical background.

JANET AND THE LEGACY OF RIBOT

Janet read Ribot while still at the Ecole Normale, and the example of a philosopher-turned-psychologist must have been an inspiration for the young *normalien*.[52] He avidly read Ribot's monographs as they appeared, and they obviously made a great impression, for many years later Janet recalled them as the works of Ribot that he liked the best, "perhaps as a *souvenir de jeunesse*."[53] Janet praised Ribot highly as the initiator of scientific psychology in France and considered himself to be following in the path blazed by his predecessor. It is not surprising, therefore, that Janet's psychology resembles that of Ribot in a number of ways. Nevertheless, Ribot was not the only source of Janet's psychology, and Janet was not afraid to disagree with Ribot on several central issues.

At the most speculative level, Janet was, like Ribot, a philosophical monist. He thought that mind and motion were two aspects of the same reality. They only seem distinct because we know them through such different means—the senses, in the case of matter and motion, consciousness in the case of mind.[54] He did not believe that thinking substance had a separate existence, and in this sense was opposed to the Cartesian dualism of his uncle. On the other hand, he was attracted more to the dynamic idealism of Leibniz than to the mechanical rationalism of Spinoza.[55] This made his brand of monism more acceptable to spiritualists, who also respected Leibniz, than the Spinozist variety, championed by Taine, Ribot, and Espinas. Still, Janet interpreted his monism to mean that psychology and physiology were not two different sciences, but merely two different approaches to the same phenomenon.[56]

For this reason among others, Janet was as convinced as Ribot that psychology should be studied according to "the method of the natural sciences."[57] This meant the objective rather than the subjective method. Like Ribot, Janet did not dismiss the data of introspection completely; on the contrary, he thought that only introspection can reveal the nature of consciousness, since we cannot observe directly the thoughts of others.[58] Nevertheless, both agreed that self-observation had two limitations. First, contrary to popular and philosophical belief, one can be mistaken about the contents of consciousness. Janet found this to be true especially in the mentally ill, who rationalized their behavior in ways that obscured its true origins, but the problem extends to normal individuals as well.[59] More importantly, introspection by definition cannot reveal those aspects of psychological life that escape consciousness—in other words, the testimony of consciousness cannot illuminate the unconscious.[60] To overcome the limitations of self-observation, one must have recourse to the study of others.

Most of Janet's observations were carried out on the mentally ill in the context of clinical psychiatry, first in Le Havre and then in Paris at the Salpêtrière. But Janet's psychological theory was explicitly intended to apply to normal psychology, too. This implied that psycho-pathology could illuminate the workings of the ordinary mind, and, like Ribot, Janet invoked Bernard's dictum that "the laws of illness are the same as those of health and that there is in the former only the exaggeration or the diminution of certain phenomena that could already be found in the latter."[61] Mental illness constitutes a "natural experiment" that enables one to observe how altered conditions affect the workings of the mind.[62] Disease changes the variables in ways that would be impossible or immoral for the experimenter, and it reveals mental phenomena that reason and the observation of normal individuals might never suggest. Janet was so fond of the statement that "disease decomposes and analyses memory better than psychology"—meaning philosophical psychology—that he included it in two of his works.[63]

Like Ribot, Janet had a sense of the complexity of the psychological domain that made him scornful of the simplicity of philosophical psychology. The theory of the self illustrates this complexity. Spiritualist philosophers assumed it to be simple and indivisible. Indeed, as we have seen (chapters 1 and 3), these properties constituted one argument for the soul's immortality. But the idea of the self, Janet insists, is the culmination, not the foundation, of consciousness. "The idea of the self (*le moi*) is, in effect, a very complicated psychological phenomenon that includes memories of past acts, the notion of our situation, our powers, our body, even our name, which, uniting all these dispersed ideas, plays a large role in the knowledge of the personality."[64] This echoes Ribot's theory that self-consciousness is the highest and most complex of mental phenomena. And below self-consciousness exist many layers of psychological and physiological phenomena, few of which are accessible to introspection. This is why the objective method is so necessary.

There are a number of other ways in which Janet borrowed directly or indirectly from Ribot. Ribot and Janet both placed more emphasis on the will than on the intellect, although their analyses of will were significantly different. The loss of will power became one of the defining features of "psychasthenia," one of Janet's two major classes of neurosis.[65] Moreover, Janet placed action above contemplation in his hierarchy of psychological functions, in part because he thought that action required the coordination of more mental elements (perceptual and motor) than abstract thought, but also because the will was more directly involved in action.[66]

Janet was especially indebted to Ribot for the theory of different memory types. Ribot had developed a theory of partial memories that held that the recollections of the various senses were relatively independent and that the memories of one sense—for example, hearing—could be lost without affecting those of another.[67] In his early works, Janet claimed that the different personalities exhibited in multiple consciousness tended to divide up along different sensory systems, so that whereas his patient Lucie was a visual type, having abandoned most tactile sensations, her alter ego Adrienne was a tactile type, taking over and communicating with the world through precisely those senses forgotten by Lucie. Janet used this to explain why Lucie had no knowledge of Adrienne.[68]

Finally, Janet adopted the evolutionary framework Ribot brought to French psychology, although he generally stressed the evolution of psychological life in the individual rather than the species. Nevertheless, in his hierarchy of mental functions, Janet subscribed to "Ribot's law," assuming that these functions were disturbed by mental illness in inverse order of their complexity and recency of acquisition—the most recent and complex were impaired first, the earliest and simplest only in the advanced stages of disease.[69] Furthermore, Janet used as his criterion in constructing this hierarchy not "our artistic or moral preferences," the traditional criteria he attributed to philosophers, but "the conditions of life, health, and illness. . . . The theories of evolution have taught naturalists that there is an order of perfection in beings, that one must discover it and take account of it in descriptions."[70] Hence the hierarchy of mental functions ultimately reflects the order in which adaptive responses were developed during the course of evolution.

Yet despite these broad areas of agreement, Janet differed from Ribot in a number of crucial respects. One of these was in their respective attitudes toward academic philosophy. Ribot, it will be recalled, was uncompromisingly hostile. He rejected philosophical psychology in toto, because it did not use experimental methods and because it made metaphysical assumptions. Janet tried to paint Ribot as a liberal arbiter who avoided conflict with philosophers of other schools, but he drew this conclusion from Ribot's policy as editor of the *Revue philosophique,* not as the author of the prefaces to *English Psychology* and *German Psychology.*[71] Only when philosophical theories wandered too far away from observation and into metaphysics did Janet refuse to consider them.[72]

Indeed, Janet repeatedly stated that his aim was to join philosophy and medicine to create a unified psychology. As Charcot put it in his preface to *The Mental State of Hystericals,* "M. Pierre Janet wished to unite as completely as possible medical studies with philosophical studies; it was necessary to bring together these two kinds of knowledge and these two educations in an effort to analyse clinically the mental state of a patient."[73] Often Janet phrased this in a slightly different way, by saying that he wanted to unite psychology and mental pathology. He addressed two of his works to both "doctors and psychologists."[74] But his definition of psychology in these and other cases never excluded philosophy. He was happy that the Third International Congress of Psychology dropped the adjectives "physiological" and "experimental" from its title:

> This signifies, in my opinion, two things: first, that there are not two psychologies, that there is no psychology other than that which takes account of facts, and secondly, that this psychology in no way has the childish pretension of eliminating reasoning and system, of forgoing the conceptions of the great philosophers, and that on the contrary it reaches out to all of good will and to all disciplines, whatever they may be.[75]

In a passage in which he quarreled with the Nancy school's extensive use of the concept of suggestion, he cried, "is it well to take no account of the distinctions established for centuries by the philosophers, between the diverse psychological phenomena, sensation, imagery, association of ideas, judgment, will, personality, etc.?"[76] Janet did take account of those distinctions, even if his clinical studies led him to modify many of them.

Methodologically, Janet differed from Ribot not so much in theory as in practice; whereas Ribot could only summarize the observations and experiments of others, Janet actually undertook firsthand psychological research. Nevertheless, Janet's more direct experience led him to concentrate on a number of specific techniques that gave his psychology its distinctive character. One of these was the psychological case method, which involved extended and intensive observation of individual illnesses, and above all, talking with and listening to the patient:

> We believe that we should, before all, know well our subject in his life, his education, his disposition, his ideas, and that we should be convinced that we can never know him enough. We must then place this person in simple and well-determined circumstances and note exactly and on the spur of the moment what he will do and say. To examine his acts and words is as yet the best means of knowing men, and we find it neither useless nor wearisome to write down the wandering speeches of a lunatic.[77]

Janet would explore the responses of his patients to his questions and to situations in which they found themselves. Ribot tended to abstract the findings of psychiatrists from their particular context and to present them as "facts" to be generalized through induction into laws. Janet also generalized, but he had a greater respect for the individuality of cases, and he came to place more and more emphasis on this form of observation.

He was not opposed to more direct manipulation of patients, however, especially in his earliest works, and hypnotism was the means through which he experimented on many subjects. Certain forms of hypnotic trance presented only very limited psychological phenomena, and by isolating and varying these through suggestion, Janet could identify the components of mental life. Indeed, he claimed that hypnotism allowed one at last to carry out in practice the thought experiment of Condillac with his living statue, into which he introduced sensations and ideas one by one.[78] Hypnotism, however, had the disadvantage of displacing normal consciousness, and in order to study the workings of the subconscious during normal consciousness, Janet stumbled onto the method of automatic writing, borrowed from the practice of mediums who transcribed the sayings of the spirits of the departed while in a trance. In certain highly suggestible patients, Janet was able to induce responses to questions of which the conscious mind was not aware. Typically, the subject would be distracted by another colleague, who would engage her (the patients were usually female) in conversation. Meanwhile, Janet would approach the subject from behind and whisper questions into her ear. To his initial surprise, he got answers from the handwriting on paper. He first used this method with Lucie, and it became one of the principal tools by which he developed his theory of double personality. He used it to discover what was going on in parts of the mind not accessible to conscious reflection.[79]

Another way in which Janet's methods differed in degree from those advocated by Ribot involved their respective attitudes toward quantitative methods. Despite the fact that Ribot himself employed the pathological rather than the experimental method, he was in principle enthusiastic about German psychology, especially in the first half of his career.[80] Janet, on the other hand, although he made use of quantitative methods in his laboratory, remained fundamentally skeptical about the value of psychometry.[81] "I am convinced that it is through the study of these natural experiments, more than through mathematical theories and measurements that we will achieve a better understanding of our intelligence and our action."[82] In part, Janet considered the application of quantitative methods premature. "Psychology is not yet advanced enough to admit of many precise measurements."[83] He thought that psychology was still in the stage of natural history, in which phenomena needed to be observed and classified before quantitative methods could be fruitfully applied. At such a stage in its development, the employment of overrefined measures could only hurt. "It is useless and even dangerous to use a microscope to do large-scale anatomy; one runs the risk of not knowing what one is looking at."[84]

One of Janet's attempts to expose the misuse of quantification was an article on "The Measure of Attention and the Graph of Reaction Times," presented at the International Psychological Congress at Munich in 1896 and published in *Neuroses and Fixed Ideas*. Presented in the land of Wundt, it tried to deflate the enthusiasm for measurement by showing an instance in which experimenters had failed to make sure that they were in fact measuring what they claimed to be measuring. A number of researchers had investigated the nature of attention by measuring reaction time to visual or auditory stimuli. Janet undertook similar research on some of his hysterical patients, and some of the results were predictable: his patients had difficulty concentrating on the test, and their reaction times were slow. But other patients scored phenomenally well, even when they did not pay attention to the test. For these patients, attention and reaction time were inversely, not directly, related. At the extreme, one patient experienced an ecstatic attack during the test, but continued to respond to the stimuli with reaction times faster than normal, healthy individuals. Janet concluded that simple reactions to simple stimuli were most likely to become automatic, and thus removed from conscious attention, so that a fast reaction time was as likely to be a measure of automatic response as of conscious attention. In short, the test did not measure what it purported to measure, and quantitative psychologists had been seduced by the search for numbers into abandoning their critical faculties.[85]

The greatest methodological difference between Janet and Ribot, however, is that Janet refused to invoke physiology to explain psychological phenomena. This may seem odd for one who was convinced that physiology and psychology were two sides of the same coin. Indeed, Janet considered it probable that a physiological theory of mental process would eventually be developed.[86] Nevertheless, he considered physiological theories of mind premature, for two reasons. First of all, until psychological analysis makes some progress, one does not know what the explanandum is. ". . . One must seek by means of precise and experimental analysis of the trouble itself to specify its nature, in order to facilitate later the interpretation of the cerebral lesions that will be discovered."[87] In a purely clinical and pragmatic sense, diagnosis will always involve the analysis of psychological factors, simply because that is how the symptoms manifest themselves. More importantly, research into the anatomy and physiology of the brain had not yet reached the point at which any meaningful connections could be drawn between the brain and human thought and behavior. "In my opinion, these anatomical hypotheses are only the pure and simple translation of observed psychological facts into another language, and they will become of interest only when histological research is able to establish the nature of the modification produced in these parts of the cerebral cortex."[88] It is the physiological theories, not the psychological theories, of mental illness that Janet accuses of being "metaphysical," "philosophical," and "deductive." This, it may be remembered, was exactly the argument used by Paul Janet in response to Ribot.[89]

Moreover, unlike Ribot, who readily conceded the possibility of purely physiological adaptive reponses, Janet argued on methodological grounds that any act that normally requires conscious thought must be assumed to contain conscious elements, even if the observed subject is not aware of them. "When we know that a complicated phenomenon, like the movements of anger or the gestures of prayer, can exist in us only with a set of conscious emotions and ideas, we do not have the right to assume that the exact same gestures are produced during catalepsy without being directed and unified by some sort of consciousness."[90] Janet held to the Cartesian criterion that language and intelligent action are the signs of consciousness, and he inferred the latter wherever he found the former. He even extended this reasoning to habit and instinct; to the extent that they were adapted to the specific context in which they occurred, such acts must be conscious in some sense.[91]

All this points to a vast extension of the domain of consciousness, far beyond what Ribot, physiologists, psychologists, and even philosophers would accept. Janet extended consciousness beyond habit and instinct to include even the regulation of the body's physiological functioning.[92] This extension is in fact a consequence of Janet's monism, and it is significant that Janet drew different conclusions from the same premises as Ribot, who distinguished between physiological phenomena, which had only one aspect, and psychological phenomena, which had two. Janet defended Ribot against the charge of epiphenomenalism, but it is clear that the two held disparate views on the domain of psychology.[93] Janet's monism is in this respect closer to that of Espinas, who also extended consciousness into the remotest regions of biology. Seen in this light, Janet's "psychological automatism," which meant that even automatic acts are psychological—that is, conscious—in nature, contains an implicit critique of Ribot, since for the latter automatic meant physiological.

From this interpretation of monism, stressing the extent of consciousness rather than its dependence on physiology, Janet drew a number of other consequences that differentiate him from Ribot. For instance, he concluded that every action required the intervention of an idea, and conversely that every idea tended to express itself in action. This ideo-dynamism was not unique to Janet by any means, nor was it foreign to

Ribot, but Janet incorporated it as a central feature of his theory. He insisted that wherever there was action, there must also be motor images. Hence the loss of movement might be due to the inability to recall motor images rather than to the physical impairment of the nervous system. This theory allowed him to give a psychological explanation of hysterical paralysis. More radically, Janet denied the common distinction between sensory and motor images—that is, the theory that there is one set of images that results from the input of the senses and another set associated with the motor activity of the body. He found it more economical to assume that there was only one set of images that served both as the terminus of the afferent impressions and as the origin of the efferent instructions.[94] At the same time, he held that there were as many different types of images as there were senses. This also helped him explain hysterical anesthesia, since he could attribute it to the inability to recall a particular class of tactile images.

Another consequence of Janet's interpretation of monism is that the emotions play a much different and less important role in his psychology than they do in Ribot's. We have seen in chapter 2 that Ribot gave great importance to emotional life among psychological phenomena. For instance, he thought that emotions constituted the essence of the will, and that without emotion, action was impossible. Janet thought instead that ideas could determine action directly, without the intervention of emotions. He therefore downplayed the role of emotions in voluntary action.[95] Furthermore, whereas Ribot thought that the emotions provided the ultimate basis of our sense of self, Janet thought that the sense of self was based on the synthesis of a large number of elements, of which emotions were only one part. Hence, alterations in the sense of the self were more likely to be related to the power of synthesis than to the emotions.[96] Indeed, for Janet, emotions were above all disruptive, shattering syntheses and leading to dissociation. They were effective mainly during moments of weakness, and they could cause the onset of neurotic symptoms. Finally, Ribot thought that emotions were basically the manifestations of organic states, whereas Janet, after flirting with James's theory that emotions were the reports of physiological conditions in the body, developed in *Obsessions and Psychasthenia* the idea that sentiments were related to purely psychological and intellectual states.[97]

Janet interpreted many neurotic phenomena as the inability to recall certain classes of psychic images. But one might well ask how one can simply "forget" certain classes of images if there is no organic lesion and hence, under Janet's own monistic assumptions, the images must still be conscious in some sense? To explain this, one must return to Janet's theory of consciousness. If consciousness is very extensive, it is not uniform. Unlike the Cartesians, who make of consciousness an all-or-nothing phenomenon, Janet held it to be capable of several degrees.

Sensation, the lowest degree, includes the psychological aspect of sensory input in its most raw, unassimilated form—unadorned elementary images. Sensation, though conscious, need not involve the discrimination between self and other. Hence there can be sensations that do not reach the self, and that therefore cannot be predicated of the subject "I."[98] Perception, the next level of consciousness, unifies sensations into discrete objects and distinguishes between subject and object. It synthesizes the messages of the senses, but it does not involve general ideas. The highest level of consciousness is intelligence, which requires judgment, the ability to discern an abstract relation between two terms provided by perception. As we have seen, Janet believed that the idea of the self was one of the most complex achievements of mental life, and as such, involved intelligence. The individual was not always capable of the effort required to sustain such an idea. It was therefore in theory possible for a sensation to be conscious, in that it existed as an image in the mind, without being attached to the idea of the self, and therefore present to self-consciousness. For this reason, Janet acknowledged an entire class of phenomena Ribot did not—subconscious ideas.

Moreover, Ribot and Janet differed in their analysis of the process by which ideas combined. Ribot was more or less associationist; he generally thought that all ideas could be generated through the mechanical operation of contiguity and resemblance. Janet on the other hand, held that association can only reproduce a link between sensations or perceptions that has already been made in the past. It is a "conserving" rather than a creating activity, and it is the basis of automatic actions.[99] Sensations are initially linked by another process, "synthesis," that not only creates new associations but generates higher mental phenomena as well. Although synthesis is a fundamental feature of Janet's psychology, he is not very specific about the nature of this activity. It "unites more or less numerous given phenomena in a new phenomenon differing from the elements."[100] It goes on constantly, assimilating the elements of sensation into a representation of the present situation and adapting the organism to the environment.[101] It is prior to the self, since the self is one of its products. Above all, it is not reducible to association. Janet never tires of repeating that synthesis is creative, that it is not a mechanical product of properties of the elements on which it operates.[102] Both association and synthesis are important, but they are completely distinct. In *Obsessions and Psychasthenia,* Janet distinguished further levels of synthetic activity to create a hierarchy of psychological phenomena based on the number and complexity of elements synthesized.[103]

Synthesis requires a certain amount of energy, which Janet called "psychological force" and which, like synthesis, has no clear antecedent in Ribot's psychology. The quantity of psychological force available to the individual determines his ability to synthesize incoming sensations and to respond to the environment. Psychological force varies among individuals, but normally it is sufficient to maintain the idea of the self and to put one into contact with reality.

If psychological force enables one to synthesize one's inner and outer impressions and to deal adequately with the world, then it is obvious that the ultimate cause of mental illness is a lack of psychological force. Janet at first called this condition *la misère psychologique* and later coined the term *psychasthenia* to describe it.[104] Janet believed that one's reservoir of psychological force was ultimately derived from one's fundamental constitution, and as such was largely inherited from one's parents. Disease could also lower one's psychological force. Moreover, one might have a small quantity of psychological force and still get by, provided the demands of one's environment do not exceed one's capacity to respond. Finally, Janet came more and more to stress the role of traumatic emotions in lowering one's psychological force. Violent emotions disrupt existing syntheses and drain one's energy, leaving one open to other pathological phenomena.

Janet came to distinguish two broad classes of mental disorder, based on how one responded to the weakening of one's psychological force—hysteria and psychasthenia proper. In hysteria, the patient responds by abandoning whole classes of sensation, giving up the effort to synthesize all stimuli and concentrating on a few considered most important or easiest to assimilate. The rest are no longer assimilated to the idea of the self. This "narrowing of the field of consciousness" results in the disintegration of or dissociation (*désagrégation*) from the self of many sensations and explains the anesthesia, paralysis, and amnesia of hysterics, who have abandoned less important stimuli—tactile images, muscular impulses, and memories—in order to be able to deal with the more pressing needs of visual and auditory impressions.[105]

But the problem is not limited to the loss of certain sensations, because these sensations are not completely lost. The senses still function, impressions are received, and memories survive. They are no longer available to the self, but they continue to exist, and they impinge on behavior in several ways. Fixed ideas can remain in the mind and determine several kinds of pathological phenomena, including hysterical contractures and attacks, the latter of which Janet considered the repetition of a forgotten traumatic event.[106] Unassimilated into the self, such ideas cannot be coordinated with the totality of other sensations and hence can appear at inappropriate moments, triggered by purely automatic association. Moreover, such ideas can in certain situations become hallucinations, because the patient does not associate them with the self and cannot resist them, thereby giving them the force of objective reality.

Most dangerously, however, the work of synthesis continues to operate on these dissociated phenomena, integrating and developing them outside the control of the self. Fixed ideas develop in richness and complexity. Hysterics may act in quite sophisticated ways, responding without knowing it to phenomena of which they are not aware. In this way Janet explained the phenomena of automatic writing and hypnotism. If severe enough, this unassimilated synthetic activity could lead to the formation of a second idea of the self, resulting in split personality. This is why Janet defined hysteria as *"a form of mental disintegration characterised by a tendency toward the permanent and complete doubling* (dédoublement) *of the personality."*[107]

Although hysterics suffered from a lack of psychological force, Janet came to reserve the term *psychasthenia* (etymologically, "lack of psychic force") for another class of neurotics that today would be called obsessive-compulsives. Psychasthenics are usually obsessed by some idea that both fascinates and repels them. They cannot seem to rid themselves of it, even though they recognize its falsity or repulsiveness. But more important in Janet's opinion are the complaints of inadequacy and difficulty in dealing with the world that such patients generally have. Janet thought that psychasthenics also suffer from a lack of psychological force, but for some reason this weakness does not lead to the complete dissociation of certain classes of stimuli. All the sensations and memories are available, but the patient has great difficulty in synthesizing them. Psychasthenics are not strong enough to avoid the formation of fixed ideas, but such ideas are not completely subconscious, as in hysteria, and, unlike hysterics, psychasthenics can resist them. Janet placed great importance on his finding that, unlike hysterics, psychasthenics are not suggestible, either through hypnotism or through distraction.[108] Comparing hysteria and psychasthenia, therefore, Janet found that "the reduction of consciousness seems to proceed geometrically in the former, through a reduction in the number of functions retained, and dynamically in the latter through a reduction in the force and perfection of all the phenomena."[109]

JANET AND ACADEMIC PHILOSOPHY

Despite the fact that Ribot presented an obvious model for a young academic philosopher who wanted to study scientific psychology, and despite the fact Janet first substituted for and then succeeded Ribot at the Collège de France, Ribot's influence on Janet should not be exaggerated. We have just investigated several ways in which Janet's psychology differs from that of Ribot—Janet's use of purely psychological concepts, his extension of the domain of consciousness, and his distinction between automatic and synthetic activity are the most significant. But even in those areas in which the two agreed, similarity need not imply influence. Janet's early interest in psychology was fostered by his uncle rather than by Ribot, and he never studied under Ribot. Janet read widely in psychological literature, and he gave his allegiance unreservedly to none. To the extent that he can be said to have had a *maître,* given that his formative years were spent in the provinces, it was Charcot rather than Ribot. Ellenberger writes that Janet should not be considered merely a disciple of Charcot, since he "came to the Salpêtrière, not as a student, but as an experienced collaborator."[110] This is true, but it is also true that, after some initial hesitation, Janet had opted for Charcot over Bernheim long before he arrived in Paris, and he never ceased to defend the work done at the Salpêtrière against the criticisms of the Nancy school. Charcot was a more natural choice as a mentor, because although Ribot also advocated the conjunction of psychological analysis with mental pathology, he was not in a position to bring about that union himself. As a neurologist and psychiatrist, Charcot was in such a position, and it is understandable that Janet would seek him out in preference to Ribot for the training he needed.

For the same reason—namely, that Charcot and his school were conducting actual research on patients, whereas Ribot could only summarize it—Janet quoted the work of the Salpêtrière far more frequently than that of Ribot. There are very few places in which Janet relies on Ribot exclusively for a fact or even an interpretation. In fact, far from Janet being at first a student of Ribot and later achieving his independence, as Sjövall suggests, it seems more likely that Janet's debt to Ribot was retrospectively enlarged as his career progressed.[111] This was for two reasons. First of all, since Janet ended up succeeding Ribot rather than Charcot, historians and Janet himself have tended to write the story from the teleology of this outcome, erecting analogies between Ribot and Janet into influences. It is significant that, whereas in the introductions to his earlier works Janet attributed the pathological method to Charcot, in *Obsessions and Psychasthenia* he calls it the method of Ribot. This change of indebtedness may have something to do with the fact that *Obsessions and Psychasthenia* was the first work Janet published after assuming Ribot's chair. Janet may also come to have rely more heavily—if still somewhat indirectly—on Ribot as the evolutionary framework of his psychological theory became more pronounced. Ribot had thought out the implications of evolution for psychology as Charcot had not.

If there are overlaps between what Janet may have acquired from Ribot and what the school of the Salpêtrière had to offer, there are also overlaps between the medical tradition and the psychology of academic philosophy. One reason for this is that a number of spiritualist concepts had entered psychiatry through contacts between spiritualist philosophers and sympathetic psychiatrists earlier in the century.[112] Even where those concepts had received other, more biological interpretations, they remained a part of the medical tradition. For example, the distinction between physiological and psychological approaches to mental phenomena cannot be reduced to the opposition between medicine and philosophy, for this was an issue within the medical community as well. Although by the beginning of the 1880s the pendulum had swung toward physiological approaches, shortly thereafter it began to swing back, at least in some minds, and Janet thought that Charcot had arrived at a purely psychological theory in many areas by the end of his life.[113] Conversely, Janet opposed what he considered to be the premature recourse to physiological explanation he found in many philosophers, even of the spiritualist variety. We have seen, for example, how Rabier in particular embraced the distinction between conscious and physiological and indeed how the Cartesian tradition could actually abet such a distinction. Neither tradition, medical or philosophical, was completely homogeneous, and if in general one can still assert that the philosophical community was more psychologically oriented and the medical community more biologically oriented, there was enough diversity and discussion for each side to recognize itself in at least part of the other.

The distinction between automatic and synthetic activity was not unique to the tradition of French academic philosophy, either. Goldstein has found it enunciated by the psychiatrist Jules Baillarger in 1845, and it could be found in many later medical writings.[114] Taine used it in *On Intelligence,* and Richet used a very similar distinction in an article that Janet quotes extensively in *Psychological Automatism.* Richet distinguishes between a "superior force" that directs the intellect and constitutes the self, and another self, completely automatic, that "regularizes and coordinates willed movements."[115] Richet even describes this second self as a form of "psychic automatism," as opposed to a "somatic," or purely physiological, automatism.[116] Nevertheless, immediately after having posited this superior force for the purposes of illustration, Richet denies it once again, maintaining that the superior force can be reduced to the equilibrium of tendencies in the mind.

Despite antecedents like this and others in the medical tradition, it is more likely that Janet first came to appreciate the opposition between automatic and synthetic activity in academic philosophy and then found it confirmed in the medical literature he subsequently encountered. One reason for this hypothesis is simply that the philosophical influence was the earlier. The philosophy of the syllabus always distinguished between sensory and intellectual operations, placing association among the former and judgment among the latter. If this general similarity were not enough to account for the distinction between automatic and synthetic activity, however, there is in addition the fact that Paul Janet used precisely these terms to characterize the distinction between association and judgment.[117]

Another important element of Pierre Janet's theory that came from the philosophical tradition is his emphasis on the theory of judgment, "the principal point of contemporary psychology, that which divides psychologists the most today."[118] In one sense, judgment is simply the highest level of synthesis, above sensation and perception. But Janet also stresses that judgment is qualitatively different from either sensation or perception, which both result in images. Judgment, on the other hand, produces ideas of relation, which cannot be reduced to images.[119] Janet makes judgment the basis of his theory of the will, since he maintains that "voluntary acts are precisely those that are determined by judgments and ideas of relation."[120] If in this sense judgment is a class unto itself, a subset of all synthetic activity, in many passages it becomes the type of synthetic activity. Janet often uses the opposition between judgment and association to describe the distinction between synthesis and automatism. His earliest *experimentum crucis* to determine the existence of unconscious mental activity was an effort to find out if posthypnotic suggestion could extend to subconscious judgment rather than acts that could be explained by automatic association.[121] In this way, philosophical definitions of mental processes played a crucial role in the development of one of his central concepts.

This distinction led him to distance himself from associationist psychology, as his review of Binet's first book, *The Psychology of Reasoning (La psychologie du raisonnement)*, illustrates. Binet admired British psychology in his early days, and in this book he tried to prove "that the fundamental element of mind is the image; that reasoning is an organization of images, determined solely by the properties of images, and finally that it suffices for images to be given for them to organize themselves and for reasoning to follow with the fatality of a reflex."[122] Janet objected that a judgment involves two different things: first, sensations and images, the "materials," as it were, and second "the idea of relation that unites them."

> Undoubtedly there is a reason that causes B to follow A in a mind, but as long as this reason is not known by me, as long as it does not present itself to me as the consciousness of a relationship between A and B, there is in my mind a simple juxtaposition of terms by association, there is not a judgment. . . . There is thus something new in my mind when I conceive this relationship, . . . and the mechanical association of terms was only the preparation for it.[123]

Synthesis is not a simple product of the properties of the elements combined. It is a separate operation that discerns those properties by its own independent power. Thus understood, Janet's psychology comes close to reinstating the concept of faculty that Ribot had tried to eliminate with his associationist analysis of mental phenomena, and despite the fact that Janet objected to hypostasizing classes of phenomena into faculties, he used the term extensively to refer to the mind's capacity for synthesis.

At first glance, it would seem that another of Janet's principal innovations with respect to Ribot, the extension of the domain of consciousness, owed little to the tradition of academic philosophy, which was predominantly Cartesian. We have seen how Rabier embraced the distinction between physiological and conscious, and how on the other hand Espinas had extended consciousness considerably. Janet read Espinas, quoting him precisely on this issue.[124] Nevertheless, one of the main reasons Janet gave for attributing a form of consciousness to subconscious acts was precisely the Cartesian criterion for intelligent action and Janet's insistence that "reasoning without consciousness makes absolutely no sense," both of which were arguments used by spiritualists against the notion of unconscious cerebration propagated by epiphenomenalists.[125]

Moreover, when Janet rethought the implications of his discovery of subconcious intelligent actions, he concentrated on the distinction between sensation and self-consciousness, and he relied above all on Maine de Biran to support the theory that there can be sensation that is not attached to the sense of the self.[126] All the writers who have discussed Janet's sources have pointed out his debt to Maine de Biran, but few have noted how odd this allegiance was in some ways on the part of someone interested in scientific psychology. For whatever Maine de Biran's influence on a few psychiatrists such as Baillarger and Jacques Joseph Moreau de Tours in the 1840s, in the 1880s he was the darling of the neospiritualists. As we have seen in chapter 1, Ravaisson made Maine de Biran the basis of his opposition to Cousinian eclecticism, praising him for asserting that inner experience gave one access to substantial reality, not merely phenomenal appearances. Ravaisson and Lachelier found inspiration in Maine de Biran to reassert the legitimacy of metaphysics, and Bergson found in him one source of his concept of intuition, the immediate grasp of the inner self that was so opposed to the reason that applied to the outer world. Many of the similarities between Janet and Bergson that have been pointed out by numerous authors have their source in the fact that both read Maine de Biran.

It is probable that Janet came to Maine de Biran through his contact with neospiritualist authors, or more likely still through his uncle Paul, who also had a keen interest in the early nineteenth-century philosopher. If Pierre Janet were to make Maine de Biran safe for science, however, he would have to read him in a way that avoided the consequences of the neospiritualists. And in fact, Janet repeatedly stressed that Maine de

Biran was acquainted with the science of his day, that he visited asylums and encountered magnetizers, to bolster the idea that he engaged in objective and not just subjective psychology.[127] Janet even unearthed a document to prove that Maine de Biran attended an exhibition of hypnotism. Janet asserted that Maine de Biran "merits, more than is generally believed, to be considered a precursor of scientific and experimental psychology."[128] In this way Janet drew upon an alternative source within the academic tradition and forged a positive concept—affective consciousness—from the ideas of the very person claimed by the neospiritualists as their inspiration.

The trace of academic philosophy may also be seen in Janet's conception of method. Despite the greater specificity of his investigative technique, his general conception of scientific method differs very little from that taught in the lycée. This should not be surprising, given that Janet not only taught lycée philosophy for many years; he also wrote a textbook of which the philosophy of science constituted fully half.[129] Janet generally characterized his method as "psychological analysis" (*analyse psychologique*). This expression first appeared in an article of 1887 in which he described the use of automatic writing, which he hoped would provide "a method of psychological analysis" to investigate the nature of unconscious phenomena.[130] In *Psychological Automatism,* Janet compared his work with cataleptic patients to Condillac's thought experiment with the living statue to which sensations were added one by one. Janet thought the way Condillac chose "to analyze the human mind" was "an excellent scientific method," and indeed, Janet considered himself to have found a means of doing in practice what Condillac had only done in theory.[131]

In both cases, analysis meant to separate mental phenomena into their elemental parts as a means of understanding them. Once he had resolved them into their elements, he could isolate them, vary them systematically, and thereby discover the relations between them. Despite the overtones of chemistry and empiricism suggested by this characterization of analysis, the term was also used in eclectic psychology. Cousin had engaged in the "analysis of thought," and eclectics "analyzed" psychological phenomena in their classrooms and in their textbooks. Analysis had in fact been a standard part of philosophical method since Descartes enjoined his readers to break their problems into their smallest parts and to work on each separately. Janet's innovation was to try to carry out his analysis on living patients, using hypnotism and automatic writing. He became more cautious, however, during the course of his career, and he came to use the less theatrical methods of extended observation and controlled questions to perform his "psychological analysis."

After he had analyzed phenomena through observation and experimentation into their component parts and fundamental relations, he then assumed these elements and relations to be general and tried to deduce other phenomena from them. This part of method he called "synthesis" (related to but not to be confused with his use of that term in the expression "synthetic activity"), in the geometrical sense of working from principles to consequences, or verification, since it was intended to see if the consequences derived actually agreed with the facts.[132] Although Janet may have been echoing Spencer's extensive use of analysis and synthesis, these two terms were also a standard part of the syllabus.

Another way in which Janet's psychology reflects academic philosophy is in its acknowledgment of the hypothetical character of its generalizations. Janet constantly treated his theories as "simple résumés, symbols that represent more or less well the momentary state of a question."[133] In this sense, they were not intended to be so much explanatory theories as empirical definitions, which are "nothing but general ideas, résumés which should comprise only the greatest number of possible facts."[134] Such definitions and theories are always provisional, giving only an "approximate synthesis of these phenomena still so poorly known."[135] They are constantly subject to revision, "since there is a constant increase of known facts, which renders [them] soon too narrow."[136]

This modesty stems from his theory of hypothesis, which he defined as "a procedure by means of which the human mind *goes beyond sensory observation* and *adds* to the facts known by the senses *some notion that cannot actually be sensed* and which perhaps never will be."[137] This definition of hypothesis is essentially that of Rabier, and like Rabier, Janet thought that hypothesis intervened at several stages of the experimental process—in experimentation itself, to suggest lines of research, in classification, and even in "the most rigorous induction."[138] Janet held that even after one had isolated an invariable antecedent using Mill's experimental methods, the supposition that this antecedent was the real cause remained hypothetical.[139] Such a view implied that any scientific law, even one that had been verified, was nevertheless a hypothesis.

One final but crucial area in which hypothesis enters touches the central area of his research—the hypothesis that unconscious acts are nevertheless psychological rather than physiological. The notion of a psychological unconscious is of its very nature hypothetical, since it cannot be observed directly, either by oneself or by someone else. Nevertheless, it is manifested by the intelligence of the acts and words produced without conscious awareness. Janet argued that such an inference is no different from any indirect measure in science—for example, the chemist who determines the composition of the stars from their spectral analysis, or the physicist who measures temperature by the height of a thermometer.[140] He also argued that a physiological interpretation of the unconscious is even more hypothetical. But he nevertheless admits that his own position is hypothetical.[141]

This admission of the hypothetical character of scientific laws is one reason why Janet tried to distinguish sharply between facts and their interpretation. Like Rabier, Janet's recognition that laws are hypothetical did not shake his belief in the givenness of observation. He always published the findings from the psychological laboratory at the Salpêtrière in two volumes, one containing his systematic interpretation and one containing only "observations," the case histories from which his conclusions were drawn. He thought that his observations would continue to have relevance even if his theories were proved incorrect.[142] Of course, in its extreme form, such an impartial observation is impossible, and Janet himself recognized this. "No doubt it is always a little hypothetical to choose in the midst of innumerable details those that one believes interesting and to neglect the others, but without this hypothesis no exposition is intelligible."[143] This remark was intended to apply only to the selection of phenomena recorded in his notes that found its way into his texts, but it could be applied to the recording of the notes themselves. Moreover, in at least one instance, the accuracy of Janet's notes was called publicly into question by Bernheim, who believed that Janet had misquoted him.[144] Nevertheless, Janet held onto the notion that one's observations should be as theory-neutral as possible.

In addition to his methodological borrowings from the syllabus, Janet's notion of the content of psychology was also fairly traditional. In the introduction to the second edition of *Neuroses and Fixed Ideas* (1904), Janet listed the most interesting topics to be discussed from the point of view of psychology: they included the sensory organs, elementary movements, attention, memory, ideas, will, emotion, personality, and "the laws of social hierarchy."[145] This is a fairly standard enumeration of the major topics of eclectic psychology.[146] Janet's psychological analysis of a mental patient would "pass in review all her psychological faculties, and determine exactly what, in each group of phenomena, deviates from the normal law."[147] Hence, even though he brought new methods to bear, he addressed familiar questions. He directed the categories of philosophical psychology outward rather than inward, applying them to the "objective" rather than the "subjective" method.

Moreover, in determining what is normal, Janet also draws implicitly upon the philosophical definition of a human being. Often he uses a functional and even evolutionary standard of normality, as if the only criterion were adaptation to one's environment. In this sense, Janet would appear to deny universal standards by which to evaluate mental functioning. But Janet made no attempt to define normality quantitatively, as Binet was doing at exactly the same period. In many ways, Janet's picture of the human mind is remarkably traditional. Whereas Ribot's individual is driven by his body and passions, Janet's is ideally dominated by conscious synthetic activity. Janet's name is associated with the theory of split personality, and his claim that the individual can harbor more than one self set him at odds with many academic philosophers. Nevertheless, unity of consciousness remained the ideal and the definition of normality for Janet. The healthy mind is master of its contents, which only escape self-consciousness if they are insignificant or have properly integrated automatic mechanisms to deal with them. Janet resisted even purely psychological theories, such as that of Bernheim, which made irrational phenomena like suggestion normal features of the mind. And if he made action rather than thought the summit of his hierarchy of mental functions, he found the source of action in an intellectualist rather than an emotivist theory of will.

CONCLUSION

Pierre Janet tried to build an empirical psychology using the methods of pathological psychiatry. He was inspired in this project by Ribot, Charcot, and others who objected to the purely subjective methods of philosophical psychology. But in the crucial areas that defined his originality—his purely psychological approach, his extension of consciousness, and his conception of synthetic activity—he drew on the resources of the philosophical tradition in which he was first educated. As a result, he was able to practice a psychology in which empirical did not necessarily mean biological nor explanatory mechanical.

This approach made him immensely popular with his philosophical colleagues. They saw in him a vindication of their arguments against associationism and biologism from someone who used the very tools of science. They seized upon him and elevated him quite quickly to the highest position in the academic world, and they looked to him as a model of the cooperation of philosophy and science when they sought others to take his place.

6

Metaphysician: Emile Durkheim's Science of the Moral

Thus renovated, the theory of knowledge seems destined to unite the opposing advantages of the two rival theories, without incurring their inconveniences. It keeps all the essential principals of the apriorists; but at the same time it is inspired by that positive spirit which the empiricists have striven to satisfy.
—Emile Durkheim, *The Elementary Forms of the Religious Life*

EMILE Durkheim (1858–1917), Janet's classmate at the Ecole Normale, followed in the footsteps of Espinas rather than Ribot. Durkheim shared Espinas's distaste for humanistic, literary culture, and advocated its replacement by a culture based on science. Early in life Durkheim set as his goal the establishment of an empirical science of society employing the methods and assumptions of the natural sciences, and he often cited Espinas as both precursor and inspiration. He also borrowed from Espinas some of his central concepts, such as social realism and collective representations. Durkheim agreed with his predecessor that the moral needs of the Third Republic could best be served by a scientific understanding of the social origins of morality. On a professional level, Espinas paved the way for Durkheim by making it possible for an advocate of positivist sociology to gain entry into the French university system. In all these ways, Durkheim's debt to Espinas is greater than has generally been recognized.[1]

Nevertheless, Durkheim differed in crucial ways from Espinas. Espinas thought science and metaphysics were linked, whereas Durkheim, like Ribot, sought to maintain a distinction between them. Durkheim backed away from the organic analogies that were the explanatory center of Espinas's sociology, and Durkheim's obsession with the conditions of experimental proof was in part an attempt to counteract the uncontrolled analogizing of his elder colleague. Espinas hoped to make sociology scientific by attaching it to biology, whereas Durkheim thought that sociology could only achieve positivity by becoming autonomous.

Similarly, Espinas stressed the links that bound society to the rest of nature, whereas Durkheim emphasized the break between social phenomena and other levels of reality. Finally, while Espinas tended to identify morality with technology as simply another need of the social organism, Durkheim distinguished the two and emphasized the obligatory character of morality. In short, Durkheim reasserted the autonomy and centrality of the "moral," in its equivocal meaning as spiritual and ethical. These differences, especially his methodological rigor and his insistence on a purely sociological analysis, parallel in significant ways the differences of Janet from Ribot. They also constitute the principle reasons why Durkheim and not Espinas is generally considered to be the founder of scientific sociology in France.

What accounts for Durkheim's divergence from the direction in which Espinas had pointed sociology? The formative influences on Durkheim were many and complicated. They include his early work with historians at the Ecole Normale, his reading of Comte and Renouvier, as well as his encounter with experimental psychology and social science in Germany. But academic philosophy had an equally important effect on Durkheim's development. Durkheim studied under and read a variety of academic philosophers who criticized the associationism and biologism of Espinas and Spencer, philosophers who argued for the irreducibility of concepts to sensations and for the irreducibility of scientific method to analogy. Durkheim accepted important elements of the philosophical critique of his predecessors, incorporating them into his conception of sociology while retaining the goal of a natural science of society. This attempt to reconcile the methods of the natural sciences with the insights of nonpositivist philosophy defines much of Durkheim's uniqueness.

As important as the specific borrowings from individual philosophers, which have been documented by a number of scholars, was the structure of academic philosophy as a discipline. Any attempt to reexamine Durkheim in his intellectual context, as several writers have recommended, must address philosophy at the disciplinary level.[2] That is what this chapter attempts to do. As shown in chapter 4, by the time Durkheim entered the profession during the 1880s, academic philosophy was becoming much more diverse and open to different schools of thought. Along with changes in the larger political and ideological climate, this openness gave those advocating a scientific approach to society a better chance of gaining access to elite positions. The criteria for advancement were shifting from doctrinal orthodoxy to professional competence. Indeed, philosophers were increasingly encouraged to elaborate original positions and to defend them ably rather than to conform to a preexisting school of thought. Durkheim made use of this situation, staking his claim to a position in Paris on the originality and quality of his work as much as on its content and approach. In this context, even Durkheim's denunciations of philosophy and his insistence that he was doing something entirely different could be an asset in the philosophical field of the day. I argue that academic philosophers recognized and rewarded this strategy; indeed, the widespread belief that Durkheim was kept away from Paris by philosophical opposition is in large measure a founding myth begun by Durkheim and his followers and perpetuated uncritically by subsequent scholars.

Academic philosophy was also important for Durkheim's sociology because it still claimed the subject matter of the human sciences, and it still claimed the right to judge scientific method. These claims allowed its practitioners to borrow from and intervene in a number of related disciplines. Despite his assertions that he restricted himself exclusively to sociology, Durkheim used the continuing connections among philosophy, psychology, and the social sciences to intervene actively in all these disciplines. This is generally recognized in relation to philosophy. Durkheim made no attempt to disguise the fact that his science of morals was designed to replace philosophical approaches to ethics, and his sociology of knowledge was aimed at another traditional topic of philosophy—the origin of the categories. It is also acknowledged in relation to the other social sciences, where Durkheim's sociological imperialism attempted either to absorb or remold neighboring disciplines such as history and law. Durkheim's attempts at intervention have been less remarked in psychology, though they are just as real. Durkheim constructed a psychology that complemented his particular notion of the sociological domain, and he argued against psychologists and psychiatrists whose findings threatened his sociology.

In short, Durkheim constructed his sociology by borrowing from and contributing to a number of different disciplines and schools of thought. As he did this, he portrayed his own position as a middle way between opposite extremes. In sociology he presented himself as midway between social organicism and methodological individualism. In psychology he steered a course between psychophysical reductionism and traditional introspectionism. In philosophy he portrayed his emergent realism as midway between materialism and spiritualism. Despite a deserved reputation for systematic and rigorous thought, Durkheim shows the traces of the eclectic tradition, recognizing truth in divergent schools and trying to reconcile them through a synthetic union. If these divergent elements are integrated by Durkheim's powerful intellect into a system of considerable consistency, the integration is not seamless. Durkheim's crypto-eclecticism helps explain why commentators—starting with his contemporary critics—have had such difficulty categorizing his thoughts, portraying him as either positivist, neo-Kantian, or idealist.[3]

This tendency to seek the middle way does not mean that Durkheim accepted the right of all schools of thought to go their separate ways. Although, as scholars such as Besnard and Schmaus have pointed out, Durkheim tolerated a considerable diversity of approach within his *Année sociologique* team, his tolerance had definite limits.[4] Durkheim, like Cousin, felt that since he had selected the correct elements from all schools, other schools were no longer necessary. As with Cousin, so with Durkheim the ability to choose from all schools could coexist with the willingness to exclude those who did not meet his minimum requirements. His battle with Gabriel Tarde and his dismissal of René Worms provide examples of the exclusionary tactics of which he was capable.

This chapter has three main purposes. First, I examine Durkheim's career until he achieved an academic appointment at the Sorbonne in 1902 to show the extent of his philosophical support. Then I compare and contrast Durkheim's sociology as it emerged during the same period with that of Espinas, both to emphasize the extent to which Espinas constituted a model for Durkheim and to specify the differences that require explanation. Finally I look at the sources of Durkheim's innovative positions in his interaction with philosophy, psychology, and the social sciences. This interaction involved a complex set of borrowings and interventions through which Durkheim constructed a position that drew from many sources but was reducible to none. Although the question of whether Durkheim's position shifted during the course of his career is an interesting one, I do not enter into that debate here.[5] On the one hand, I accept that there is enough continuity to warrant treating Durkheim's works as a whole. On the other hand, if, as I hope to show, Durkheim's position was from the very beginning of his career an attempt at reconciling many diverse trends, it should not be surprising that the balance of elements might shift as he struggled to work out the implications of his fundamental insights, nor that different readers might interpret his writings in very different ways.

CAREER

As with most other educated Frenchmen, Durkheim's connection with philosophy began in the secondary education system. He took the philosophy class at the Collège d'Epinal in Eastern France, which enabled him to pass the baccalaureate in letters in 1874.[6] As a promising student, he went to Paris to prepare for the entrance exams for the Ecole Normale Supérieure. He studied at the lycée Louis-le-Grand, where his professor was Thomas-Victor Charpentier, the eclectic who also taught Pierre Janet. Georges Davy reports that Charpentier took an active interest in his pupil, encouraging him when, as with most candidates for the

Ecole Normale, he did not get in the first time he tried.[7] Once he did gain entrance, Durkheim studied with Léon Ollé-Laprune, whose Catholic spiritualism he disliked, and with Emile Boutroux. Although he never mentioned it, he may also have encountered Elie Rabier, since Rabier replaced Ollé-Laprune for a year while Durkheim attended Normale. In any event, he took care to read the textbook Rabier published soon thereafter, since he quoted it more than once in the course of his writings.[8] Moreover, in several respects Durkheim's discussion of method echoes the kinds of concerns evident in Rabier's philosophy of science (see the following section). However, Emile Boutroux was the teacher who most impressed Durkheim. Durkheim had great respect for Boutroux as an intellect and as a person, dedicating his first major work, the *Division of Labor in Society (De la division du travail social)*, to him and remaining on cordial social terms with him throughout his life.[9] Durkheim was a serious and imposing youth who acquired a reputation for dialectical skill. These qualities earned him the nickname "The Metaphysician" among his classmates.[10]

When Durkheim left the Ecole Normale and passed the *agrégation* in 1882, he became a lycée professor, first at Puy, then at Sens, and finally at Saint Quentin. He was apparently an excellent teacher, methodical and clear, as students and administrators both agreed.[11] Lachelier considered him "one of the most serious of our young professors of philosophy. . . ."[12] He taught this course for four years, long enough to master the discourse of the syllabus. Notes of his philosophy course at Sens taken by one of his students, André Lalande, reveal a thorough and careful presentation still heavily influenced by the neospiritualism of the Ecole Normale.[13] His teaching was sufficiently orthodox that he did not get into trouble with provincial clerics, even though his students reportedly called him "Schopenhauer" because he so frequently cited that philosopher.[14]

Durkheim's lycée teaching was interrupted by a sabbatical during the academic year of 1885–86, funded by a fellowship from the French government. Durkheim attended classes in Paris for part of this period, and during the first half of 1886 he traveled to Germany, where he worked in the psychological laboratory of Wilhelm Wundt. He also encountered the historical schools of jurisprudence and economics. The articles he wrote upon his return about his experience in Germany, as well as the other review articles that had begun to appear in the *Revue philosophique* even before his departure, marked him as an astute observer and as a vigorous champion of the scientific approach to morality. These qualities brought him to the attention of Louis Liard, who in 1887 appointed him to the University of Bordeaux—where, as we have seen, Liard himself had taught—as a *chargé de cours,* primarily to assume the course in pedagogy initiated by Espinas, who had just become dean of the Faculty of Letters. He was given this position despite the fact that he had not yet started, much less completed, his dissertation and so did not yet have his doctorate. Espinas may have recommended Durkheim to replace him, and he must have heartily approved of the addition of the words *science sociale* to the title of the course.[15] He certainly welcomed Durkheim to Bordeaux in a speech that celebrated the event as a triumph for social science.[16]

Although Durkheim later made much of the fact that this position allowed him "to devote ourselves early to the study of social science and, indeed, to make it the substance of our professional occupations," his connections with official philosophy were by no means severed by this appointment.[17] His classes in social science included a large number of philosophers, who were preparing for examinations in philosophy, not sociology. Durkheim helped students prepare for the *licence* and *agrégation* in philosophy throughout his stay in Bordeaux.[18] He also taught pedagogy, educational psychology, and psychology. These courses obviously required him to stay abreast of developments in psychology, and they are probably one reason for the quite detailed knowledge of contemporary psychological theory Durkheim displayed in his sociological works. In addition to teaching, Durkheim also remained in the milieu of academic philosophy through his friends, colleagues, and professional organizations. His closest associates at Bordeaux were Octave Hamelin, a metaphysician whose work has been described as a mixture of Renouvier and Hegel and whose posthumous *Système de Descartes* Durkheim edited, and Georges Rodier, a historian of ancient philosophy.[19]

It was in this philosophical milieu that Durkheim brought forth and defended his sociological treatises. *The Division of Labor* was Durkheim's doctoral dissertation in philosophy, which he defended in 1893 before a committee consisting of Janet, Waddington, Boutroux, Marion, Séailles, and Victor Brochard (a neocriticist).[20] Durkheim was provocative at his defense, assuming the demeanor of a scientist in a den of philosophers and claiming that his work "had not departed from absolute mechanism," that it was "a purely scientific thesis."[21] At the same time, however, he had claimed in his preface that there was no barrier between science and ethics, and he had tried to demonstrate that the division of labor was not only a fact but a moral obligation in modern society. It is therefore not surprising that his committee reacted strongly, and that their questions centered on the relation of his functional description and mechanical explanation of ethical rules to the properties they considered essential to any moral theory—liberty and abstract duty. Any work on moral theory that did not take into account these two features of moral action would necessarily be inadequate in the eyes of a spiritualist or neo-Kantian philosopher. Thus, Marion thought that Durkheim's thesis was "not refined enough (*pas assez fine*) to reach ethics," Janet objected that he had confused function with duty, and Waddington felt that he had dealt only with "the inferior regions of morals."[22] According to one report, Janet became so agitated that he pounded his fist on the table.[23] Although to some extent Durkheim and his committee were talking past each other, there was also an important and relevant point implicit in these criticisms. By using an exclusively objective and mechanical method, Durkheim had missed the specificity of the moral domain, in the opinion of the committee.

Despite their objections, the committee passed Durkheim, not because they agreed with his thesis, but because it was well done and original, and he had defended it well. As in the case of Espinas, professional competence overcame doctrinal disagreements. All accounts of the defense, from both friend and foe, stressed Durkheim's command of his subject, his ability to respond to questions, and his self-assurance. One witness predicted, "*He* will be a master."[24] Nor was this estimation limited to his defense. We have already seen that Lachelier thought highly of him, and other philosophers, even those most diametrically opposed to his ideas, grudgingly admitted his mastery of dialectic, his ability to give logical force to his empirical arguments.[25] The sheer intellectual power of his work earned him the respect of academic philosophers.

This respect was confirmed by the two major works that followed *The Division of Labor: The Rules of Sociological Method* (*Les règles de la méthode sociologique,* first published in the *Revue philosophique* in 1894) and *Le suicide* (1897). These books generated much controversy but also cemented Durkheim's reputation as a formidable thinker. However much his reviewers disagreed with him, and we will return to the substantive criticisms in the next section, they never failed to mention the boldness and rigor of his works. Thus, for example, Léon Brunschvicg and Elie Halévy, two young neocritical philosophers who reviewed *The Division of Labor* for the *Revue de métaphysique et de morale,* stated that "the conception of M. Durkheim is so audacious and so original, it overturns so completely our habitual ways of thinking, that one should not be astonished at the intensity of the opposition it will encounter."[26] This respectful attitude characterized all of the many interactions between the *Revue de métaphysique et de morale* and Durkheim, despite what one might expect given the journal's general orientation.[27] Gabriel Tarde, Durkheim's chief rival in sociology, thought that "there are few sociological truths so useful to examine as the errors of M. Durkheim, however manifest they may be. And we must thank him for having expressed them with in such a clear, intrepid manner. They were in the air, they were asking to be incarnated in a logical and vigorous mind; it is fortunate that they found his."[28] Marcel Bernès, a lycée philosopher who taught a course in sociology at the University of Montpellier, was critical of Durkheim's definition of social facts in *The Rules,* which he thought illustrated a prevalent error in contemporary sociology. "Above all, what I reproach M. Durkheim for is that in wanting to express social facts in a very precise manner and generally with great logical clarity, he does not always render the true physiognomy of social reality."[29] In other words, Durkheim's logic is sound, but his characterization of the social world is wrong.

Durkheim's forceful, bold approach won him admirers among students as well as professors of philosophy. He gained a number of supporters for his vision of sociology from students in philosophy at the Ecole Normale and at the University of Bordeaux. To an even greater extent than Ribot, and unlike Espinas and Janet, Durkheim cultivated his disciples carefully, encouraging young philosophers to undertake dissertations in topics that could be treated sociologically and helping them to find university positions. His considerate and frequent correspondence with such young talents must have flattered them, and the opportunity he provided in his *Année sociologique,* launched in 1898, allowed them to contribute to sociology in an institutional environment in which most of them could not hope to address sociology in their official work.[30] This cultivation of support among young philosophers sometimes paid off in unexpected ways. The early reviews of Durkheim's work in the *Revue de métaphysique et de morale,* which in principle ought to have been opposed to his sociology, were largely favorable, since they were written by precisely such young supporters.[31]

Despite this respect and support, Durkheim had to wait for his own appointment to a position in Paris. As we have seen (chapter 3), the Faculty of Letters requested the creation of a *cours complémentaire* in sociology in the fall of 1893.[32] Because the position was eventually defined as the history of social theory rather than as empirical sociology, it went to Espinas, who had more seniority, instead of Durkheim. Espinas's appointment both helped and hurt Durkheim's future chances. On the one hand, Espinas lobbied for Durkheim and blocked at least one attempt by outsiders to appoint Tarde to a position in sociology there.[33] On the other hand, Espinas's position was similar enough to sociology to make it more difficult to justify another position. Paul Janet had already tried this in his report to the General Council of Faculties. He argued that the two positions were entirely distinct and had nothing to do with each other.[34] A subcommittee seemed to agree with this reasoning, but before the issue reached the full council, the Faculty of Letters withdrew its request, citing the course created by the Comte de Chambrun as a reason.[35] The Faculty of Law, which had competed with the Faculty of Letters for a position in sociology, also decided to wait to see the results of Espinas's course before requesting its own course in sociology. From these decisions, one can infer that although the professors of the Sorbonne recognized the distinction between sociology and the history of sociology, and although they were willing to push for both, they did not consider the subject sufficiently important to push too hard, at least for the time being.

Even if there had been a position, Durkheim had competitors. Tarde had written a number of well-received books on sociology and was a serious contender for a position in sociology, though he was by training a jurist rather than a philosopher.[36] Moreover, his brand of sociology was more attractive to antipositivist philosophers than Durkheim's. Unlike Durkheim, Tarde admitted the data of introspection and found a place for indeterminism within a framework that was nevertheless scientific. Bernès had written an extended critique of Durkheim's methods and had introduced a course in sociology at his provincial faculty.[37] His approach also included introspection and contingency along with such neospiritualist themes such as the conventional character of science and the creative nature of human action. Espinas confided in a letter to Hamelin his fear that the "ultra-metaphysicians" would try to get Bernès rather than Durkheim into a position at the Sorbonne.[38] However, Espinas thought he could counter any such attempt: "From this side I can do something and I will do it. Bernès is not comparable to Durkheim."[39] Espinas thought that he could use the superior quality of Durkheim's work as a shield against the imposition of an inferior candidate, even if that candidate were ideologically preferable. He also used the weapon of academic credentials on Durkheim's behalf. Espinas convinced Tarde's extramural supporters that the Sorbonne would never stand for an outsider to be chosen over Durkheim.[40] These incidents illustrate the extent to which professional competence had become an overriding consideration in university appointments. One may also cite as competitors Jean Izoulet, whose works on social philosophy drew heavily on the language of science and attracted great enthusiasm in the highest Republican circles, and René Worms, who had founded a journal and several professional organizations in sociology.[41] Finally, less closely linked with philosophy but still concerned with sociological issues were anthropologists (in the nineteenth-century sense, which implied physical anthropology and the study of racial characteristics) such as Charles Letourneau and criminologists such as Alexandre Lacassagne.[42] It is hindsight to assume that Durkheim's difficulties in securing a position in such a crowded field represent more than typical intellectual competition.

Morever, if Durkheim cultivated certain connections that could advance his career and his science, he ostentatiously ignored others. He shunned the sociological organizations established during the 1890s by René Worms, organizations that brought together a large number of sociologists. He did not publish in the *Revue internationale de sociologie* founded by Worms in 1893, preferring to let his articles appear in the *Revue philosophique* until he established his own journal. Nor did he become a member of the Société Sociologique de Paris or the Institut International de Sociologie, two institutions also founded by Worms. This prevented him from becoming better known in Paris among a wider public outside the walls of the Sorbonne.[43]

In part, as Karady suggests, this must have been calculated.[44] Durkheim banked on his elite credentials and would not risk them by associating with the nonacademic and/or second-rate writers whom Worms allowed to contribute to his *Revue internationale de sociologie*. He preferred to cultivate a narrow audience of elite academics rather than a wider audience of the educated public. At the same time, as Laurent Mucchielli argues, Worms himself was a sufficiently well-credentialed competitor that Durkheim may not have wanted to raise the status of his rival's journal by publishing in it.[45] But Durkheim also failed to build bridges to elite academics in other disciplines who would certainly vote on any future appointment. History is a good example. History was one of the largest disciplines in the Faculty of Letters, but instead of appeasing the historians, Durkheim alienated them with an imperialistic rhetoric that attempted to subordinate history to sociology. In fact, the only discipline with which Durkheim maintained regular and cordial relations was philosophy. It is true that philosophers could help him advance his career, but in light of his willingness to alienate others who could help him, it is hard to resist the conclusion that at least part of Durkheim's continued association with philosophy was voluntary.[46]

For all these reasons—hostility, yes, but also limited funds, other priorities, competitors, and maladroitness—Durkheim had to wait until 1902 to obtain a position at the Sorbonne. Izoulet was named to a new chair in social philosophy at the Collège de France in 1897, and Tarde assumed the chair of modern philosophy at the same institution in 1901. Durkheim thus had to wait longer for an elite appointment than some of his contemporaries and longer than the average elite academic philosopher at this period, but the difference—twenty years rather than 17.5—hardly constitutes evidence of a conspiracy against him, despite a historiographic tradition to the contrary (table A.3).[47] And it was with the approval of the academic philosophers on the General Council that he was appointed as a *chargé de cours* at the Sorbonne. However, it was as a professor of education, not of sociology, that Durkheim came to Paris, and it was not until 1913 that his chair (he had been made *titulaire* in 1906) was renamed "Education and Sociology."[48] Institutional inertia and intellectual opposition were therefore not negligible factors even after Durkheim's arrival in Paris.

ESPINAS AND DURKHEIM

Although modern historians of sociology often omit Espinas entirely from their accounts, Durkheim himself in his reconstructions of the history of social science always made Espinas an important precursor. In his first course at Bordeaux, he claimed that Espinas was "the first to have studied social facts in order to make a science of them and not in order to secure the symmetry of a grand philosophical system."[49] In 1895, he called Espinas the first of the academic philosophers to be attracted by sociology.[50] Much later, he claimed to have gotten his social realism "in direct line from Comte, Spencer, and Espinas, whom I knew well before knowing Schaeffle."[51] He willingly deferred to Espinas's candidacy for the chair of social economy. Durkheim and Espinas do not seem to have been personally close, even at Bordeaux, where both taught from Durkheim's arrival in 1887 to Espinas's departure in 1893. They grew more distant with the passage of time, in part because Durkheim renounced the kind of biological analogies in which Espinas indulged, and in part because Espinas seems to have become suspicious of the party spirit he saw developing among the Durkheimians.[52] Moreover, Durkheim apparently did not want Espinas to publish in the *Année sociologique*. But he directly criticized Espinas only once, and that was in response to an article in which Espinas had intimated that Durkheim had abandoned social realism.[53] For the most part, he allowed his protégés, notably Simiand and Bouglé, to undertake the task of dismantling social organicism.[54]

One reason for Durkheim's circumspection was professional gratitude. Espinas had taken some of the edge off the opposition to sociological positivism within the academy, making it easier for his successor to find a place. And as we have seen, Espinas lobbied on behalf of his younger colleague at crucial points in the latter's career. Just as important as the professional assistance was the intellectual inspiration. There is no evidence that Durkheim read Espinas at the Ecole Normale, but as he himself said, he knew him well before his first published review article in 1885 on Albert Schaeffle's *Structure and Life of the Social Body (Bau und Leben des sozialen Körpers)*.[55] Already in 1884, Durkheim had made himself known to Ribot by going to his office to protest Espinas's article on the *agrégation de philosophie,* which characterized young Normalian philosophers as neo-Kantian dilettantes. Durkheim assured Ribot that they were "evolutionists, not at all criticists and they are troubled that one of the most authoritative of their own attacks them. . . . What is certain is that Durkheim appears nourished by *Animal Societies*."[56] Both institutionally and intellectually, Espinas was the closest thing to a model or paradigm Durkheim had available to him as he began his career.

Among the opinions that Durkheim shared with Espinas was, first of all, a disdain for humanist literary culture and the type of philosophy that was tied to it. Durkheim's attacks on classical education and the literary mind run throughout his career. He had been uncomfortable with what he considered the superficial literary brilliance of the Ecole Normale during his years there, and he disliked his literature professors.[57] In his report on philosophy in the German universities in 1887, he found that French philosophy still had "too many links with literary art."[58] In *The Division of Labor*, Durkheim argued that the age of general humanist culture was past, that modern society "feels an ever more pronounced distance from the dilettante and even from these men who, too taken with an exclusively general culture, refuse to let themselves be completely caught up in professional organization."[59] In his lectures on the history of French pedagogy, Durkheim reserved his greatest contempt for Renaissance humanism, which elevated literature and aesthetics above a concern for logic, truth, and even morality.[60]

Much of this diatribe against humanist culture and literary philosophy was part of the common coinage of positivist polemic of the period, and it is familiar to the reader from the sections on Taine and Renan in chapter 1. Very few of Durkheim's ideas in this direction were original, and not all can be said to have been derived from Espinas. But Espinas had given recent and vigorous expression to them in his piece on the *agrégation* in philosophy, which Durkheim read, and in 1895, when Durkheim undertook his own article on the *agrégation* and on the teaching of philosophy, he echoed many of the specific points raised by Espinas a decade earlier.[61] Like Espinas, Durkheim found academic philosophy infected with literary preciosity, causing philosophers to avoid facts and solid reasoning in favor of rhetorical effect and novelty. Like Espinas, Durkheim analyzed the situation as the result of Cousin's efforts to create an official philosophy as a unifying force in postrevolutionary France. Like Espinas, Durkheim thought that Cousin's institution had served its purpose, but that society had changed to the point that such an institution was no longer effective. What society needed today, Durkheim thought, were individuals trained to clear, sober thought by the discipline of the natural and social sciences. Philosophy could play a role in producing such citizens by adopting the spirit of science and teaching the principles of science. Thus, Durkheim proposed to drop metaphysics and the history of philosophy from the syllabus, leaving method, psychology, and ethics. These would be taught from the perspective of the social sciences, emphasizing facts, not systems. And in order to have professors who could teach such a syllabus in the proper spirit, Durkheim, like Espinas, would require future professors of philosophy to have substantial scientific training.

Much of the problem with philosophy was a question of method.

Both Durkheim and Espinas condemned the deductive method, because it was based on principles that were themselves unprovable and, in their opinion, ultimately subjective. Such a method encouraged the idea that morality was something abstract, universal, eternal, and unchanging, not necessarily connected to society and therefore not subordinate to the demands of society. Like Espinas, Durkheim advocated an empirical approach to the study of morals. As he stated in the preface to *The Division of Labor,* "This book is pre-eminently an attempt to treat the facts of the moral life according to the method of the positive sciences."[62] That method is inductive rather than deductive or "geometrical." Durkheim was not opposed to the use of quantification, but he resisted the notion that science should consist of reasoning about abstract concepts not derived from phenomena, and he repeatedly condemned not only philosophy but other disciplines such as political science and political economy as examples of what sociology should not be.[63] Social science should be a natural science, deriving its concepts from systematic observation and comparison.

Because sociology is a natural science, Durkheim also agreed with Espinas that the methods and vocabulary of the other natural sciences could be useful in sociology. His early work, especially *The Division of Labor,* is filled with concepts drawn from biology, and such distinctions as organ/function, morphology/physiology, and normal/pathological played a crucial role in the development of his sociology.[64] This use of biological concepts extended to the organic analogy, which Durkheim used both to illustrate and to structure his arguments. One need only glance at his references in *The Division of Labor* to Edmond Perrier, Espinas's rival in the theory of the organism as society, to see the extent to which Durkheim drew upon biological analogies to support his work.[65] Durkheim could have derived these principles from Espinas, and it is ironic that he chose to cite not the person he considered his forerunner in social science, but a naturalist whose credentials in biology would lend greater weight to an example in sociology. The distinction between the normal and the pathological, which allowed both Espinas and Durkheim to ground prescriptions for social health in the needs of society, depends on conceiving of society as a kind of organism.[66]

Durkheim also followed Espinas in the extensive use he made of concepts and analogies drawn from psychology. This may seem surprising, given Durkheim's reputation for dismissing psychological explanations of social phenomena. ". . . Every time that a social phenomenon is directly explained by a psychological phenomenon, we may be sure that the explanation is false."[67] But this is not the same as adapting concepts from psychology and applying them in sociology. Durkheim claimed not to have known about developments in experimental psychology until after he left the Ecole Normale, and he complained that his knowledge remained superficial thereafter.[68] This statement cannot be taken at face value, however, because there is evidence that he read Ribot's theses as well as his works on English and German psychology while still at Normale.[69] In any event, *The Division of Labor* contains extensive reference to developments in psychology, and in *Suicide* Durkheim cited a vast literature in psychiatry in his efforts to combat a psychopathological explanation of the phenomenon. Psychological metaphors tended to replace biological ones in his later works, but it is important to note that the former do not exclude the latter. Hence it is entirely consistent with a biological conception of society to introduce psychological terms in the study of collective mentality.

With the tools provided by the method and examples of the natural sciences, Durkheim constructed a theory of society that also owed much to Espinas. First and foremost, he owed much, as he acknowledged, to Espinas's formulation of social realism. In order for sociology to be a science, society must have an existence that surpasses the individuals who inhabit it. This was one source for Durkheim's insistence that social facts be defined by their externality to, and constraint upon, the individual. It was also one source for his definition of morality as "the fundamental conditions of social solidarity."[70] Like Espinas, Durkheim dismissed the notion that there were transcendental standards of moral action. Instead, he maintained that moral rules represent the constraints needed to hold society together, and these rules change with the society. "Whatever it [ethics] is at any given moment, the conditions in which men live do not permit its being otherwise, and the proof is that it changes as conditions change, and only then."[71] Durkheim also shared Espinas's conviction that the demystification of moral principles would not lessen their efficacy. Both thought that demonstrating the reality and power of society and the foundation of ethics in the needs of society would increase the individual's respect for the rules that made social life possible. Both believed that social science would support the French Republic by demonstrating that the existing state of things was a natural product of social forces, one that could be altered only slowly.

Finally, it will be recalled that Espinas spoke of society as "an organism of ideas," constituted by the collective representations that bind it together. Durkheim borrowed his concept of collective representations directly from Espinas, specifically citing *Animal Societies* when he described how they arose in crowd situations.[72] Two points follow from this filiation. One is that Durkheim did not necessarily use the term *representation* in a Kantian sense. The other is that Durkheim's shift from using "collective conscience" to "collective representations" and from an organic to a psychological model does not in itself necessarily mean that he abandoned Espinas's model of society. It may be more accurate to say that the model itself was so flexible that it could lead to very different pictures, depending upon which parts of it were emphasized.

In a number of ways, then, Durkheim developed themes present in Espinas's work. Some, like the collective conscience and collective representations, were derived directly from Espinas. Others Durkheim may first have come across in other writers. Nevertheless, Espinas embodied a number of ideas that Durkheim found attractive and that served as a powerful inspiration in the early part of his career.

However, there are an equal number of ways in which Durkheim's sociology differs from that of Espinas. They may be summed up by saying that whereas Espinas stressed similarity, Durkheim emphasized distinctness. This difference between them is evident first of all in their respective attitudes toward philosophy and its relation to social science. As argued in chapter 3, Espinas thought that science inevitably presupposed a metaphysics, and therefore that the two were ultimately inseparable. Durkheim's position was closer to that of Ribot: science should be kept distinct from metaphysical questions, and making that distinction is the condition of any scientific progress. Durkheim repeatedly insisted that sociology should be "neither spiritualist, nor pantheist, nor evolutionist, but simply the science of morals."[73] For Espinas, sociology could not be itself without also being naturalist.

Although Durkheim recommended sweeping changes in the teaching of philosophy, he did not want to eliminate it, as Espinas seems to have wanted to do. Precisely because science and philosophy were distinct, philosophy would always retain a legitimate role in human thought and in French education. For Espinas, science generated its own metaphysics, and hence it was unnecessary to have any other form of speculation. Similarly, Espinas seems to have thought, at least in the early part of his career, that science generated its own morality. It was therefore unnecessary to teach morality in the school. Durkheim, on the contrary, believed that society had to inculcate its beliefs into the individual, and he thought that morality was an essential part of education. He therefore thought that the philosophy class was the perfect place to culminate moral education by providing students with the facts of moral life.[74]

Durkheim's philosophical enemies differed significantly from those of Espinas. Espinas fought against all forms of dualism, spiritualist and especially neo-Kantian. Durkheim was comfortable with spiritualism and neocriticism, even though he insisted that they accept the phenomenal determination of the moral realm. The axis of his sympathies was not naturalism/dualism, but rationalism/mysticism. Durkheim decried what he considered to be the "renascent mysticism" of his age.[75] This included the intuitionism of Bergson, the Catholic-tinged philosophies of action of Ollé-Laprune and Maurice Blondel, and the traditionalism of Charles Maurras and Maurice Barrès. His condemnation also covered less extreme manifestations of the new spiritualism, whenever they denied the relevance of science for moral phenomena. In principle this might stain most of his spiritualist contemporaries, but in practice Durkheim was tolerant of anyone who recognized the rule of reason. This is how he could be on such intimate personal and intellectual terms with a pure metaphysician like Octave Hamelin.

Another area in which Durkheim differed from Espinas was in his conception of method. Immediately after proclaiming in *The Division of Labor* that ethics should be studied according to the method of the positive sciences, Durkheim added the following:

> The moralists who deduce their doctrines, not from an a priori principle, but from some propositions borrowed from one or more of the positive sciences like biology, psychology, sociology, call their ethics scientific. This is not the method that we propose to follow. We do not wish to extract ethics from science, but to establish the science of ethics, which is quite different.[76]

Laws and models derived from other sciences can not be applied directly to sociology. Espinas had assumed that the laws of evolution and of the formation of aggregate individuals must be the same in the human as in the animal world, since nature is essentially one. Having constructed his model for the latter, he could apply it with certainty to the former. This is why, after writing *Animal Societies,* Espinas felt no need to continue his empirical research into human societies. Durkheim did not allow himself this assumption; he was forced to study human society directly. If he nevertheless borrowed concepts from other sciences, which he certainly did, such concepts could only be considered hypotheses, which had to be verified by their usefulness in explaining specifically social phenomena.

Durkheim's concern with the control and verification of hypotheses derived by analogy is just one aspect of his attention to method. In both his empirical works and in *The Rules,* Durkheim tried to articulate and follow a method that would be objective, systematic, and demonstrative. His famous injunction to "treat social facts as things" was an attempt to get sociologists to distance themselves from preconceived notions— whether borrowed from common sense, philosophy, or other sciences— and behave toward social phenomena as the biologist would toward plants and animals.[77] His attention to definition and to taking one's definition from phenomena was designed to ensure that the investigator did not arbitrarily privilege some phenomena over others. And his concern in establishing the conditions of proof was an effort to give his findings demonstrative force. Durkheim did not think that resemblance could be an explanatory principle in itself. He dismissed the attempts of those such as Spencer—and by implication Espinas—to prove points by piling up supportive evidence.[78] Durkheim's efforts at scientific rigor may not always have been successful. Both contemporary and subsequent critics have found fault with his enunciation of the principles of scientific method, his practice of them in his own works, or both. Nevertheless, Durkheim's level of sophistication in demonstration was far greater than that of Espinas, and far greater than that of most of his contemporaries.

If Durkheim's concern with the distinctness of the social led him to distance himself from the organic analogy and to emphasize the importance of method, it also made a difference in the way he used psychology. Espinas had made much use of the analogy of consciousness: we can only understand the consciousness of others on the analogy of our own. Espinas applied this principle not only to the consciousness of animals, but to the collective conscience as well. In this way, he continued the spiritualist conception of the moral sciences as sciences of introspection at the same time that he added a comparative dimension. Durkheim denied himself this principle, holding that science must be completely objective, even when dealing with consciousness, individual or collective. He rejected the testimony of introspection, whether of one's own consciousness or the reports of others, as inherently unreliable.[79]

Durkheim also differed from Espinas in crucial aspects of his theory of collective representations. Espinas was an associationist and an empiricist. He thought that concepts could be derived from sensation by a gradual process of repetition and association. Collective representations evolved in the same way, through the gradual accretion of individual contributions to a representation as it was exchanged among members of a society. Durkheim's views on concept formation are not entirely clear or consistent. For one thing, he used the term *representation* to refer to a variety of mental phenomena, from sensations to perceptions to concepts.[80] At times he used the model of the exchange and fusion of individual representations among members of a society to describe the emergence of collective representations.[81] This would seem to imply a gradual process of fusion and merging. However, he also described concepts as emerging from the combination of sensations with entirely different and irreducible properties. To illustrate this, he employed the analogy of individual atoms combining to produce molecules with different properties.[82] In whatever manner they were conceived to arise, collective representations always had for Durkheim a form and content that were irreducible to those of individual representations. His critique of utilitarianism in *The Division of Labor* was designed to show that moral rules were not a simple outgrowth of individual wants or sentiments, and the sociology of knowledge he began to develop in Primitive Classification ("De quelques formes primitives de classification") and later elaborated in *The Elementary Forms of the Religious Life (Les formes elémentaires de la vie religieuse)* was premised on the idea that logical categories are irreducible to sensations.

There was another way in which Durkheim's theory of morals differed from that of Espinas. As discussed in chapter 3, Espinas tended to collapse morality and technology. On the one hand, he considered the practical arts to be part of the collective conscience, and on the other, he treated morality as the practical art of maintaining the social organism. Durkheim resisted this assimilation, distinguishing between moral rules, which contained a sanction imposed by society, and practical rules, which involved no sanction other than the automatic problems of failing to perform effectively. He stressed the obligatory character of moral rules and the aura of sacredness that surrounded them. This is true even when he argued that the division of labor was a moral force and proposed, in the preface to the second edition to his first book, a system of occupational associations to reintegrate society.[83] The point of his important critique of Spencer was that the act of exchange in itself was not moral and could not provide the basis for a moral order. It was the obligatory rules surrounding that act, making it possible and regulating its conditions, which constituted the moral dimension of economic activity.[84] Even though moral rules must correspond to the material activity of society, they are not identical with it. This distinction became more pronounced during the course of Durkheim's career, until by the time he wrote *The Elementary Forms,* he opposed moral to economic activity as sacred to profane.[85]

Finally, if Durkheim tried to maintain a distinction between science and metaphysics, he did not always strictly observe this distinction, because he clearly thought that his sociology supported some metaphysical conceptions of reality and undermined others. In *The Rules,* he suggested that "sociology, as it becomes more specialized, will furnish more original materials for philosophical reflection."[86] Specifically, sociology features "an idea which might well be the basis not only of a psychology but of a whole philosophy—the idea of association."[87] The idea of association is the idea that a whole exhibits properties not present in the parts taken separately. One may call it a form of emergent realism. The sources of this emergent realism will be discussed in the next section, but the important point here is that Durkheim differed from Espinas, who was a monist. For Espinas, collective representations were not different in kind from individual representations, and society was an organism in much the same way as individuals.

From this discussion it can be seen that Durkheim accepted parts of the tradition represented by Espinas and the sources on which he had drawn—Ribot, Spencer, and the other English psychologists and philosophers cited in *Animal Societies.* Nevertheless, Durkheim rejected significant parts of that legacy. He rejected the continuity of science and metaphysics that Espinas espoused; he demanded stricter conditions of scientific proof; and he insisted on the autonomy and irreducibility of the social realm. The next section explores some of the sources for this change in direction.

DURKHEIM AND ACADEMIC PHILOSOPHY

The sources of Durkheim's sociology are many and complicated. Any effort to reduce them to one or two specific thinkers fails to recognize the breadth of Durkheim's erudition and the independence of his thought. Durkheim cited specific individuals as sources on several occasions, among them academic philosophers, and these debts should not be ignored. At least as often, however, he referred broadly to "spiritualists," "eclectics," "the disciples of Kant," et cetera, as either inspiration or opposition. This tendency to speak in terms of schools rather than individuals reflects his training in philosophy, where ideas rather than individuals mattered. Durkheim was also happy to cite multiple sources for or examples of his ideas, and the very multiplicity of these instances provided confirmation of the validity of the notions in question. But while he consulted a variety of schools of thought, he was completely satisfied with none of them, and he was quick to criticize what he considered to be the shortcomings of his colleagues and rivals. His own position was inspired by many sources but reducible to none. While this is a definition of originality, it is also a reflection of the state of academic philosophy, which encouraged the creation of original positions in an increasingly diverse field. Finally, Durkheim consistently tried to portray his sociology as either a middle way between extremes or a synthesis of opposites. This too is a legacy of the eclectic tradition, which sought to reconcile the best elements of opposed systems. In the following survey of some of the sources of Durkheim's ideas and the interventions he made in the set of overlapping disciplines related to philosophy, I am less concerned in capturing the origin of every specific concept than to demonstrate a style of picking and choosing, modifying, and intervening in a wide variety of disciplines and schools of thought. This style, characteristic of Durkheim, is perhaps the most important legacy of the discourse of academic philosophy.

Durkheim gave credit for some of his most important ideas to spiritualist academic philosophers. For example, "I owe it first of all to my teacher, M. Boutroux, who, at the Ecole Normale Supérieure, repeated often to us that each science must be explained by 'its own principles,' as Aristotle put it: psychology by psychological principles, biology by biological principles. Very penetrated by this idea, I applied it to sociology."[88] Durkheim also credited spiritualist philosophers with opposing the reduction of psychological to biological phenomena.

> There is, then, no temerity in affirming that, from now on, whatever progress is made in psycho-physiology will never represent more than a fraction of psychology, since the major part of psychic phenomena does not come from organic causes. This is what spiritualist philosophers have learned, and the great service that they have rendered science has been to combat the doctrines which reduce psychic life merely to an efflorescence of physical life.[89]

Indeed, Durkheim insisted on the "hyperspirituality" of social phenomena, since they consisted of collective representations that sprang from individual representations, which were, Durkheim agreed, "spiritual" in character.[90] He complained repeatedly and apparently sincerely that his sociology had been misconstrued as materialist.[91] In his preface to the second edition of *The Rules* (1901), he protested that "whereas we had expressly stated and reiterated that social life is constituted wholly of collective 'representations,' we were accused of eliminating the mental element from sociology."[92] Of course, if some critics, such as Brunschvicg and Halévy, thought he was a materialist, others agreed that he had reified society into a kind of ideal substance.[93] Tarde accused him of "scholastic ontology," of creating "transcendental social facts that M. Durkheim exalts to the rank of the rescucitated Ideas of Plato."[94] Charles Andler charged him with "mysticism" for asserting the reality of social facts independent of their individual manifestations.[95]

However, Durkheim also disagreed with many aspects of spiritualist philosophy. As Lukes points out, if Durkheim agreed with Boutroux's insistence on the irreducibility of one level to another, he did not follow Boutroux's further claim that each higher level contained a higher degree of contingency.[96] Durkheim bitterly opposed any philosophy that introduced contingency into the phenomenal realm, denouncing such philosophies as "mysticism" and the negation of science.[97] The hyperspirituality of the social realm did not preclude its determinism, and Durkheim was firm on introducing the principle of determinism into the social domain. He insisted, especially in his early work, that social phenomena could be explained "mechanically." This meant that the causes of social facts were to be conceived as efficient causes on the model of physical push and pull rather than as final causes dependent on conscious intentions and goals.[98] It was this mechanism rather than materialism to which spiritualist philosophers objected. Or more accurately, it was this mechanism that *constituted* materialism for most neospiritualists, since they defined mechanism in the human sciences as the attempt to model human action on the causality of the material world.[99]

Neospiritualists, following Ravaisson and Lachelier, defined metaphysics in terms of creative activity rather than substance. Mind exists wherever there is such creative activity, and to the extent that the mind constitutes reality by interpreting it and giving it objectivity, reality is penetrated by mind. Durkheim abhorred this interpretation of metaphysics and of reality. In his piece on the *agrégation de philosophie,* he cited a prize-winning essay from the competition of 1890 in which the author, a successful candidate for the *agrégation,* claimed that one could find certitude in psychology and sociology only *"through an explanation of these phenomena that suits us.* . . . The greatest sociologist will be the one who, realizing the illusion of trying to give an exactly scientific interpretation of these facts, would abandon this scientific direction of the desire to comprehend, and who would make his science serve his needs for action."[100] Of this passage Durkheim lamented, "Is *that* what is called philosophizing? Is *that* the idea that is formed in our classes of science, of objective and impersonal truth?"[101] And yet, as Durkheim admitted, this essay had won the prize. Moreover, it is a good example of neospiritualist thinking. It may be that Durkheim's predeliction for describing social facts in terms of representations rather than actions comes in part from the fact that the neospiritualists had claimed action as the site of metaphysics and contingency.

However, it may have been Boutroux's interest in Kant that first led Durkheim to Renouvier, the other nonpositivist philosopher for whom Durkheim had the greatest respect.[102] We have seen in chapter 4 how Renouvier stressed the a priori character of the categories, the irreducibility of concepts to sensations, the impossibility of explaining higher phenomena in terms of lower, and the foundation of morality in duty rather than interest. All these concepts reappeared in Durkheim's sociology, and significantly, these are all ways in which Durkheim differed from Espinas. It is therefore likely that Renouvier's vigorous critique of Spencer left an impression on Durkheim. Renouvier was not the only critical philosopher Durkheim read, however; he also perused the neocritical theses of former *normaliens* such as Auguste Penjon and Pierre Nolen, he followed *La critique philosophique* and his knowledge of Kant was based on reading him in the original German.[103]

Nevertheless, it would be a mistake to think of Durkheim as a neocriticist, as some have proposed to do.[104] Durkheim made use of a number of Kantian distinctions, but he also found more positivistic ways of expressing the same dichotomies when it suited his purpose. For example, despite his debt to Boutroux and Renouvier for the distinction between levels of reality, he also cited the mechanistic metaphor of chemical synthesis to illustrate this idea. Moreover, he disagreed with certain fundamental parts of critical philosophy. He did not like the way in which Kant tried to derive a universal criterion of moral action from the idea of the person, and he ridiculed Kant's attempts to deduce specific moral duties from the categorical imperative.[105] He also questioned the whole premise of critical philosophy, since he did not think one could deduce the a priori conditions of knowledge through any sort of transcendental analysis. He followed Comte in arguing that one could truly understand logic and method only by studying the history of science. Epistemology was for Durkheim an object of sociological inquiry. This was one reason why he became so interested in the social origins of the categories in his later works. Indeed, he often seems to have raised Kantian problems only to give them a Comtean solution. Some of Durkheim's criticisms of Kant are similar to those made by Renouvier. But if Durkheim borrowed from Renouvier, he was not enough of a neocriticist to please Renouvier's collaborator Pillon, whose reviews of Durkheim castigate him for denigrating the role of innate moral ideas and for assimilating society to an organism.[106] For Pillon, Durkheim retained too much of the social organicism he and Renouvier had criticized in Spencer's work.

Before passing to Comte, however, we should not neglect the influence of eclecticism and the tradition of syllabus philosophy. Durkheim taught the syllabus for most of his career, and the impact of its structure and its designers can be seen throughout Durkheim's work. For example, Durkheim cited Paul Janet more than once to support his contention that one must begin with observation in the study of moral rules.[107] The terms *moral fact* and *social fact* were part of eclectic philosophy. They were at the heart of Cousin's *méthode psychologique* and continued to figure prominently in the syllabus. In this sense, Durkheim could present himself as *really* doing what the eclectics merely *claimed* to be doing. Indeed, one can speak of a certain nostalgia for eclecticism in Durkheim's quarrel with philosophers—at least the eclectics accepted in principle the relevance of fact to moral questions, unlike the newer "mystical" spiritualists, who divorced the moral from the factual domain. Durkheim invoked Janet's support in opposing "the growing tendency of metaphysics to leave its natural limits and to invade every part of the course [of philosophy]."[108]

Durkheim's methodological prescriptions can also be seen as a variation on a standard theme of academic philosophy. Although its scope was limited to one science, *The Rules* was essentially a textbook treatment of method, the outlines of which would be familiar to anyone who had been through lycée philosophy. It is true that Durkheim opposed the superficial and a priori way in which scientific method was taught in the philosophy class.[109] He thought that one should learn science by doing it, and he represented his own book as the result of his practical experience rather than as an abstract, secondhand set of prescriptions.[110] Nevertheless, the general organization of the book closely follows the standard treatment of scientific method in the syllabus: definition of the object (chapter 1), observation (chapter 2), classification (chapter 4), isolation of causes (chapter 5), and demonstration (chapter 6). Compare this order with the elements of scientific method in the syllabi (Appendix B). Durkheim's general conception of scientific method, with its blend of almost Baconian empiricism and Cartesian rationalism, resembles that found in any philosophy textbook.[111] Even a number of particular points, traditionally considered to be peculiarly Durkheimian, have antecedents in the kind of philosophy of science that was traditional to academic philosophy.

A good example is Durkheim's well-known emphasis on the need for external, easily identifiable characteristics to serve as indexes of social phenomena. However controversial the specific indexes he chose—externality and constraint—the general idea of objective indexes was a commonplace of the theory of classification taught in the lycée. Textbook philosophy generally incorporated a discussion of artificial versus natural classification, the former using a superficial or even externally imposed marker to aid the researcher in finding the objects of study, the latter employing characteristics inherent in the objects. Textbooks included discussions of the advantages and disadvantages of different systems of classification. They generally agreed that at the beginning of any study one must select an external, even arbitrary characteristic of the object in question as a provisional means of identifying and classifying phenomena.[112] This is necessary, as Durkheim points out, because "at the very beginning of research, when the facts have not yet been analyzed, the only ascertainable characteristics are those external enough to be immediately visible."[113] As one learns more, one can begin to assess the relative importance of various attributes and move from an artificial to a more natural basis of classification. Because this was such an elementary part of academic philosophy, Durkheim felt aggrieved that his critics claimed he was making constraint the essential characteristic of social phenomena, when in fact it was only a working principle of practical classification.[114] On the other hand, it was also a commonplace of textbook philosophy that external indexes were not necessarily essential and that one could be misled by one's choice of index. Critics therefore complained that Durkheim abused the principle of external characteristics to privilege an a priori characterization of social facts.[115]

Durkheim's discussion of causality is another example. Although he has been portrayed by some as a positivist in the narrowest sense, concerned only with the correlation of phenomena, he was in fact a "rationalist" and an essentialist, as Schmaus has argued.[116] For Durkheim the link between cause and effect was logical, not just empirical, deducible from the essential nature of phenomena, and he specifically criticized Mill on this point. But so did most French academic philosophers.[117] On the other hand, Durkheim took great pains to assess Mill's four methods of experimental proof for their applicability to sociological research. These too had become a standard part of the syllabus, discussed in most philosophy textbooks of the day.[118] Far from being original or controversial, Durkheim used a fairly standard vocabulary of methodological discourse. This is not a trivial point, since interpreters have argued at length about whether Durkheim was epistemologically closer to Comte or to Kant.[119] His actual roots are much more mundane, in the largely eclectic *méthode* of academic philosophy, with its sometimes incoherent blend of empiricism and rationalism. Indeed, the rhetorical force of *The Rules* came in part from the fact that Durkheim was championing a model of science with which academic philosophers largely agreed.

This is not to say that critics could not find anything to argue with in Durkheim's description of method. Brunschvicg and Halévy, for example, thought that Durkheim failed adequately to relate higher and lower levels, thereby differing from most forms of mechanism, which attempted precisely to connect the higher to the lower. They also thought he abused the procedure of abstraction and generalization by providing only general causes of general facts, rather than constructing a model that could explain both general trends and particular events. They concluded that Durkheim's sociology was "a sterile classification of forms."[120] Durkheim's realism found critics from a wide range of schools, positivist to spiritualist, who thought he was claiming too much for the ability of science to penetrate "reality."[121] Most criticisms, however, concentrated not on his description of method in the natural sciences but rather on his application of this method to the study of human action. In particular, they centered on Durkheim's "objectivism," his exclusion of the subjective method from the study of morals and his refusal to consider intention as a cause of social action. Bernès felt that this was the source of all the many problems with Durkheim's method.[122]

The irreducibility of the psychological to the biological and of "intellectual" to "sensitive" operations was a feature of eclectic philosophy as well. Durkheim's sociology of knowledge basically agreed with the eclectic description of higher mental functions. Durkheim simply offered a sociological explanation of those functions. Of course, he consistently opposed the *méthode psychologique,* which was grounded in the validity of introspection. He also thought that despite their claim to begin with facts, eclectics and other philosophers did a poor job of surveying and analyzing the facts, rushing to metaphysical conclusions before systematically investigating the phenomena.

When Durkheim first approached Comte, it was with the words of antireductionist and antievolutionist philosophers firmly in mind. Ribot and Espinas had read Comte, but for a variety of reasons they felt that Spencer better represented the cutting edge of the scientific human sciences. The philosophical critique of Spencer had created a context in which Comte's antireductionism warranted a second look. For Durkheim, Comte offered insights into a scientific way of accounting for the qualities the philosophers had defended. Durkheim never failed to acknowledge his debt to the man who coined the term *sociology*. Just after he credited Boutroux with the idea explaining each science by its own principles, he added, "I was confirmed in this method by reading Comte. . . ."[123] Like Comte, Durkheim insisted that social phenomena constituted an autonomous realm that had to be explained in its own terms, but like the founder of positivism, Durkheim distrusted the evidence of introspection.

Nevertheless, Durkheim also disagreed with Comte on several equally important points. The least of these was the fact that Comte was rather the philosopher than the founder of scientific sociology. It is true that Durkheim claimed in a number of places that Comte had not gone beyond philosophical generalities in his social theory, but Durkheim excused his precursor in a typically Comtean fashion by asserting that it was necessary for the idea of sociology to have been formulated by a positivist philosopher before others could turn it into an empirical enterprise.[124] More serious was Comte's contention that the object of social science was humanity in general rather than particular societies. Durkheim, agreeing with Espinas on this point, thought that humanity was an abstraction that could not be the object of a science of society.[125] Related to this was Comte's assumption that social evolution followed a single path, so that the historical development of all societies could be described by the same formula. Durkheim thought that this underestimated the diversity of societies and of their patterns of change.[126]

Durkheim's philosophical background provided him with a powerful critique of associationism, evolutionism, and reductionism that set the terms for any future sociology. If he was to provide a scientific explanation for moral rules and logical categories, he would have to account for the characteristics of these phenomena that had been elucidated by philosophers. His philosophical background also provided him with a range of alternative solutions to this problem, from which he selected those elements that seemed to fit the vision of sociology he was developing. Durkheim's major contribution to philosophy, the "principle of association" or emergent realism that would combine the irreducibility of levels of reality with the determinism of each level by its own laws, can be seen as an attempt to reconcile the demands of the conflicting philosophies of the day.[127] He combined the determinism of Comte with the realism of spiritualist philosophy, and in so doing he was able to claim that he had gone beyond the limitations of both.

Of course, Durkheim was not alone in claiming the middle ground. Monism, the belief that mind and matter are but two aspects of the same substance, made the same claim and offered another solution to the same problem. Monism has already been discussed in relation to Espinas, Ribot, and Pierre Janet. It was a doctrine that was capable of a variety of interpretations, depending on whether one stressed the deterministic material aspect (e.g., Spinoza) or the dynamic spiritual aspect (Leibniz). Many positivists advocated a Spinozist interpretation as a means of accounting for psychological phenomena without granting mind a separate substance outside of nature. Others, such as Janet, Tarde, Renouvier, and Durkheim himself in his early lycée lectures, inclined more toward Leibniz, emphasizing the omnipresence of mind in the material world.[128] Regardless of how one interpreted monism, it provided a way of avoiding the thorny problem of where mind comes from, and because it appeared in the writings of a sociologist such as Tarde, it had definite implications for social science. For monists, no new level of reality emerges, because the traits of the higher level are present in the lower levels as well. Thus for Tarde the social domain did not present any fundamentally new phenomena, and it was not necessary to postulate a separate social reality. For an emergent realist, the traits of the higher level are truly new—sui generis, to use one of Durkheim's favorite expressions. The social realm must be explained in part by concepts that have no counterpart in psychology or biology. Durkheim and Tarde were therefore incompatible at a most fundamental level—in their conceptions of reality. This helps explain their bitter opposition and Durkheim's attempts to combat monism by describing it as a kind of materialism or epiphenomenalism.[129]

DURKHEIM AND THE HUMAN SCIENCES

As the example of monism versus emergentism shows, philosophical issues spilled over into the human sciences. In addition to the general difficulties in keeping philosophical problems distinct from scientific issues, this spillage occurred in part because of the continued overlapping of these disciplines in late nineteenth-century France. The methodological problems of science could easily slip into the ontological problems of metaphysics, and vice versa. Moreover, academic philosophy still claimed the right to discuss issues of scientific method, and the human sciences of sociology and psychology claimed objects that had been part of philosophical discourse. It is not surprising, therefore, that philosophers intervened in the human sciences, nor that they borrowed ideas from the human sciences when this suited their purposes. Durkheim did the same, plundering a number of different disciplines for both methodological and substantive suggestions and trying to rearrange those disciplines when they challenged his notion of sociology or violated his ideas about science.

Psychology provides one example of such a discipline. In a sense, Durkheim needed no justification for intervening in psychology, because as a philosopher he could claim it as his own. But because he claimed to respect the distinction between philosophy and science, he could not tell psychologists what to do without violating his own insistence that science entailed specialization. Despite the fact that, or perhaps precisely because, Durkheim consistently opposed the use of psychological explanations for social facts, scientific psychology was important to him in several ways. First of all, it was a model of an experimental science of moral phenomena. Durkheim was very impressed by his visit to the psychological laboratory of Wundt; he felt that "however specialized these studies may be, nothing is more likely to awaken in young minds the love of scientific precision."[130] Durkheim invoked the example of the new psychology at every opportunity to show what he had in mind for the study of society. Discussing in *The Rules* the need for sociology to become an objective science of facts, Durkheim claimed that "the reform needed in sociology is at all points identical with that which has transformed psychology in the last thirty years" (i.e., since about 1864). The origin of scientific psychology is to be dated from "when it had finally been established that states of consciousness can and ought to be considered from without, and not from the point of view of the consciousness experiencing them."[131] Scientific psychology had dispelled the simplistic notion, derived from the introspectionist method, that psychic life is simple: "Today it is generally recognized that psychical life, far from being directly known, has on the contrary profound depths to which introspection cannot penetrate, to which we attain only gradually by indirect and complicated paths like those employed by the sciences of the external world."[132] When Durkheim argued in "Individual and Collective Representations" for the autonomy and irreducibility of social phenomena, he found confirmation in parallel developments in psychology: "During the last ten years [i.e., since about 1888] . . . interesting efforts have been made to constitute a psychology that would be properly psychological, without any other epithet. The old introspectionism was content to describe mental phenomena without explaining them; psychophysiology explained them, but by leaving aside as negligeable, their distinctive traits; a third school is in the process of forming that undertakes to explain them while leaving them their specificity."[133] Scientific psychology served as a legitimation of Durkheim's approach to social phenomena.

Durkheim also borrowed concepts from psychology to characterize social phenomena. This makes sense if, as he believed, society is basically "a psychic individuality of a new sort."[134] As we have seen, he described the formation of collective representations and the mechanism of their coercive power in terms largely drawn from associationist, physiological psychology through Ribot and Espinas. While he insisted that the laws of collective ideation were different from those of individual thought, he nevertheless described collective representations as things that could associate among themselves like ideas in the mind.[135] Like Ribot, Durkheim borrowed from German, British, and French sources for his psychological concepts. The majority of the psychological references in *The Division of Labor* are to British sources. He cited Alexander Bain on the type of difference that attracts and leads to friendship, Maudsley on the force of representations, and Spencer on a variety of topics. Most of these citations are positive rather than critical. The references to German psychology are fewer, but they are also important to his case. As part of his argument that civilization does not lead to happiness, Durkheim cites the evidence of Wundt, Weber, and Fechner to the effect that only moderate states of stimulation are pleasurable.[136] Durkheim also invokes Wundt as support for the idea that progress does not necessarily entail the immutable fixing of specialized social roles, as Comte and Spencer thought. Social roles become more specialized with the division of labor, but individuals have a choice in which roles they adopt, and social roles can change rapidly with changes in society. As an analogy, Durkheim cites a passage in which Wundt argues that nerve cells in the brain can substitute for others that have been destroyed.[137] This supports his general argument that as society progresses, "function becomes more and more independent of structure."[138]

He paid attention to developments in French psychopathology as well. He was clearly aware of the debates over the nature of hypnotic suggestion. In his lectures on moral education, he cited the philosopher Jean-Marie Guyau and the psychologists Alfred Binet and Victor Henry on the suggestibility of children, a trait that he encouraged teachers to use to instill in students a respect for discipline.[139] Durkheim's definition of an act of imitation in *Suicide* was essentially that of a suggested idea: *"There is imitation when the immediate antecedent of an act is the representation of a like act, previously performed by someone else; with no explicit or implicit intellectual operation which bears upon the intrinsic characteristics of the act reproduced intervening between representation and execution."*[140] He excluded cases in which people imitated others to conform to social norms and those in which representations were exchanged in a crowd. This left only what Pierre Janet would call psychological automatism. Durkheim cites Janet's experiments on dissociated suggestions in his article on "Individual and Collective Representations."[141] He would use this knowledge to intervene in the debate over crowd psychology.

Durkheim read and used the results of research in psychology, but he was not a passive consumer. He actively selected those psychological theories that supported his sociology and opposed those that presented a problem for it. One area in which Durkheim intervened was in the subject of psychological heredity. Ribot, it will be recalled, wrote his thesis on the topic, and he tried to establish that many psychological traits, some quite specific and involving higher mental function, were passed along from parent to child. In *The Division of Labor* and *Suicide,* Durkheim argued against most such heredity, maintaining that only very general capacities were inherited.[142] His point was that social phenomena such as the suicide rate and higher functions such as moral ideas, categories, and logic came from society, not from the individual. To make this point, he had to argue that heredity could not explain these phenomena.

He also had to argue that individuals did not have the resources to develop these ideas for themselves, as Mill, Pierre Janet, Rabier, and the associationists held. For such thinkers, higher mental functions were not part of the individual's biological constitution, but the individual constructs these through interaction with the environment, either through association or through a combination of experience and synthetic activity. Durkheim argued against such theories, especially in his work with Mauss on classification.[143] They even went so far as to claim that scientific psychologists had not much done work on the subject, limiting themselves to lower functions or assuming that higher functions could be explained by the same processes as lower functions.[144] Durkheim and Mauss cite only Münsterberg as a psychologist with a more sophisticated theory, but they cannot have been unaware of the implications of Binet's pioneering work on intelligence, which he began to publish in the 1890s. Their failure to discuss such works more likely reflects an a priori decision that individual psychology could not explain collective representations.

Another area of psychology in which Durkheim intervened was the debate over the existence of unconscious psychological phenomena. As I have argued at greater length elsewhere, Durkheim's article on "Individual and Collective Representations" was in part a contribution to this debate, which had a long history in French philosophical and psychological discourse.[145] Not only positivists such as Ribot but eclectics such as Rabier held that ideas were psychological phenomena only as long as they were conscious; outside consciousness, they were merely displacements in neural matter. Pierre Janet opposed this view, arguing that mental phenomena outside consciousness were psychological. Durkheim agreed with Janet, although unlike Janet, he argued further that representations were not strictly connected to their biological substratum. Not only did Janet, as a monist, not follow the argument this far, but few scientific psychologists or nonspiritualist philosophers did, either.[146] Durkheim's psychology as sketched in this article looks less like any contemporary psychological theory than like his own sociology. This of course was his point, and he turned the argument around by maintaining that those who reduced sociology to the individual should also, to be consistent, reduce mental life to biology.[147] In this way, Durkheim used contemporary psychological theory's rejection of reductionism as support for his conception of social realism.

However, another aspect of this argument shows just how selective Durkheim's reading of psychology could be. Although he argued for the irreducibility of psychological phenomena, he ignored another component of most contemporary psychology—the irreducibility of thought to association and the need to postulate a synthetic activity in the mind to account for the phenomena of consciousness. As we have seen, Janet was on the forefront of efforts to revise psychology in this direction, and he was not alone. Durkheim, however, made no mention of this aspect of contemporary thought. His model for the interaction of both individual and collective representations was that of entities that mutually attract, repel, or combine, not that of a conscious activity that synthesizes elements into a greater whole. Even though he rejected the reduction of collective to individual representations, he preferred to conceive both on a mechanistic model rather than to introduce an element that might give ammunition to the neospiritualist philosophers of action. As several commentators have pointed out, this left him with a psychology centered on content rather than on process.[148]

Durkheim took on not just psychology but psychiatry as well. In *Suicide*, he devoted considerable attention to psychopathological theories of self-destruction.[149] His overall strategy was to argue that mental illness could not account for all suicides or for the suicide rate. But he also challenged interpretations that led psychiatrists to conclude that mental illness was a sufficient cause of suicide in many individual cases. He distinguished four possible theories of suicidal psychopathology, and his strategy varied with the theory. The first he discussed was monomania, the idea that people can have localized insanities affecting only one idea or intellectual faculty. This Durkheim dismissed by claiming that most researchers had rejected the idea that mental functions could be so rigidly localized and that therefore this idea had few supporters. This claim, for which Durkheim gave no references, was nevertheless substantially true by the time he wrote.[150] The second theory held that suicide was always the result of other forms of insanity—specifically mania, melancholy, obsession, or compulsion. Durkheim rejected this theory by pointing out that these forms of insanity are defined by the fact that the reasons for the suicide are delusional. However, many people kill themselves who are not suffering from such delusions; in fact, they may commit suicide for perfectly good, or at least understandable, reasons. Hence, insanity is not a necessary concomitant of suicide, and the researcher should resist the temptation to assume that a person is insane because he commits suicide.

The third theory linked suicide with neurasthenia, which was still very much part of current psychiatric theory, as the earlier chapter on Janet attests. Durkheim did not challenge the theory itself. Indeed, he gave an almost textbook description of it, as Nye has shown.[151] Rather, he sought first to show that suicide rates do not vary among different social groups in proportion to variations in insanity. For example, he argued that since more women than men inhabit asylums, one would expect more women than men to kill themselves, but this is most certainly not the case. Durkheim reserved a final section for theories that linked suicide with alcoholism. He dealt with this theory as he had with the previous, by showing that suicide rates did not vary with alcoholism rates. Without denying that forms of insanity could contribute to individual propensity to suicide, Durkheim argued that the statistical lack of correlation should lead to a reduction in the causal importance attributed to insanity in explaining suicide, even in individual cases. The rest of his book was designed to show that social causes were primary and individual causes secondary. Without questioning the basic categories of psychiatric theories, Durkheim intended to put into question their interpretation of many acts as explained exclusively by individual pathology.

Finally, Durkheim had to take account of and turn to his own purposes the growing French preoccupation with crowd psychology in the 1890s, a subject that straddled the boundary between psychology and sociology.[152] Tarde, Gustave LeBon, and others wrote on the topic, and Tarde in particular tried to explain crowd psychology through imitation, which he described as a purely individual and psychological phenomenon.[153] Moreover, most theorists of crowd behavior characterized it as irrational, emotional, and less intelligent than the behavior of individuals outside crowds. Indeed, as Apfelbaum and McGuire have shown, Tarde and Le Bon described crowds as hypnotized subjects, devoid of autonomy and open to suggestion from a charismatic leader.[154] They also stressed the propensity of crowds for criminal, destructive, and even revolutionary action.

All this posed a problem for Durkheim, since his model for the rise of collective representations through social interaction and the sharing of representations resembles in obvious ways his rivals' descriptions of crowd behavior. But for Durkheim, collective representations are a major source of morality and social solidarity; hence he could not afford for the social situations in which they arise to be disparaged as subrational or amoral. In *Suicide,* published not long after the appearance of major works on crowd psychology, Durkheim reserved an entire chapter to address Tarde's theory of imitation, which included a description of crowd behavior. Durkheim began in typical fashion by taking the logical high road, maintaining that the common definition of imitation confused three different situations: (1) the emergence of a unison of thinking and feeling in a social group; (2) the tendency to conform to social norms; and (3) the automatic, unthinking reproduction of an act that has occurred in one's presence (ibid., 108–10). He denied that the first situation was an imitation, but not because he rejected the model of suggestion. That is, he did not try to redescribe crowd behavior as a reasonable deliberative process in which individuals retain their autonomy. Among other problems, this would leave the phenomenon at the level of individual psychology. Rather, he argued that what happened in such situations was the creation of a new idea, not the imitation of an existing one. He drew upon the same mechanism he used in *The Division of Labor* to describe the combination and fusion of individual representations into a collective representation.

> Once aroused in my consciousness, these various representations combine with one another and with the one that constitutes my own feeling. Thus a new state is formed, which is no longer mine to the same degree as its predecessor, which is less tainted with particularism and which a series of repeated elaborations analogous to the foregoing will more and more free from all excessive particularity. (ibid., 110)

Durkheim denied that the unification of the crowd resulted from each individual member imitating some one member of the crowd—that is, a leader. He turned the argument around by claiming that the leader is as often the product of the crowd as the instigator.

Situation (2) was easier for Durkheim to dismiss. If someone copies an action because it conforms to a group norm, one is respecting the authority of society, not imitating blindly—"*to act from respect or fear of opinion is not to act by imitation*" (ibid., 112–13). So Durkheim was able to confine imitation to an automatic reaction without the intervention of reflection. Then he proceeded to demonstrate that although there were indeed episodes of suicidal contagion, they had at best a marginal significance and could not account for the overall suicide rate in a given society. Throughout this whole argument, Durkheim did not deny the basic psychological processes at work. He accepted the concepts of suggestion and collective frenzy that appeared in crowd psychology, because in many ways they were essential to his theory of the origin of collective representations. Nevertheless, he attempted to minimize the importance of certain of these phenomena and to reinterpret others in a way that would vindicate his own sociology.

By borrowing from and adding to psychological theories of the day, Durkheim constructed a picture of the human psyche that made his sociology both possible and necessary. The general outlines of that picture are as follows: Individuals inherit very general capacities for sensation and perception, for emotional reaction, and for the association of these primary elements. Individuals also inherit certain general instincts, such as self-protection and procreation. This biological substratum makes possible the simpler psychological phenomena, and the combination of inherited capacities and simpler functions constitutes what Durkheim calls the "organico-psychic" constitution of the individual. The simpler or lower psychological functions are not reducible to their biological substratum, but neither can they by themselves lead to the higher capacities such as judgment, classification, logic, and morality. These are given shape and made possible only through society. They are created in the process of social interaction and instilled in the individual through education. Thus, just as Durkheim tried to remove morals from the domain of philosophy, he attempted to take the study of the higher mental functions away from both philosophical and scientific psychology. As he and Mauss stated in "Primitive Classification," such questions, "which metaphysicians and psychologists have discussed for so long, will be finally liberated from the repetitions in which they linger, the day when they are posed sociologically."[155] Hence Durkheim's psychology leaves room for and indeed requires sociology to complete the explanation of human mental functioning.

This willingness to borrow, rearrange, criticize, and construct across disciplinary lines is also evident in relation to other disciplines. Perhaps the first of these disciplines Durkheim encountered was history, which he studied at the Ecole Normale under Gabriel Monod and Fustel de Coulanges.[156] Durkheim followed their courses and produced papers for them, and he retained from their example a conviction of the need for methodical empirical inquiry. Moreover, he saw in their work not merely a chronicle of contingent individual events, but an effort to understand in history the effect of social institutions.[157] This is most obvious in Fustel de Coulanges's major work, *The Ancient City,* which examined the relationship between the family, the city, and religion in classical civilization. Nevertheless, Durkheim did not become a historian, and he thought that sociology had to transcend the historical point of view to arrive at a truly scientific conception of itself. Historians, he often complained, too often remained enmeshed in the particular, the unique, and the contingent. As a result of their narrow focus, they denied themselves the resources of the comparative method and could not see the general laws that underlay the events they reconstructed. In this sense, the sociological spirit that Durkheim thought had begun to penetrate history but that needed to be extended, would make history at once more scientific and more "philosophical" in the Comtean sense, and Durkheim more than once portrayed sociology as lying between philosophy and history.[158] This axis corresponds to the conflicting poles of his education at Normale, and in many ways his sociology can be seen as an attempt to reconcile these two opposites.

Political economy was also an object of Durkheim's attention, if only because it claimed to study the laws of a portion of social activity—the production and distribution of wealth—and because its mathematical laws gave it a greater appearance of scientificity than any other human science. Nevertheless, Durkheim had little good to say about it. He considered political economy ideological, because it began with abstract notions of value not derived from experience and made assumptions about human nature that were neither derived from, nor verified by, empirical investigation to see if people really behave that way.[159] He also thought economics failed to recognize the reality of society by reducing it to a set of interacting individuals.[160]

Law was of crucial importance to Durkheim because he held that it was an index of morality and hence one of the best ways to study society. In his first major work, Durkheim argued that laws with repressive sanctions—penal law—represent the collective conscience and those with restitutive sanctions—for example, administrative and commercial law—the division of labor.[161] Hence by measuring the proportions of the two types of law, one can assess the relative preponderance of the collective conscience and the division of labor as sources of social solidarity. Durkheim admitted that this distinction was not one generally used by jurists; indeed he warned that "we cannot use the distinctions utilized by jurisconsults. Created for practical purposes, they can be very convenient from this point of view, but science cannot content itself with these empirical and approximate classifications."[162] According to Durkheim, this practical orientation, which included a too narrow focus on the exegesis and interpretation of legal documents, prevented legal scholars from attaining a truly scientific point of view. As for political economy, law was also condemned for being excessively individualistic.[163]

All of these disciplines suffered from the fact that they were partial and abstract. "In reality, all these special sciences—political economy, comparative history of law and of religions, demography, and political geography—have been up to the present conceived and applied as if each formed an independent whole, whereas to the contrary the facts with which they are concerned are only diverse manifestations of the same activity—collective activity."[164] Only the Germans escaped this tendency to some extent. Durkheim was much impressed by the historical schools of jurisprudence and economics he encountered on his trip abroad. In the report he published upon his return, Durkheim wrote that "only the German moralists see in moral phenomena facts that are at once empirical and *sui generis*."[165] These scholars all pursued an empirical investigation of moral phenomena without introducing metaphysics. Moreover, in the case of the historical jurists and economists, the Germans had broadened the traditionally narrow focus in their disciplines by relating their matter to social institutions and seeking systematic relations between them. Wundt, whose *Ethics* appeared in 1886, was one of the first to treat moral facts as objects of scientific analysis.

Students of Durkheim have differed over the significance of the German influence. Some have found it decisive, pointing to the general German tendency to exalt the role of the state and to hypostasize the *Volkseele*.[166] Others, following Durkheim's own indications, have concluded that his German sojourn clarified and confirmed rather than inspired most of his major ideas.[167] It should be recalled that Durkheim found fault with the German approach. In "The Present State of Sociological Studies in France" ("L'état actuel des études sociologiques en France"), he argued that if the French tended to be too analytic and reductionist, the Germans were insufficiently endowed with these qualities. "If the German mind is more sensitive than ours to the complex in social things, on the other hand, owing to its poor analytical ability, it has found it difficult if not impossible to submit such a complex reality to scientific analysis."[168] This is why German theorists claimed that the study of society can lead only to the discovery of empirical generalizations, not scientific laws. Durkheim thought that the French Cartesian spirit kept a healthy faith in the ability of the human mind to find rational laws for all phenomena. The proper attitude for sociology would therefore be a combination of the best aspects of the French and the German approaches.

From elements of this wide variety of sources, Durkheim constructed his unique vision of sociology and promoted it in a crowded field of competitors. As part of this promotion, Durkheim consistently tried to portray his sociology as either a middle way between opposite extremes or a synthesis of the partial truths of other approaches. We have already seen how he attempted to distance himself from the social organicism of Spencer and Espinas while retaining a commitment to an empirical method and to the autonomous reality of the social. By portraying these men as predecessors, Durkheim managed to convey a sense that this phase in the development of sociology was a thing of the past, even as he acknowledged his indebtedness to it, and he rarely bothered even to mention the few remaining social organicists, such as René Worms. Durkheim also distanced himself from another form of biological sociology—those schools who sought the explanation of social phenomena in the biological constitution of the individual or race. The "anthropological" school, represented by Charles Letourneau, continued a long nineteenth-century tradition of physical anthropology grounded in racism. At the end of the century, this tradition still retained scientific respectability. Georges Vacher de Lapouge put forth a theory of "anthroposociology" that was a modified form of social Darwinism. He held that the class struggle was a form of natural selection that furthered the evolution of the species. Finally, the "criminological" school, represented by Lacassagne, had moved away from the theories of the Italian criminologist Cesare Lombroso, who maintained that criminality was biologically determined and inheritable. Still, Durkheim considered their theories insufficiently free of biologism.[169] Durkheim's attacks on psychological heredity were designed in part to counter such theories by denying heredity the specificity needed to explain social facts.

But Durkheim's most formidable rival was Tarde. As we have seen, many of Durkheim's positions were either formulated or articulated in opposition to Tarde. Durkheim's insistence on purely sociological explanations opposed Tarde's methodological individualism and psychologism; Durkheim's emergent realism contrasted with Tarde's monist nominalism; Durkheim's rationalism differed markedly from Tarde's insistence upon the element of chance and contingency in human invention. This last contrast led to Durkheim's charge that chance was a deus ex machina, a nonexplanation that rendered science impossible by proliferating the unexplained instead of trying to explain it.[170] Several accounts of this debate are available, but two important features have received little attention.[171] One is that the polemics between Tarde and Durkheim crossed the boundaries of sociology to invade psychology and philosophy as well. However much both sides tried to keep their arguments couched in terms of method and social phenomena, they were not able to prevent spillage into other domains. For example, Durkheim's psychological discussion of the emergence of representations by fusion and chemical synthesis in the human mind not only reinforced his conception of the relation between sociology and psychology, it also undercut Tarde by attacking the latter's portrayal of the mind as a society of nerve cells, each with its own idea.[172] And this excursion into psychology led into a flirtation with metaphysics, as both Tarde and Durkheim hinted that their opponent's monism or emergent realism was incompatible with science while simultaneously insisting that metaphysics was irrelevant to science.[173] In a context in which philosophy, psychology, and sociology were still not completely differentiated, the temptation to cross, or simply deny, boundaries occasionally proved irresistable, especially when doing so would either enhance one's own position or undermine that of another before the interdisciplinary audience of late nineteenth-century French academics.

Second, despite the inevitable tendency of a debate to dichotomize positions, Durkheim always retained a desire to take the middle ground wherever possible. Thus, he usually portrayed himself as combining the elements of truth in the nominalism of Tarde and in the realism of the biological sociologists, while avoiding their shortcomings. With Tarde, Durkheim acknowledged that social phenomena should not be reduced to biological explanations. Unlike Tarde, he maintained that psychological explanations were inadequate as well. With social organicists, he acknowledged that society was a sui generis reality, not reducible to the interactions of individuals. Unlike social organicists, he denied that society could be explained by the direct application of biological models. With the anthroposociologists, Durkheim insisted upon a deterministic approach to social phenomena. But unlike the anthroposociologists, he denied that heredity and race could explain significant portions of social life. "Beyond the ideology of the psycho-sociologists and the materialist naturalism of socio-anthropology, there is room for a sociological naturalism that sees in social phenomena specific facts and that undertakes to explain them while religiously respecting their specificity."[174]

But while he could acknowledge partial truths in a variety of positions, he was still capable of ruling certain positions out of bounds if they did not meet minimal standards of method or fundamental assumption. He considered Tarde's approach the negation of science and passed over Worms in silence. In this way he was not all that different from Victor Cousin, as Bernès pointed out in a response to Durkheim's article on the teaching of philosophy in the lycée. Durkheim argued that Cousin had constructed eclectic philosophy to create social solidarity among Frenchmen by instilling them with common values in a society that no longer enjoyed religious unity. Cousin had been willing to enforce conformity to his philosophy among academics because of the social function academic philosophy served. The price paid was the reduction of academic philosophy from a science to ideology. With the collapse of the Cousinian project in the Second Empire, academic philosophers claimed the freedom of science for themselves. This led to a variety of different positions and destroyed the unity of philosophy in the lycée as a new generation of professors taught their different doctrines to lycée students. Durkheim considered this situation to be pathological. Like Cousin, he thought that the teaching of philosophy in the lycée should serve a social function. Durkheim thought that function should be "to accustom the generations that follow us to see human and social things as objects of science, that is as natural things."[175] For this purpose, Durkheim recommended some changes in the syllabus. Psychology and ethics would be retained, though they would be treated from a more scientific and less philosophical point of view. Philosophical discussions of ethics were not only useless but potentially subversive, since "among these theories there are some that are not sufficiently animated by a sense of duty and as a consequence cannot be taught to very young people without danger."[176] Logic too would stay, but it would be more geared toward the study of scientific method. With these changes, philosophy could once again fulfill a useful function by preparing students to deal seriously and scientifically with the social problems of the day. Durkheim proposed to eliminate those parts of the syllabus that did not serve this purpose—metaphysics and the history of philosophy—and above all, he wanted to correct what he saw as a dangerous dismissal of science on the part of "mystical" professors and students.

In his response, Bernès denied that the teaching of philosophy was in a state of crisis. What Durkheim interpreted as pathological, feverish activity, Bernès described as a healthy sense of independent and critical thinking in lycée professors and students. A certain amount of divergence of opinion is inevitable, natural, and desirable in modern life. "Society is not and cannot be, like an organism, an equilibrium among diverse functions or between all these functions and the ambient milieu. Its strength comes only from diversity. . . ."[177] Bernès accused Durkheim of not being that different from Cousin. "The last general objection that I will address to M. Durkheim is that he does not differ as much as he thinks from the Cousinian idea of philosophical education, and that he attaches too much importance to the uniformity of this education."[178] He also thought that Durkheim overestimated the certitude of science and its value in preparing students to live in society. What Durkheim called mysticism was merely "a sense of the relativity of science, of the need for an education of the will, and of the value of strong convictions."[179]

While Durkheim was careful to restrict his remarks in this piece to the teaching of philosophy in the lycée, implying that academic freedom should be the rule at the university level, his program for the lycée is consistent with his hopes for sociology as a science. He clearly hoped that the spirit of science would pervade society as more and more academic disciplines adopted the scientific point of view. The rise of sociology was evidence of this progress, just as it contributed to that progress by showing that moral ideals were capable of scientific solutions. Moreover, through his courses in pedagogy, he was in a position to try to influence the future teachers of France to adopt a more scientific view of their methods and their mission. Just as a combination of judicious method (neither too empiricist nor too deductive), logical rigor, and careful definition of the social realm (neither biological nor purely artificial) would alone lead to progress in sociology, so these qualities, diffused through society, would lead to progress in French society as well. In this sense, the intellectual and institutional projects of Durkheim's sociology were not that different from each other or from the project of Victor Cousin.

Conclusion

Was Durkheim a latter-day eclectic? He himself would have bridled at the thought. Although he might cite Paul Janet approvingly on occasion, he condemned the eclectic school quite harshly as devoid of science.[180] He often employed the word "eclectic" in a derogatory sense, as a juxtaposition of elements without logical connection.[181] But he also used it neutrally when characterizing the state of German philosophy, and favorably when describing Wundt's effort to reconcile the empirical and the speculative methods in the study of morality, which he thought admirable.[182] Without employing the word, he could regret that Taine, whom he greatly respected and generally followed, had "rather juxtaposed than logically united the two tendencies [empiricism and rationalism] that he tries to reconcile."[183] Hence it was not the attempt to examine and reconcile opposites to which Durkheim objected in eclecticism; it was the failure to approach them methodically and to join them logically. Durkheim clearly thought that he had done so.

His commentators have argued ever since about whether he succeeded. I think that in many ways he did. Durkheim was able to combine the method of the natural sciences with a sense of the autonomy of the social in a system of great consistency and power. However, this success came at a price. By insisting on a scientific method that was completely rationalist and deterministic, he was unable to recognize, much less account for, the element of conventionalism in scientific concepts or of contingency in social action. And by insisting on a social realism on the model of psychological realism, he could only define social facts as a kind of thing—collective representations—that raised as many questions as it answered. As a result, Durkheim's conception of society wavered on the edge of metaphysics even as he sought to found a science. By providing a historically specific constellation of philosophical positions and a particular model of philosophical practice, academic philosophy contributed to both the strengths and limitations of Durkheim's sociology. Attempting to construct a science of society by choosing from the materials available to him, he inevitably reflected the thought of his day, even as he tried to surpass it.

Conclusion

> For these educational institutions did not begin to exist the day
> the regulations that define them were written; they have a past of
> which they are an extension and from which they cannot be sepa-
> rated without losing a great part of their signification.
> —Emile Durkheim, *The Evolution of Pedagogy in France*

SHORTLY after the turn of the twentieth century, both Janet and Durkheim attained elite institutional positions in Paris that enabled them to
propagate their theories and methods with relative security, reducing
their dependence on philosophy. Moreover, both had effectively initiated
empirical practices that put their arguments beyond the reach of philosophical objections, since philosophers neither could nor cared to muster
an opposing array of empirical counterexamples. Of course, such empirical pretensions brought them into conflict with others who had equal
claims to "objectivity." Janet and Durkheim both had to contend with
specialists within their own disciplines and with practitioners of other
disciplines, such as neurology, physiology, history, and economics. They
did not always win such battles, or if in retrospect their arguments seem
intellectually valid, their opponents could often retreat behind their own
institutional ramparts. The success of the new disciplines was only partial. Janet became increasingly isolated from the psychiatric community,
and Durkheim never realized his imperialistic ambition of coordinating
the various social sciences under the banner of his sociology.[1]

Both, however, inspired a new generation of researchers who continued the process of building the human sciences. Durkheim was more
successful, fostering a substantial group of researchers around his journal, the *Année sociologique,* which largely defined French sociology in
the early twentieth century. At least half of these social scientists had
backgrounds in philosophy, more than three times the number of any
other discipline.[2] Janet was not as successful at building a school of
followers. But if he was not a direct teacher of the next generation of
French psychologists, he was still influential, and above all representative of a tendency among French psychologists to combine philosophical
with medical training.[3] Hence it is not too much to say that Janet and

233

Durkheim established paradigms that helped solidify their respective disciplines.

The purpose of this book, however, has not been to write the history of the human sciences after Janet and Durkheim; rather, I have tried to throw new light on the origins of the human sciences in France by placing some of the founding figures in a context that has not been adequately explored or appreciated—academic philosophy. By doing so, I hope to have shown that academic philosophy played a significant role in shaping the new disciplines that emerged from its bosom. As the institution within which the first social scientists were educated and pursued their careers, academic philosophy offered a method, a set of solutions, and a panoply of advocates against which social scientists had to contend. Moreover, some of the specific features of French academic philosophy showed up in the ideas of psychologists and sociologists. The spiritualism of the philosophical tradition manifested itself in the insistence of both Janet and Durkheim on the irreducibility of higher to lower levels of reality. The scientism of French academic philosophy reappeared in a conception of method that borrowed from the logic of the philosophy textbook. And the eclecticism of the academic field found expression in an effort to reconcile the spirituality of moral phenomena with the demands of positive science. The influence of philosophy on the human sciences made itself felt through a complex process in which unacknowledged assumptions, conscious borrowing, active opposition, and attempted persuasion all played a role. A rich rhetorical situation engaged philosophers and social scientists in an ongoing debate that had profound consequences for the human sciences in France. Because it was created largely by the eclectic spiritualists, this rhetorical situation constitutes the eclectic legacy.

Academic philosophy also contributed to another feature of the emerging discourse of the human sciences in France—the pronounced tendency of sociologists to borrow from psychology. Both Espinas and Durkheim used the emerging discipline of scientific psychology in many ways: as a model of scientific method, as an authority on the individual human capacities upon which any sociology must rest, and as a source of concepts that could be adapted to the description and explanation of social phenomena. The collective conscience is only the most obvious manifestation of this tendency, which pervaded the work of early sociologists even when they tried to differentiate sociological from psychological explanations. Sociologists could borrow so easily from their psychologist colleagues in part because academic philosophy embraced both disciplines, allowing intellectual and institutional proximity to reinforce each other. Indeed, for most of the nineteenth century, one cannot speak of different disciplines. Psychology and sociology were different parts of the encompassing syllabus of philosophy, and just as philosophers did not hesitate to intervene in the human sciences, so sociologists felt few compunctions about kibitzing the developments in scientific psychology.

This is not to say that academic philosophy determined every feature of the new human sciences. A more comprehensive study of the history of these disciplines would have to consider the impact of a wide variety of other disciplines as well as their institutional settings. Even within philosophy, by the beginning of the Third Republic the French academy was less a place where successors repeated the doctrines of their predecessors than a forum within which contenders competed by developing original positions and defending them competently before their peers. Social scientists could—indeed, were required to—articulate creative solutions to the question of how to constitute positive moral sciences. Nor were all social scientists obliged to come up with identical solutions; within each new discipline different approaches were offered. But those competitors with solid philosophical credentials and persuasive philosophical arguments were most likely to gain access to the elite positions of the French university system.

The debate between philosophy and the human sciences did not end with the institutionalization of empirical research programs. At the level of fundamental assumptions, theory, and method, philosophers continued to intervene in the human sciences, and social scientists like Durkheim continued to enter the lists in defense of their theories. The intellectual sparring did not abruptly come to a halt, even though it became increasingly possible for psychologists and sociologists to ignore philosophical objections when it suited their purpose.

Yet just as the human sciences had changed under the impact of philosophical criticism, so did philosophy change as a result of its confrontation with the human sciences. Academic philosophy largely abandoned the claim to be a science of observation, redefining itself as critique or metaphysics and employing exclusively logic or intuition. In this way philosophy made itself inaccessible to the encroachments of science, although in the process it relinquished territory that had historically belonged to it.

Philosophers also responded by radicalizing the ideality of science, undercutting the privilege of the "objective" method by rethinking the nature of facts, moral or otherwise. We have seen that during the last third of the nineteenth century, academic philosophy increasingly emphasized the hypothetical character of science. Moreover, under the impact of Kantian philosophy, mathematics came to be viewed as something that was not abstracted from observation, as Comte and Mill seemed to think, but constructed by the mind according to its own necessity. Nevertheless, these idealistic theories about hypothesis and mathematics coincided with a continued acceptance of the givenness of

observation. For much of the nineteenth century, philosophers continued to make a sharp distinction between inductive and deductive sciences, between natural and mathematical sciences.

One of the most significant developments in turn-of-the-century
French philosophy of science was to apply the ideality of mathematics
to the natural sciences, thereby putting into question the nature of facts
and largely erasing the distinction between facts and their interpretation.
This was implicit in the work of Emile Boutroux, but it came to fruition
in the writings of scientists such as Henri Poincaré and Pierre Duhem
as well as philosophers such as Léon Brunschvicg and Emile Meyerson.[4]
One effect of such changes in the philosophy of science was to undermine the consensus that had existed between philosophers and social
scientists about what constituted an empirical science. This had been
one area in which the two sides could agree, and it had been one of
the ways in which social scientists vindicated their competence before
philosophical judges. They might have disagreed over whether the methods of the natural sciences should be applied to moral phenomena, but
at least they both understood what those methods were, and philosophers could be forced to acknowledge that even if the assumptions of
the project were questionable, at least they were being applied according
to the canons of scientific method. This ground of consent was removed
by the spectacular developments in the philosophy of science.[5]

The account of the history of the human sciences in France offered
here results from what may be called a rhetorical approach. It is an
approach to intellectual history that holds that actors and audiences
reciprocally constitute and affect each other through the medium of
argument. It is this approach that led me to investigate the space—
institutional and intellectual—in which arguments about the human sciences took place, and that investigation led me to posit the existence of
an eclectic legacy. It is this approach that enabled me to attend to the
inconsistencies within texts without undermining the possibility of cir-
cumscribing meaning, to ground meaning in historical context without
reducing thought to social class, and to accept the discontinuities in
conceptual discourse without constructing chasms of incommensurability. If the account has proved persuasive, the approach might be applied
to other areas of intellectual history.

For example, it is beyond the scope of this work to examine the
implications of conventionalism and other innovations in the philosophy
of science for the human sciences. It is also impossible to examine the
reasons for the relative stagnation of both philosophy and the human
sciences in France after the First World War, although this is a problem
that has not received adequate attention. Yet it seems to be the case that
philosophy, psychology, and sociology all suffered crises in the 1930s,

which led to the introduction of Hegel in philosophy, Freud in psychology, and Marx in social theory.[6] The similar fates shared by these three disciplines suggest that the connections between them may have continued at some level and that a history of their relations in the early twentieth century could yield insights that remain to be explored.

Appendix A
Academic Philosophers, Notable Philosophers, and Social Scientists in Nineteenth-Century France

In support of the claim that there was a close association between academic philosophy and the academic human sciences, this appendix contains statistics about academic philosophers, notable philosophers, and academic social scientists in nineteenth-century France.

To represent the elite of academic philosophy, I drew up a list of individuals who held chairs of philosophy in one of three institutions of higher learning in Paris: the Paris Faculty of Letters (the Sorbonne), the Collège de France, and the Ecole Normale Supérieure (at which most future professors of philosophy were trained). For the Faculty of Letters, I relied on Christophe Charle's biographical dictionary, part of a series on the Biographical History of Education being published by the Institut National de Recherche Pédagogique and the Centre National de Recherche Scientifique.[1] In 1932, the Collège de France published a quatercentenary volume that listed everyone who had held a chair.[2] Similarly, in 1895 the Ecole Normale Supérieure published a centenary publication that contained a list of all professors and students.[3] I tried to use the same criteria as Charle: I included all the *titulaires* and most of the significant *suppléants,* although some of these latter were undoubtedly overlooked, since they were not always documented with the same care as the *titulaires.*

In addition to teaching professors, I wanted to include a sample of educational administrators, since in France administrators played a large role in deciding what kind of philosophy was taught, who taught it, and where. For this reason, I included among elite academic philosophers those inspectors general with a background in philosophy. The responsibilities of these officials are described in chapter 1. I have depended on the volume on the Inspection General in the Biographical History of Education just mentioned for information about these officials.[4]

Academic philosophers are not necessarily intellectually notable philosophers. To measure the notability of academic philosophers, as well as to gauge the relation of academic philosophy to the field of philosophy as a whole, I constructed an independent list of French philosophers considered notable by their peers. Before doing so, I consulted two previous efforts to count French philosophers. Paul Vogt counted contributions to the major philosophical journals in France.[5] Unfortunately for my purposes, the first such journal in

France—the *Revue philosophique*—dates only from 1876, and I wanted to compare philosophy during the early Third Republic with preceding periods. Jean-Louis Fabiani took his sample from the table of contents of an exhaustive history of French philosophy written during the 1920s, J. Benrubi's *Sources and Streams of Contemporary French Philosophy (Les sources et les courants de la philosophie française contemporaine,* 1933). Although this method is subject to the biases of the source, in fact it agrees tolerably well with Vogt's list.[6]

I used several histories of philosophy written at different times by authors of different philosophical inclinations to minimize the risk of selection bias.[7] To focus on the most significant philosophers, I included only those who rated notices of at least six lines in at least two contemporary sources. I tried to identify authors whose first significant work appeared between 1811 and 1910. For the Third Republic, the resultant sample was not significantly different from the lists of Vogt and Fabiani; however, by using earlier histories I was able to project further back into the nineteenth century with greater confidence and so test hypotheses about changes in French philosophy during the course of the century. Because of the tendency of philosophers to include psychology and sociology as part of philosophy, the list of notable philosophers also included the social scientists studied in this book, as well as most other notable social scientists in nineteenth-century France.

Having in this way constructed a list of 125 philosophers and social scientists, I divided the list into generations of twenty years by date of birth, beginning in 1791 and ending in 1870. The former date includes the generation of Victor Cousin (born 1792), the latter the generation of Emile Durkheim (1858) and Pierre Janet (1859). Because the average age of appointment to an elite academic position or first publication by a notable philosopher was around forty, the generations so constructed correspond rather well to the major political periods of nineteenth-century France. The generation of 1791–1810 achieved power and/or prominence from 1831 to 1850—that is, roughly during the July Monarchy. The next generation, including the protégés of Cousin, generally gained their elite positions during the Second Empire. The third generation attained eminence during the formative years of the Third Republic, and the last generation came of age during the 1890s and the first years of the twentieth century, during the greatest reform and expansion of the universities.

I collected information about the education, career, and philosophical orientation of the individuals on my list, using the sources just mentioned. To supplement any missing biographical information, I used a variety of sources, from personnel files at the Archives Nationales to obituaries in the leading philosophical journals to the *Grande Encyclopédie.* Readers interested in the complete list and biographical information may consult the dissertation on which this book is based.[8] Of the information, philosophical orientation is in many ways the least precise and most open to debate. The categories I used to classify philosophers are given at the bottom of table 6. They are based on the histories of philosophy I used to create my list of notable philosophers. Although each source had its own schemata, most of them recognized a spectrum from positivism to spiritualism with some sort of Kantian middle way in between. For this reason, I was able to use a simplified version of the classification found in Benrubi, the most extensive of the histories. For elite academic philosophers who did not appear in any of the histories of philosophy, I used indications from the other sources I consulted or my own best judgment about their written works.

After bringing the different classifications into alignment with each other, the sources sometimes still disagreed with each other over how to classify individual philosophers. Sometimes, indeed, the same source placed a given philosopher in two different camps. Rather than try to reduce these discrepancies to a spurious unity, I simply noted them and put a mark in each column to which an author was assigned. Dividing the total number of attributions of philosophical orientation by the number of philosophers yields a ratio I have called the ambiguity quotient, because it represents the degree of ambiguity present in the assignment of philosophers to schools of thought (tables 4 and 8). The presence of ambiguity should not be surprising. Besides the fact that judgments of philosophical orientation are ultimately subjective, it is a central argument of this book that many philosophers consciously or unconsciously borrow from schools other than the one to which they feel principal allegiance.

The following tables are illustrative rather than statistically conclusive in any formal sense. Nevertheless, they give a concise overview of general trends in French philosophy and can serve to place in a larger context the various individual cases described in this book.

Table 1.

Higher Education of Elite Academic Philosophers[a]

Year of Birth	N	None	Prov. Fac.[b]	Paris Faculties			ENS	Relig.	Other
				Letters	Law	Sci/Med			
1791–1810	15	6.7%	6.7	13.3	6.7	0.0	60.0	13.3	0.0
1811–1830	16	0.0	0.0	18.8	12.5	0.0	81.2	0.0	0.0
1831–1850	13	7.7	7.7	0.0	15.4	0.0	76.9	0.0	0.0
1851–1870	13	0.0	15.4	7.7	0.0	0.0	92.3	0.0	0.0

[a]For the definition of an elite academic philosopher, see the text accompanying this appendix. This table includes people who attended more than one of these institutions, such as the Faculty of Letters and the Faculty of Law. Percentages may therefore add up to more than 100%. Percentages are calculated by dividing the number who attended a given institution by N.

[b]Provincial Faculty of Letters. Other abbreviations: Sci/Med—Paris Faculties of Science and/or Medicine; ENS—Ecole Normale Supérieure; Relig.—seminaries.

Table 2.
Employment History of Elite Academic Philosophers[a]

Year of Birth	N	No Acad.[b]	Prov. Lyc.	Prov. Fac.	Paris Lyc.	Tot. Lyc.	Tot. Prov.	ENS	Paris Letters	CF	IG
1791–1810	15	6.7%	60.0	0.0	60.0	80.0	60.0	53.3	33.3	20.0	40.0
1811–1830	16	0.0	32.1	56.2	50.0	93.8	93.8	37.5	33.6	12.5	37.5
1831–1850	13	7.7	92.3	38.5	46.2	92.3	84.6	61.5	46.2	7.7	30.8
1851–1870	13	0.0	92.3	38.5	69.2	92.3	100.0	38.5	69.2	23.1	0.0

[a]Each entry represents the percentage of elite academic philosophers who held the indicated position at some point during their careers. In this table, all positions are positions in philosophy rather than other kinds of academic appointment. Most individuals moved through several of the junior positions on their way to elite positions, and many moved through more than one elite position. For this reason, percentages may add up to more than 100%.

[b]No Acad.—Individuals who held no junior academic position before occupying an elite position. Other abbreviations: Prov. Lyc.—Provincial Lycée; Paris Lyc.—Paris Lycée; Tot. Lyc.—Total Lycée (either Paris or Provincial); Tot. Prov.—Total Provincial (Lycée or Faculty); CF—Collège de France; IG—Inspection General.

Table 3.

Length of Apprenticeship and Service of Elite Academic Philosophers

Year of Birth	N	Yrs. Jr.[a]		Age Apptd.		Yrs. Held[c]	
		All	Non-IG[b]	All	Non-IG	All	Non-IG
1791–1810	15	18.7	10.0	38.8	30.7	18.9	22.6
1811–1830	16	22.2	16.8	41.8	35.4	19.8	23.2
1831–1850	13	18.3	13.5	42.5	40.0	14.8	12.4
1851–1870	13	17.5	17.5	40.6	40.6	17.8	17.8

[a]Average number of years between first appointment to a junior academic position (lycée or provincial faculty) and first appointment to an elite position.

[b]Inspectors General tended to be appointed much later than other elite appointments. They were selected on the basis of their experience with the system of secondary education rather than on grounds of exceptional brilliance. Hence precocity was not rewarded in Inspectors as it sometimes was in teachers.

[c]Calculated from year of initial appointment to an elite position to the year in which the individual died, retired, or left academic philosophy for another position.

Table 4.

Philosophical Orientation of Elite Academic Philosophers

Year of Birth	N	Posi-tivist[a]	Neo-crit.[b]	Spiritualist[c]				Amb.[g]
				Total[d]	Ecl.	Nonecl.[e]	Cler.[f]	
1791–1810	15	6.7%	0.0	93.3	78.6	7.1	14.3	1.00
1811–1830	16	0.0	6.2	100.0	87.5	6.2	6.2	1.06
1831–1850	13	15.4	30.8	69.2	55.6	44.4	0.0	1.15
1851–1870	13	30.8	61.5	46.2	33.3	66.7	0.0	1.38

[a]*Positivists* in the the broad sense of the term, including all those who apply the principle of determinism to human action and who consider the methods of the natural sciences to be the principal source of knowledge about human as well as natural phenomena.

[b]*Neo-criticists:* those who accept the Kantian analysis of the a priori character of certain fundamental principles of scientific knowledge, as well as the impossibility of attaining certainty in metaphysical speculation. Besides neo-Kantians of various inspirations, this group includes a number of scientists (Poincaré, Duhem, and Cournot) who also disagreed with the positivist conception of science.

[c]*Spiritualists,* including idealists (those who give precedence to ideas over spiritual activity).

[d]*Total* of Eclectic, Noneclectic, and Clerical Spiritualists. Numbers under these last three categories are percentages of Spiritualists, not of N.

[e]*Noneclectic Spiritualists:* either those who simply remained outside the eclectic fold, or those who defined themselves in opposition to it. They are distinguished from clerical philosophers by their insistence on the autonomy of philosophy and on the exclusive use of human means (reason and intuition) rather than dependence on revelation.

[f]*Clerical Spiritualists:* Priests or ministers who write on philosophical matters, as well as laypersons who write on theology. These writers subordinate philosophy to theology, and natural to divine inspiration, in varying degrees.

[g]*Ambiguity:* In many cases the sources used to identify philosophical positions did not agree among themselves. As a result, the number of positions attributed is often greater than the number of individual philosophers, and "ambiguity" is the ratio of these two numbers.

Table 5.

Percentage of Elite Academic Philosophers
Who Were Notable[a]

Year of Birth	N	Notable	
		#	%
1791–1810	15	3	23.1
1811–1830	16	6	37.5
1831–1850	13	9	69.2
1851–1870	13	10	76.9

[a]For the definition of a notable philosopher, see
the text accompanying this appendix.

Table 6.

Final Career Position of Notable Philosophers[a]

Year of Birth	N[b]	Academic Philosophy							Tot. Acad. Phl.
		Prov. Lyc.	Prov. Fac.	Paris Lyc.	Paris Fac.	ENS	CF	Other	
1791–1810	10	0.0%	0.0	0.0	0.0	0.0	10.0	0.0	10.0
1811–1830	14	0.0	0.0	0.0	14.3	7.1	0.0	7.1	28.5
1831–1850	21	0.0	0.0	4.8	19.0	0.0	4.8	14.3	42.9
1851–1870	46	4.3	6.5	4.3	17.4	0.0	6.5	2.2	41.2

Year of Birth	N	Non-acad.	Academic But Not Philosophy						Tot. Non-acad. Phl.
			Lyc.	Prov. Fac.	Paris Fac.	ENS	CF	Other	
1791–1810	10	70.0	0.0	0.0	0.0	0.0	0.0	20.0	90.0
1811–1830	14	28.6	0.0	0.0	0.0	0.0	21.4	21.4	71.4
1831–1850	21	19.0	0.0	4.8	14.3	4.8	4.8	9.6	57.3
1851–1870	46	26.1	0.0	6.5	15.2	0.0	2.2	8.7	58.7

[a]In this chart only the final career position is shown, defined as that held by the individual at retirement age or death.

[b]Information not available on 1 individual born from 1831 to 1850 and 3 from 1851 to 1870.

Table 7.
Final Career Position of Notable Philosophers Who Were
Academic Philosophers at Some Time

Year of Birth	N	Human Sciences[a]	Acad. Phil.	Other
1791–1810	5	0.0%	40.0	60.0
1811–1830	8	12.5	62.5	25.0
1831–1850	15	20.0	60.0	20.0
1851–1870	26	23.1	73.1	3.8

[a]Includes positions in public educational institutions commonly
associated with the human sciences—psychology, sociology,
philology, et cetera.

Table 8.
Philosophical Orientation of Notable Philosophers[a]

Year of Birth	N	Positivist	Neo-crit.	Spiritualist				Amb.
				Total	Ecl.	Nonecl.	Cler.	
1791–1810	10	50.0%	10.0	50.0	40.0	0.0	60.0	1.10
1811–1830	14	35.7	21.4	57.1	37.5	37.5	25.0	1.14
1831–1850	22	36.4	31.8	45.4	10.0	60.0	30.0	1.14
1851–1870	49	51.0	44.9	36.7	5.6	50.0	50.0	1.33

[a]See table 4 for definitions of categories.

Appendix B
The *Programmes* of 1832, 1874, 1880, and 1902

1832[1]

Introduction

1. Object of philosophy.—Utility and importance of philosophy.—Its relations with the other sciences.

2. Different methods that have been pursued up to the present in philosophical research.—Of the true philosophical method.

3. Division of philosophy.—Order in which the parts must be arranged.

Psychology

4. Object of psychology.—Necessity of beginning the study of philosophy with psychology.—Of consciousness and the certitude that belongs to it.

5. The phenomena of consciousness, and of our ideas in general.—Their different characters and kinds.—Give some examples.

6. The origin and formation of ideas.—Take as examples some of the more important of our ideas.

7. Give a theory of the faculties of the soul.—What determines the existence of a faculty?

8. Sensation.—Its character.—Distinguish sensation from all the other faculties, and mark its place in the order of their development.

9. The faculty of understanding, or reason.—Specific character of this faculty.

Faculties related to the general faculty of understanding:

Consciousness.
Attention.
Exterior perception.
Judgment.
Reasoning.
Memory.
Abstraction.

Generalization.
Association of ideas.

10. Activity and its various characteristics.—Voluntary and free activity.—Describe the phenomenon of will and all its circumstances.—Demonstration of free will.

11. The self; its identity; its unity.—The distinction between body and soul.

Logic

12. Method.—Analysis and synthesis.
13. Definition; division and classifications.
14. Certitude in general and the different kinds of certitude.
15. Analogy.—Induction.—Deduction.
16. Authority of the testimony of men.
17. Reasoning and its different forms.
18. Sophisms and the ways to resolve them.
19. Signs and language in their relations to thought.
20. Characteristics of a well-made language.
21. Causes of our errors and the ways to remedy them.

Ethics

22. Object of ethics.
23. The various motives of our actions.—Is it possible to reduce them all to a single? What is their relative importance?
24. Describe the moral phenomena on which rest what is called moral conscience: the sentiment or idea of duty, the distinction between good and evil, moral obligation, et cetera.
25. Merit and blame.—Punishments and rewards.—The sanction of ethics.
26. Division of duties.—Individual ethics, or duties of a man to himself.
27. Social ethics, or duties of man to others:
    1. Duties to man in general;
    2. Duties to the State.
28. Enumeration and assessment of the different proofs of the existence of God.
29. The principal attributes of God; divine Providence, and the plan of the universe.
30. Examination of the objections drawn from physical and moral evil.
31. Destiny of man.—Proofs of the immortality of the soul.
32. Religious ethics, or duties to God.

History of Philosophy

33. What method must be applied to the study of the history of philosophy?
34. In how many general epochs can the history of philosophy be divided?
35. Discuss the principal schools of Greek philosophy before Socrates.

36. Discuss Socrates and the character of the philosophical revolution of which he is the author.

37. Discuss the principal Greek schools from Socrates to the end of the Alexandrian school.

38. What are the principal scholastic philosophers?

39. What is the method of Bacon?—Analyze the *Novum Organum*.

40. Of what does the method of Descartes consist?—Analyze the *Discourse on Method*.

41. Discuss the principal modern schools since Bacon and Descartes.

42. What benefits can one derive from the history of philosophy for philosophy itself?

1874[2]

Course of Philosophy.

Object of philosophy.—Its principal divisions.—Its relations with the other sciences.

Psychology

Psychological facts.—Consciousness.—Distinction between physiological and psychological facts.

The faculties of the soul.—Sensation, intelligence, will.

Sensation.—Sensations.—Sentiments.—Instincts, penchants, passions.

Intelligence.—Exterior perception.—Interior perception.—Reason.

Ideas in general.—Classification of ideas.—The origin of ideas.—Different theories proposed on this question.

Primary ideas, axioms, and principles of reason.

Intellectual operations.—Memory.—Association of ideas.—Imagination.

Attention, abstraction.—Comparison.—Generalization.

Judgment.—Reasoning.

Signs and language.—The relations of language with thought.

Will. Instinct. Habit.

Moral liberty or free will.—Demonstration of freedom.—The principal systems that deny freedom.

Harmony of the faculties of the soul.—Unity of the principle of these faculties.—Human personality.

The spirituality of the soul.—Distinction between body and soul; their union; laws of this union.—The different systems that deny the distinction between body and soul.

Metaphysics or Theodicy

Principal elements of general metaphysics.

The existence of God.—Proofs of the existence of God.

Attributes of God and of Providence.—Refutations of objections drawn from physical and moral evil.

Logic

The foundation of certitude.—Skepticism.—Principal forms of ancient and modern skepticism.

The different kinds of certitude: sense evidence, rational evidence, moral evidence.

Method in general.—Analysis and Synthesis.

Deductive method. Definition.—Deduction.—Syllogism.—Demonstration.—Abuse of syllogism, sophisms.

Inductive method.—Observation.—Experimentation.—Induction.—Hypothesis.—Division.—Classification.

Error.—Logical and moral causes of error.

Application of method:

1. To the exact or mathematical sciences;
2. To the physical and natural sciences;
3. To the moral and political sciences (philosophy, law, political economy, etc.);
4. To the historical sciences. The testimony of men, critique of witnesses.

Ethics

General or theoretical ethics.—Principle of moral law.—Refutation of contrary or incomplete principles.

Moral conscience.—Principal phenomena of conscience.—The distinction between good and evil.—The various motives of our actions.

Duty and right.—Justice and virtue.—The sanctions of moral law.

Practical ethics and division of duties.—Duties of man to God, himself, others, family, society, and the State.

Destiny of man.—Immortality of the soul.

History of Philosophy

Systems in general.—Definition of the principal philosophical systems.

Rudiments of Greek philosophy before Socrates: Ionians, Atomists, Eleatics, Pythagoreans, Sophists.

Socrates.—Plato.—Aristotle.

Rudiments of the schools after Socrates: Pyrrhonians, Epicureans, Stoics.

Rudiments of Roman philosophy and the Alexandrian School.

Rudiments of Scholastic philosophy.

Rudiments of the philosophy of the Renaissance.

The philosophy of the seventeenth century.—Bacon.—Descartes and his principal disciples.—Malebranche.—Leibniz and Locke.

Rudiments of the philosophy of the eighteenth century in France, England, and Germany.

Philosophical Works.

Xenophon: *Memorabilia* on Socrates.
Plato: *Phaedo;*—the seventh book of the *Republic* (allegory of the cave; theory of ideas.)
Epictetus: *Manual.*
Cicero: *De Officiis;*—the two first books of *De Finibus Bonorum et Malorum* (exposition and refutation of the system of Epicurus).
Port-Royal Logic.
Descartes: *Discourse on Method.*
Pascal: *On Authority in Matters of Philosophy;*—*Interview with M. de Saci.*
Bossuet: *Treatise Concerning the Knowledge of God and Oneself* (general plan of the work; analysis of the fourth and fifth part).
Fénelon: *Treatise on the Existence of God.*
Leibniz: *Theodicy* (extracts).

1880[3]

Introduction

Science.—Classification of the sciences.—What do we call the philosophy of the sciences, of history, et cetera?—Specific object of philosophy; its divisions.

Psychology

Object of psychology: distinctive character of the facts it studies.—Degrees and limits of consciousness.—Distinction and relation of psychological and physiological facts.
Sources of information of psychology: consciousness, languages, history, et cetera—Utility of comparative psychology.—Experimentation in psychology.—Classification of psychological facts.
Sensation.—Emotions (pleasure and pain).—Sensations and sentiments.—Inclinations and passions.
Intelligence.—Acquisition, conservation, elaboration of understanding.
Acquisition: data of consciousness and of the senses.
Conservation and combination: memory, association of ideas, imagination.
Elaboration: formation of abstract and general ideas; judgment, reasoning.
The directing principles of the understanding: data of reason; can they be explained by experience, association of ideas, or heredity?
Results of intellectual activity: idea of the self, idea of the exterior world, idea of God.
Elements of esthetics: the beautiful.—Art.—Principles and conditions of the beaux arts.—Expression, imitation, fiction, and the ideal.
Will.—Analysis of the voluntary act: freedom. The various modes of psychological activity: instinct, voluntary activity, habit.
Manifestations of psychological life: signs and language.

Relations of body and soul.—Sleep, dreams, somnambulism, hallucinations, madness.

Elements of comparative psychology.

Logic

Definition and division of logic.

Formal logic.—Ideas and terms.—Judgments and propositions.—Definition.—Deduction and syllogism.

Applied logic.—Methods: analysis and synthesis.

Inductive logic.—Methods of the sciences of nature: observation, hypothesis, experimentation, classification, induction, analogy.—Empirical definitions.

Application of these methods to the psychological sciences,—to the historical sciences.—Sources of history: critique of testimony.

Deductive logic.—Method of the abstract sciences: rational definitions, axioms, deduction, demonstration.—Use of deduction in the experimental sciences.

Place of deduction and experience in ethics, law, and politics.

Nature, causes, and remedies of error.

Ethics

Speculative ethics.—Conscience, the good, freedom, duty.

Various conceptions of the supreme good: utilitarian and sentimental doctrines.

Doctrine of obligation. Duty and right.—Absolute value of the person. Virtue.—Responsibility and sanction.

Practical ethics.—Personal ethics: temperance, wisdom, courage, human dignity, and relation with inferior beings.

Domestic ethics: the family.

Social ethics: justice or respect for the law.—Rights.—Charity.

Elements of society: idea of the State. Distinction among natural law, civil law, political law.—Vote.—Obedience to the law.—Military service.—Devotion to country.

Religious ethics.—Duties to God.

Elements of political economy.

Production of wealth.—The factors of production: material, work, savings, capital, property.

Circulation and distribution of wealth.—Exchange, money, credit, salary, and interest.

Consumption of wealth: productive and unproductive consumption.—The question of luxury.—Expenditures of the State.—Taxes, budget, borrowing.

Metaphysics or Theodicy

The problem of certitude.—Skepticism.—Idealism.

Various conceptions of matter and life.

The mind.—Materialism and spiritualism.

God: his existence and attributes.—The problem of evil.—Optimism and pessimism. Immortality of the soul.

Conclusion of the course.—Role of philosophy.—Its importance from the intellectual, moral, and social points of view.[1]

Philosophical Authors

Plato: *Republic,* one book.
Aristotle: *Nicomachean Ethics,* one book.
Seneca: *De Vita Beata*.
Cicero: *De Legibus,* first book
Descartes: *Discourse on Method*.
Leibniz: *Monadology*.

1902[4]

Philosophy

N.B. The order adopted in the syllabus does not restrict the freedom of the professor; it suffices that all the questions indicated be covered.

Introduction

Object and divisions of philosophy.

Psychology

Specific characteristics of psychological facts. Consciousness.

Intellectual life.

The data of the understanding.—Sensations.—Images.—Memory and association.

Attention and reflection.—The formation of abstract and general ideas.—Judgment and reasoning.

The creative activity of the mind.

Signs: relations of language and thought.

Principles of reason: their development and role.

Formation of the idea of the body and perception of the exterior world.

Affective and active life.

Pleasure and pain.—Emotions and passions.—Sympathy and imitation.

Inclinations.—Instincts.—Habit.

Will and character.—Freedom.

---

1. The order adopted in this syllabus must not restrict the freedom of the professor, provided that all the indicated questions are treated.

Conclusion: The physical and the spiritual.—Psychological automatism.—Personality: the idea of the self.

Elements of Esthetics

Rudimentary ideas on the beautiful and on art.

Logic

Formal logic: Terms.—The Proposition.—The various forms of reasoning.
Science: Classification and hierarchy of the sciences.
Method of the mathematical sciences: Definitions.—Axioms and postulates.—Demonstration.
Method of the sciences of Nature: Experience: The methods of observation and experimentation.—Hypothesis; theories.—Role of induction and deduction in the sciences of Nature.—Classification.
Method of the moral and social sciences: The methods of psychology.—Relations of history and the social sciences.

Ethics

Object and character of ethics.
The data of moral conscience: Obligation and sanction.
The motives of conduct and the ends of human life: Pleasure, sentiment, and reason.—Personal interest and the general interest.—Duty and happiness.—Individual perfection and the progress of humanity.
Personal ethics: The sentiment of responsibility.—Virtue and vice.—Personal dignity and moral autonomy.
Domestic ethics: The moral constitution and social role of the family.—Authority in the family.
Social ethics: Law.—Justice and charity.—Solidarity.
Rights: Respect for life and individual liberty.—Property and work.—Freedom of thought.
Civic and political ethics: The Nation and the Law.—The Fatherland.—The State and its functions.—Democracy; civil and political equality.
N.B.—The professor will insist, as much in personal ethics as in social ethics, on the dangers of alcoholism and its physical, moral, and social effects: moral degradation, weakening of the race, poverty, suicide, criminality.

Metaphysics

Value and limits of knowledge.
Problems of first philosophy: Matter, the Soul, and God.
Relations of metaphysics with science and ethics.

Philosophical Authors

The professor will choose in the following list four texts that will be commented upon in class and that will serve as a basis for the exposition of the systems of philosophy to which they are related.

Xenophon: a book of the *Memorabilia.*

Plato: *Phaedo; Gorgias;* a book of the *Republic.*

Aristotle: a book of the *Nicomachean Ethics;* a book of the *Politics.*

Epictetus: *Manual.*

Marcus Aurelius.

Lucretius: *De Natura Rerum,* Book 2 or 5.

Seneca: Extracts from the *Letters to Lucilius* and the *Moral Essays.*

Bacon: *On the Dignity and the Growth of the Sciences.*

Descartes: *Discourse on Method; Meditations; Principles of Philosophy,* Book 1.

Pascal: *Pensées* and opuscules.

Malebranche: *The Search for Truth,* Book 1 or Book 2.—*Dialogues on Metaphysics.*

Spinoza: *Ethics* (one book).

Leibniz: *New Essays,* forward and Book 1.—*Theodicy* (Extracts).—*Monadology.*—*Discourse on Metaphysics.*

Hume: *A Treatise of Human Nature* (one book).

Condillac: *Treatise on Sensations,* Book 1.

Montesquieu: *The Spirit of the Laws,* Book 1.

J.-J. Rousseau: *The Social Contract* (one book).

Kant: *Foundations of the Metaphysics of Morals.*—*Prolegomena.*

Jouffroy: Extracts.

A. Comte: *Course of Positive Philosophy,* first and second lessons.—*Discourse on the Positive Spirit.*

C. Bernard: *Introduction to the Study of Experimental Medicine,* first part.

Stuart Mill: *Logic,* Book 6.—*Utilitarianism.*—*On Liberty.*

Spencer: *First Principles* (first part).—*The Study of Sociology.*

Notes

INTRODUCTION

1. Durkheim dedicated his dissertation, *The Division of Labor in Society,* to Boutroux. Pierre Janet, "Autobiography," in *A History of Psychology in Autobiography,* ed. Carl Murchison (Worcester, Mass.: Clark University Press, 1930-), 1:123.

2. Terry N. Clark, *Prophets and Patrons: The French University and the Emergence of the Social Sciences* (Cambridge, Mass.: Harvard University Press, 1973).

3. This is especially true of general histories of sociology, such as Talcott Parsons, *The Structure of Social Action,* 2d ed. (New York: Free Press, 1949); and Raymond Aron, *Les étapes de la pensée sociologique* (Paris: Gallimard, 1967). Björn Sjövall, in his book on Janet, devotes a great deal of space to Maine de Biran, but very little to Paul Janet (Björn Sjövall, *Psychology of Tension* [Stockholm: Norstedt, 1967]).

4. This assumption was most often made by sociologists and psychologists, probably for reasons of disciplinary identity. The following statement from an eminent Durkheim scholar is a good example: "Before Durkheim sociology was a provocative idea; by his professional endeavors it became an established social fact" (Edward Tiryakian, "Emile Durkheim," in *A History of Sociological Analysis,* ed. Tom Bottomore and Robert Nisbet [New York: Basic, 1978], 187).

5. For an account of this rupture in French philosophy, see Vincent Descombes, *Modern French Philosophy,* trans. L. Scott-Fox and J. M. Harding (Cambridge: Cambridge University Press, 1979), 3–32.

6. For an example of this rebellion, see Luc Ferry and Alain Renaut, *French Philosophy of the Sixties: An Essay on Anti-Humanism,* trans. Mary H. S. Cattani (Amherst: University of Massachusetts Press, 1990). See also the volume edited by Mark Lilla, *New French Thought: Political Philosophy* (Princeton: Princeton University Press, 1994).

7. See, for example, J. Benrubi, *Les sources et les courants de la philosophie contemporaine en France,* 2 vols. (Paris: Alcan, 1933); and Emile Bréhier, *Histoire de la philosophie,* 3 vols. (Paris: Presses Universitaires de France, 1932).

8. On the extended "crisis" of French philosophy, see Jean-Louis Fabiani, "La crise du champ philosophique, 1880–1914" (Thèse de troisième cycle, Ecole des Hautes Etudes en Sciences Sociales, 1980), portions of which have been published as *Les philosophes de la République* (Paris: Les Editions de Minuit, 1988). See also Victor Karady, "The Durkheimians in Academe," in *The Sociological Domain,* ed. Philippe Besnard (Cambridge: Cambridge University Press, 1983), 76–77.

9. William Logue, *From Philosophy to Sociology: The Evolution of French Liberalism, 1870–1914* (Dekalb: Northern Illinois University Press, 1983).

10. Fabiani, "La crise du champ philosophique," 68, 72; Karady, "The Durkheimians in Academe," 76–79; W. Paul Vogt, "Political Connections, Professional Advancement, and Moral Education in Durkheimian Sociology," *Journal of the History of the Behavioral Sciences* 27 (1991): 65–66. See also Philippe Besnard, "The Epistemological Polemic: François Simiand," in *The Sociological Domain,* 256.

11. The fundamental reference remains Steven Lukes's magisterial biography, *Emile Durkheim, His Life and Work: A Historical and Critical Study* (New York: Harper, 1972). See also Robert Alun Jones, "Robertson Smith, Durkheim and Sacrifice: An Historical

Context for *The Elementary Forms of the Religious Life,*" *Journal of the History of the Behavioral Sciences* 17 (1981): 199–200; Stjepan G. Mestrovic, *Emile Durkheim and the Reformation of Sociology* (Totowa, N. J.: Rowman and Littlefield, 1988); Warren Schmaus, *Durkheim's Philosophy of Science and the Sociology of Knowledge: Creating an Intellectual Niche* (Chicago: University of Chicago Press, 1994); and Dénes Némedi and W. S. F. Pickering, "Durkheim's Friendship with the Philosopher Octave Hamelin: Together with Translations of Two Items by Durkheim," *British Journal of Sociology* 46 (1995): 107–25. See also the special issue of the *Journal of the History of the Behavioral Sciences* 32 (1996) devoted to "Durkheimian Sociology in Philosophical Context."

12. Mara Meletti Bertolini, *Il pensiero e la memoria: Filosofia e psicologia nella "Revue philosophique" di Théodule Ribot (1876–1916)* (Milan: Franco Angeli, 1991); Jan Goldstein, "Foucault and the Post-Revolutionary Self: The Uses of Cousinian Pedagogy in Nineteenth-Century France," in *Foucault and the Writing of History,* ed. Jan Goldstein (Cambridge, Mass.: Blackwell, 1994), 99–115; idem, "The Advent of Psychological Modernism in France: An Alternate Narrative," in *Modernist Impulses in the Human Sciences, 1870–1930,* ed. Dorothy Ross (Baltimore: Johns Hopkins, 1994), 190–209.

13. Fritz Ringer, *Fields of Knowledge: French Academic Culture in Comparative Perspective, 1890–1920* (Cambridge: Cambridge University Press, 1992).

14. John I. Brooks III, "Analogy and Argumentation in an Interdisciplinary Context: Durkheim's 'Individual and Collective Representations,'" *History of the Human Sciences* 4 (1991): 223–59; idem, "Philosophy and Psychology at the Sorbonne, 1885–1913," *Journal of the History of the Behavioral Sciences* 29 (1993): 123–45; idem, "The Definition of Sociology and the Sociology of Definition: Durkheim's *Rules of Sociological Method* and High-School Philosophy in France," *Journal of the History of the Behavioral Sciences* 32 (1996): 379–407.

15. Emile Boutroux cited the eclectic Paul Janet's lectures on Kant in 1867 as the beginning of renewed academic interest in critical philosophy (Boutroux, "Paul Janet," *Annuaire de l'Association Amicale des Anciens Elèves de l'Ecole Normale* [1900]: 31–47).

16. Most evaluations of French academic philosophy ridicule its pretensions to be scientific. Examples of this attitude include Doris Goldstein, "Official Philosophy in France: The Example of Victor Cousin," *Journal of Modern History* 1 (1968): 259–79; Alan B. Spitzer, *The French Generation of 1820* (Princeton, N.J.: Princeton University Press, 1987); and Bréhier, *Histoire de la philosophie,* 3:578–81. All of these accounts treat the content of French eclecticism as without intellectual interest, and indeed only explicable in social/ideological terms. In this they simply repeat the pioneering critiques of Renan and Taine, to be discussed in chapter 1. A notable exception to this trend is Goldstein, "Foucault and the Post-Revolutionary Self," 101.

17. For analyses of the history and multiple meanings of positivism, as well as of Comte's influence on European thought, see D. G. Charlton, *Positivist Thought in France During the Second Empire, 1852–1870* (Oxford: Oxford University Press, 1959); Walter M. Simon, *European Positivism in the Nineteenth Century: An Essay in Intellectual History* (Port Washington, N.Y.: Kennikat Press, 1963); and Anthony Giddens, "Positivism and its Critics," in *A History of Sociological Analysis,* 237–86.

18. Simon, *European Positivism,* 3.

19. For Charlton's definition, see his *Positivist Thought in France,* 1–11.

20. In addition to the studies cited in the preceding notes, see Mary Pickering, *Auguste Comte: An Intellectual Biography,* vol. 1 (Cambridge: Cambridge University Press, 1993).

21. Indeed, Schmaus argues that "the intellectual development of Durkheim . . . can be adequately explained in terms of his intellectual concerns and the core philosophical assumptions of his research program" (*Durkheim's Philosophy of Science,* 19). I agree that

that intellectual issues can have a determining role in philosophy and science; however, I do not see a hard and fast distinction between intellectual and institutional issues.

22. On the virtues and pitfalls of philosophical classifications, see Jean-Louis Fabiani, "Métaphysique, morale, sociologie: Durkheim et le retour à la philosophie," *Revue de métaphysique et de morale* 98 (1993): 176–78.

23. On the history of the French historical profession during the early Third Republic, see William R. Keylor, *Academy and Community: The Foundation of the French Historical Profession* (Cambridge, Mass.: Harvard University Press, 1975).

24. Among biographical studies of these individuals: Lukes, *Emile Durkheim;* Claude M. Prévost, *La psycho-philosophie de Pierre Janet* (Paris: Payot, 1973); Henri Ellenberger, "Pierre Janet and Psychological Analysis," in *The Discovery of the Unconscious* (New York: Basic, 1970); and Jean J. Ostrowski, *Alfred Espinas, précurseur de la praxéologie* (Paris: R. Pichon, and R. Durand-Anzias, 1973). Théodule Ribot, the founder of experimental psychology in France, has received surprisingly little attention. Some information can be found in Meletti Bertolini, *Il pensiero e la memoria.*

25. Logue, *From Philosophy to Sociology,* ix.

26. Fabiani, "La crise du champ philosophique," 1.

27. Ringer, *Fields of Knowledge,* 1–25. This section also contains a lucid explication of Bourdieu's concept of an epistemological field, central to all analyses in this tradition.

28. For an example of the Edinburgh School, see David Bloor, *Knowledge and Social Imagery* (London: Routledge and Kegan Paul, 1976). For the internalist position, see Schmaus, *Durkheim's Philosophy of Science.* For a history and analysis of the internalist/externalist debate, see Steven Shapin, "Discipline and Bounding: The History and Sociology of Science as Seen Through the Externalism-Internalism Debate," *History of Science* 30 (1992): 333–69.

29. For examples and surveys of this approach, see Donald McCloskey, *The Rhetoric of Economics* (Madison: University of Wisconsin Press, 1985); John S. Nelson, Allan Megill, and Donald N. McCloskey, eds., *The Rhetoric of the Human Sciences* (Madison: University of Wisconsin Press, 1987); Chaim Perelman and L. Olbrechts-Tyteca, *The New Rhetoric: A Treatise on Argumentation,* trans. John Wilkinson and Purcell Weaver (Notre Dame, Ind.: University of Notre Dame Press, 1969); J. Gusfield, "The Literary Rhetoric of Science," *American Sociological Review* 41, no. 1 (1976): 16–34; Bruno Latour and Paolo Fabbri, "La rhétorique de la science: pouvoir et devoir dans un article de science exacte," *Actes de la Recherche en Sciences Sociales* 13 (February 1977): 81–95; Michael A. Overington, "The Scientific Community as Audience: Toward a Rhetorical Analysis of Science," *Philosophy and Rhetoric* 10 (1977): 143–64; Richard H. Brown, *A Poetic for Sociology: Toward a Logic of Discovery for the Human Sciences* (Cambridge: Cambridge University Press, 1977); idem, ed., *Writing the Social Text: Poetics and Politics in Social Science Discourse* (New York: Aldine de Gruyter, 1992); Richard M. Weaver, "The Rhetoric of Social Science," in *The Ethics of Rhetoric* (Chicago: Henry Regney, 1968), 187–91; and Steve Fuller, *Philosophy, Rhetoric, and the End of Knowledge: The Coming of Science and Technology Studies* (Madison: University of Wisconsin Press, 1993).

30. Examples of the former include George Sarton, *Sarton on the History of Science* (Cambridge, Mass.: Harvard University Press, 1962); of the latter, Stephen Toulmin, *Human Understanding* (Princeton: Princeton University Press, 1972); and Robert J. Richards, "The Natural Selection Model and Other Models in the Historiography of Science," in *Darwin and the Emergence of Evolutionary Theories of Mind and Behavior* (Chicago: University of Chicago Press, 1987), 559–93.

31. Gaston Bachelard, *La formation de l'esprit scientifique,* 12th ed. (Paris: J. Vrin, 1983); Alexandre Koyré, *Etudes d'histoire de la pensée scientifique* (Paris: Gallimard, 1973); Thomas S. Kuhn, *The Structure of Scientific Revolutions,* 2d ed. (Chicago: University of Chicago Press, 1970).

32. A sophisticated defense of authorial intention is found in the works of Quentin Skinner, especially "Social Meaning and the Explanation of Social Action," in Patrick Gardiner, ed., *The Philosophy of History* (Oxford: Oxford University Press, 1974), 106–26. The most radical critique of attempts to circumscribe meaning is to be found in the works of Jacques Derrida. See especially "Structure, Sign, and Play in the Human Sciences," in *Writing and Difference*, trans. Alan Bass (Chicago: University of Chicago Press, 1978), 278–93. From a different perspective, Michel Foucault has argued that the reader need respect neither the unity of the author nor the unity of the text. See, among many works, *The Archaeology of Knowledge*, trans. A. M. Sheridan Smith (New York: Harper, 1972), and idem, "What Is an Author?" in *Language, Counter-Memory, Practice*, trans. Donald F. Bouchard and Sherry Simon (Ithaca: Cornell University Press, 1977), 113–38. For a discussion of the implications of such works on the human sciences and on history, see Pauline Marie Rosenau, *Post-Modernism and the Social Sciences: Insights, Inroads, and Intrusions* (Princeton: Princeton University Press, 1992); and Goldstein, *Foucault and the Writing of History*.

CHAPTER 1. PHILOSOPHY AS SCIENCE: THE ACADEMIC TRADITION AND ITS CRITICS

1. On the professionalization of philosophy, see André Canivez, *Jules Lagneau, professeur de philosophie*, 2 vols. (Paris: Les Belles Lettres, 1965); Goldstein, "Official Philosophy," 259–79; and Jean-Louis Fabiani, "Les programmes, les hommes, et les oeuvres: Professeurs de philosophie en classe et en ville au tournant du siècle," *Actes de la recherche en sciences sociales* 47–48 (June 1983): 3–20.

2. Authors have differed over when philosophy was professionalized. Doris Goldstein maintains the traditional view that it occurred under the leadership of Victor Cousin during the July Monarchy. Fabiani seems to suggest that it did not occur until the Third Republic. Part of the disagreement is over the definition of professionalization. If one means by this the creation of a profession—an occupation requiring mastery of an intellectual subject demonstrated by possession of academic credentials and adherence to recognized standards—then it undoubtedly occurred during the July Monarchy. If one means by professionalization that the discipline is carried on primarily by those in the profession, then it was not until the Third Republic that professors achieved a dominant position within the field of philosophy. Even then, however, my own findings do not support Fabiani's claim that philosophy had become the exclusive domain of professionals (table 6). Even in the generation born between 1851 and 1870, less than half of all notable philosophers were academics.

A better support for the later date would lie in the foundation of journals of philosophy—the *Revue philosophique* and the *Revue de métaphysique et de morale*—during the early Third Republic, as well as the birth of the Société Française de Philosophie, the closest thing to a professional society the community of academic philosophers produced. On the *Revue philosophique*, see chapter 2. On the *Revue de métaphysique et de morale*, see chapter 4.

3. Antoine Prost, *Histoire de l'enseignement en France, 1800–1967* (Paris: Armand Colin, 1968), 25–26. See also Ringer, *Fields of Knowledge*, 40–45.

4. On the use of the term *couronnement*, see Fabiani, "Les programmes," 5. On the importance of the philosophy class in the general culture of the French, see Theodore Zeldin's section on "Logic and Verbalism" in his *France 1848–1945: Intellect and Pride* (Oxford: Oxford University Press, 1980), 205–27.

5. The *grandes écoles* were (and are) specialized professional schools, elite, highly

competitive, and separate from the universities. They were established in large measure by Napoleon, and they reflect the narrow, pragmatic goals he had for education.

6. Walter M. Simon, "Two Cultures in Nineteenth-Century France: Victor Cousin and Auguste Comte," *Journal of the History of Ideas* 26 (1965): 45–58; Theodore Zeldin, "Science and Comfort," in *France 1848–1945: Taste and Corruption* (Oxford: Oxford University Press, 1980), 226–70. Both authors rightly insist that the issue is very complex, given the rising prestige of science during the nineteenth century; nevertheless, this increase in prestige was accompanied by an increase in specialization, so that if the image of science was widely diffused, its practice remained narrow, especially in secondary education.

7. On the hierarchy of academic disciplines, see Jean-Louis Fabiani, "Enjeux et usages de la 'crise' dans la philosophie universitaire en France au tournant du siècle," *Annales: Economies, sociétés, civilisations* 2 (1985): 385–89. The long struggle over the proper place of science in secondary education is most fully treated in Clément Falcucci, *L'humanisme dans l'enseignement secondaire en France au XIX*^e *siècle* (Toulouse: Edouard Privat, 1939).

8. Roger Geiger gives an excellent description of the faculty system of early and mid nineteenth-century France. See his "Prelude to Reform: The Faculties of Letters in the 1860s," in *The Making of Frenchmen*, ed. Donald N. Baker and Patrick J. Harrigan (Waterloo, Ontario: Historical Reflections Press, 1980), 337–62. On the other hand, professors derived a substantial part of their income from the fees for these examinations, especially the baccalaureate (Prost, *Histoire de l'enseignement*, 225).

9. Canivez notes that professors themselves denounced the effect of the baccalaureate on the teaching of philosophy, both in the amount of time it took faculty professors to conduct them and in the constraints it placed on what secondary professors could teach (*Jules Lagneau*, 193–202).

10. Roger Anderson, *Education in France, 1848–1870* (Oxford: Clarendon Press, 1975), 225.

11. Claude Digeon, *La crise allemande de la pensée française* (Paris: Presses Universitaires de France, 1959), 364–83; Prost, *Histoire de l'enseignement*, 228–29.

12. Roger Smith, *The Ecole Normale Supérieure and the Third Republic* (Albany, N.Y.: SUNY Press, 1982), 24.

13. Anderson, *Education in France*, 229.

14. A separate *agrégation* in philosophy was established in 1825. Before that date, aspiring philosophy professors took the *agrégation* in letters.

15. Victor Karady, "Les professeurs de la République," *Actes de la recherche en sciences sociales* 47–48 (1983): 101.

16. Ernest Renan was one such prodigy. After leaving the seminary, he crammed for two years and followed the courses at the Sorbonne in preparation for the *agrégation*, in which he placed first. The philosopher Alfred Fouillée was even more remarkable; he prepared for the *agrégation* on his own, while a provincial lycée professor.

17. In many cases, the *suppléants* were promising young scholars destined for a brilliant career. Jules Simon, the liberal politican, substituted for Victor Cousin for several years beginning in 1839. This helped his academic career, which he eventually abandoned for politics. But *suppléants* were not always chosen from the best and brightest. Simon claimed that for many years, Cousin deliberately placed a nonentity in his chair, to avoid overshadowing Cousin himself and to avoid having to pay more than was absolutely necessary, since the salary came out of his pocket. Even Simon was paid a pitiful wage for his services, a fact of which Cousin was quite proud and that he used as a cautionary tale for overambitious students. Jules Simon, *Victor Cousin,* trans. Melville B. Anderson and Edward Anderson (Chicago: McClurg, 1888), 87–88, 115.

18. Louis Liard was director of Higher Education, and Elie Rabier director of

Secondary Education. Cf. Fabiani, "Enjeux et usages," 386. Liard and Rabier had been classmates at the Ecole Normale, and were the only two successful candidates for the *agrégation* in philosophy in 1869 (arrêté du 5 octobre 1869, *Bulletin administratif du Ministère de l'Instruction Publique* [1869, part 2]: 215). For more on Liard and Rabier, see chapter 4.

19. For example, from 1871 to 1899 there were forty-three governments with twenty-seven different ministers of public instruction (Benoît Yvert, ed., *Dictionnaire des ministres, 1789–1989* [Paris: Perrin, 1990], 323–32).

20. Simon, *Cousin*, 97–98.

21. Isabelle Havelange, Françoise Huguet, and Bernadette Lebedeff, *Les inspecteurs généraux de l'Instruction publique: Dictionnaire biographique, 1802–1914* (Paris: Institut National de Recherche Pédagogique, 1986), 5.

22. Terry Clark characterizes this phenomenon in terms of patrons and clusters in his *Prophets and Patrons*, 67–68. Weisz, however, finds no evidence for the existence of clusters ("The Republican Ideology and the Social Sciences," in *The Sociological Domain*, 116). Whether one agrees with the specific notion of a cluster or not, the fact of institutional politics is a general one, and can with slight modifications be applied to philosophy, since many of the actors in Clark's work were themselves philosophers.

23. The following material on Cousin's life and legacy is taken from Simon, *Cousin;* Canivez, *Jules Lagneau*, 1:144–77; Paul Janet, *Victor Cousin et son oeuvre* (Paris: Calmann Lévy, 1885); and Goldstein, "Official Philosophy." See also Spitzer, *The French Generation of 1820*, 71–96, and Goldstein, "Foucault and the Post-Revolutionary Self," 101–7.

24. Victor Cousin, *Défense de l'Université et de la philosophie*, 3d ed. (Paris: Joubert, 1844), 65.

25. Goldstein, "Official Philosophy," 262; Logue, *From Philosophy to Sociology*, 17–19.

26. His *Défense de l'Université* was a response to one such attack.

27. The number of professors grew from about 60 to about 300 over the course of the nineteenth century; see Zeldin, "Logic and Verbalism," 209, and Fabiani, "Les programmes," 15, note 26.

28. On Franck, see Logue, *From Philosophy to Sociology*, 20–29.

29. It was Napoleon who initiated the use of "lycée" to designate state-supported secondary schools. He also recognized another class of public schools, the "collège," which was funded by municipalities rather the national government. During the Restoration, the word "lycée" seemed too closely associated with Napoleon and the Revolution. For this reason, the lycées became "collèges royaux." During the Second Empire, the name "lycée" was revived, and the word "collège" again reserved for municipally funded secondary schools. According to Logue, Franck's first assignment was in a municipal collège.

30. Especially Franck's preface to the *Dictionnaire des sciences philosophiques*, 2d ed., ed. Adolphe Franck (Paris: Hachette, 1875), v–xii. The first edition appeared in 1843–44.

31. Logue, *From Philosophy to Sociology*, 37–41; Goldstein, "The Advent of Psychological Modernism," 202. See also Thomas M. Telzrow, "The Watchdogs: French Academic Philosophy in the Nineteenth-Century—The Case of Paul Janet" (Ph.D. diss., University of Wisconsin, 1973).

32. See chapter 4.

33. On Ravaisson, see Havelange et al., *Les inspecteurs généraux;* Henri Bergson, "Notice sur la vie et les oeuvres de M. Félix Ravaisson-Mollien," in *La pensée et le mouvant* (Paris, 1934).

34. "Programme" can be and has been translated as either "syllabus" or "curriculum." In French it is used to designate both the list of topics for a single course—what

in English is usually called a syllabus—and the set of courses offered by an institution or required for a degree—the usual sense of "curriculum" in English. Since in this book I am focusing on the topics covered in a single year-long course, I have opted for "syllabus."

35. It is true that toward the end of the century, professors began to take more and more liberty with the syllabus. Moreover, as the universities became centers of research less tied to the secondary system, the syllabus was proportionately less constraining. But for the period in which the first generations of French social scientists were formed, the syllabus can still be used as an indicator of what constituted official philosophy. For complaints about the effect of the baccalaureate on lycée philosophy, see Canivez, *Jules Lagneau*, 1:193–202.

36. Small changes could nevertheless be significant. Fabiani discusses the implications of some of these changes ("Les programmes," 4–9). Janet reports that the order of ethics and metaphysics changed a number of times. In general, clerics, when they did not want metaphysics banished altogether from the syllabus, insisted that it come before ethics, since God was the foundation of morality. The eclectics, on the other hand, preferred ethics to come before metaphysics, for two reasons: ethics together with logic and psychology could be seen as a unit constituting the *science de l'homme*, whereas metaphysics was essentially the science of God; and philosophy should start with the most familiar and work its way up to the least (Janet, *Victor Cousin*, 325–26).

37. Cf. Goldstein, "Official Philosophy," 266.

38. This is Janet's argument (*Victor Cousin*, 314–33).

39. Victor Cousin, *Du vrai, du beau, et du bien*, 20th ed. (Paris: Didier, 1878), ii. In his preface to the edition of 1853, he claimed that this book is the best summary of his own philosophy. It was based on his lectures at the Sorbonne during 1817–18, and was first published in 1837 from notes taken by students. In the later editions, Cousin toned down some of the idealist and pantheist rhetoric of his original statements, which were influenced by his recent discovery of German philosophy. Although Cousin may legitimately have changed his views, his rewriting of the past came under criticism from his opponents.

40. Cf. Cousin, *Du vrai, du beau, et du bien*, 10, 435–36.

41. Janet, *Victor Cousin*, 372.

42. See also the article "Science," by Th. H. Martin, *membre de l'Institut* and rector of the Académie de Lyons.

43. Cousin, *Du vrai, du beau, et du bien*, 438.

44. E.g., ibid., 2, 9–10, 19, 32–34, 259, 341, 344, 365.

45. See Paul Janet, *Principes de métaphysique et de psychologie*, 2 vols. (Paris: Delagrave, 1897), 1:4.

46. Charles Jourdain, *Notions de philosophie*, 9th ed. (Paris: Hachette, 1865), 3; Paul Janet, *Traité élémentaire de philosophie à l'usage des classes* (Paris: Delagrave, 1879), 3; see also another influential textbook, Elie Rabier's *Leçons de philosophie*, 2 vols. (Paris: Alcan, 1884–86), 1:6.

47. Janet, *Traité*, 3.

48. Ibid.

49. Franck, "Philosophie," *Dictionnaire*.

50. Janet, *Traité*, 11.

51. Jourdain, *Notions*, 9.

52. Cousin, *Du vrai, du beau, et du bien*, 3.

53. Janet, *Traité*, 11.

54. Jourdain, *Notions*, 11–12.

55. "Psychologie," *Dictionnaire*.

56. This was not a problem among early eclectics, when the problem of the uncon-

scious was not an issue. After Hartmann's *Philosophy of the Unconscious* (1867) made it a topic of lively debate, however, eclectics were divided over the question of whether absolutely unconscious mental phenomena could exist.

57. Jourdain, *Notions*, 15–16.

58. On the innovations of Laromiguière and Maine de Biran, see George Boas, "The Fortunes of Ideology," in *French Philosophies of the Romantic Period* (New York: Russell and Russell, 1964), 23–69.

59. Jourdain, *Notions*, 19–20.

60. Ibid.

61. See Jourdain, *Notions*, 110–12; and Janet, *Traité*, 334–36.

62. Jourdain, *Notions*, 119; Janet, *Traité*, 365.

63. Jourdain, *Notions*, 131.

64. Janet, *Victor Cousin*, 321.

65. Martin, "Science," *Dictionnaire*.

66. Cousin, *Du vrai, du beau, et du bien*, 32.

67. Ibid., 41–43.

68. Martin, "Science," *Dictionnaire;* Cousin, *Du vrai, du beau, et du bien*, 438–39.

69. Cousin, *Du vrai, du beau, et du bien*, 436.

70. Ibid., 438.

71. "This the philosophy that we teach rests neither on hypothetical principles nor on empirical principles" (ibid., 33).

72. Janet, *Traité*, 503–4.

73. Martin, "Science," *Dictionnaire*.

74. Martin placed metaphysics among the rational sciences; Janet considered it mixed.

75. Martin, "Science," *Dictionnaire*.

76. Jourdain, *Notions*, 100; Janet, *Traité*, 302–6.

77. Hence Jourdain defined *la morale* as the "branch of philosophy that has as its object the direction of our active faculties" (*Notions*, 241). Cf. Franck, "Morale," *Dictionnaire*.

78. For example, in arguing for the existence of universal and necessary moral principles, Cousin claims that "we have done like the physicist or chemist who submits a compound body to analysis and reduces it to its simple elements. The sole difference is here that the phenomenon to which our analysis applies is in us, instead of being outside us. Otherwise the procedures employed are exactly the same; there is here neither system nor hypothesis; there is only experience and the most immediate induction" (*Du vrai, du beau, et du bien*, 341).

79. For this reason, Janet decided to reverse the usual order of presentation and put practical before theoretical ethics. See Janet, *Traité*, 569–70.

80. Cousin, *Du vrai, du beau, et du bien*, 367–70.

81. Franck excludes them ("Morale," *Dictionnaire*); Janet includes them (*Traité*, 571–72, 616).

82. Jourdain, *Notions*, 302–3; Janet, *Traité*, 711–12.

83. Cf. Jourdain, *Notions*, 269.

84. Janet, *Traité*, 759.

85. Cousin, *Du vrai, du beau, et du bien*, 58.

86. Ibid., 30–31; cf. Simon, *Cousin*, 44–47.

87. Janet, *Victor Cousin*, 324.

88. Ibid., 281.

89. ". . . what I recommend is an enlightened eclecticism which, judging with equity and even benevolence all schools, borrows from them what is true, and neglects what is false. . . . Human thought is immense. Each school has only considered it from its

point of view. This point of view is not false, but it is incomplete, and, in addition, it is exclusive. It expresses but one side of the truth, and rejects all others" (Cousin, *Du vrai, du beau, et du bien*, 10).

90. Although even this achievement came under attack. Hippolyte Taine denounced Cousin's scholarship as mere erudition rather than true science, and Janet admitted that eclectics had tended to retreat into history at the expense of dogmatic philosophy. See Taine, "M. Cousin érudit et philologue," in *Les philosophes classiques du XIXe siècle en France*, 6th ed. (Paris: Hachette, 1888; reprint, Paris: Slatkine, 1979); Janet, *La crise philosophique: MM. Taine, Renan, Littré, Vacherot* (Paris: Baillière, 1865), 2–5.

91. See, for example, Janet's discussion "Of Progress in Philosophy," *Traité*, 827–29.

92. Collected in Janet, *La crise philosophique*.

93. Canivez, *Jules Lagneau*, 1:173.

94. Janet, *Victor Cousin*, 338.

95. Of course, because of their identification with positivism, their tenure at these institutions was still precarious. See the next section.

96. This criticism was raised at least as early as 1828 by A. Marrast, who was associated with Laromiguière. See Bréhier, *Histoire de la philosophie*, 3:580. Cousin, however, explicitly acknowledged that philosophy preceeded the history of philosophy, that his criterion was provided by the results of the *méthode psychologique*, although this position may well have been a response to Marrast (*Du vrai, du beau, et du bien*, 10).

97. Félix Ravaisson, *La philosophie en France au XIXe siècle* (Paris: Imprimerie Impériale, 1868), 20.

98. Ravaisson mentions Jouffroy and Garnier, but later it could also be found in Janet.

99. Ibid., 275. Ravaisson's use of the term *positivism* is, according to Bréhier, inspired by Schelling rather than Comte (*Histoire de la philosophie*, 3:871).

100. Janet, *La crise philosophique*, 7.

101. Charles Adam, *La philosophie en France* (Paris: Alcan, 1894), 348.

102. Pierre Leroux, "Réfutation de l'éclectisme," in *Oeuvres (1825–1850)*, 2 vols. (Geneva: Slatkine Reprints, 1978), 2:277.

103. Auguste Comte, *Cours de philosophie positive*, 2 vols, ed. Michel Serres, François Dagognet, and Allal Sinaceur (Paris: Hermann, 1975), 1:853–54. Comte noted that this chapter was written from 24 to 31 December 1837.

104. When one has well established, as a logical thesis, that all our knowledge must be founded on observation, that we must proceed sometimes from facts to principles, and sometimes from principles to facts, and a few other similar aphorisms, one knows method much less clearly than one who has studied, in a somewhat profound manner, a single positive science, even without philosophical intention. It is for having misunderstood this essential fact, that our psychologists are led to take their reveries for science, believing they have understood the positive method for having read the precepts of Bacon or the *Discourse* of Descartes. (ibid., 1:35)

105. Following in this interpretation Lucien Lévy-Bruhl, *La philosophie d'Auguste Comte*, 4th ed. (Paris: Alcan, 1921), 349.

106. Comte, *Cours de philosophie positive*, 1:38–39.

107. Simon, "Two Cultures," 46.

108. Biographical details from Henri Peyre, *Renan* (Paris: Presses Universitaires de France, 1969), 10–18; Charlton, *Positivist Thought in France*, 86–126.

109. Renan's notes on Cousin's *Du vrai, du beau, et du bien*, which he read during this period, have been published as "Remarques sur le Cours de 1818 de V. Cousin," in *Etudes philosophiques: De l'Ecosse à V. Cousin*, ed. Jean Pommier (Paris: Editions A.-G. Nizet, 1972). Renan did not pursue a career in academic philosophy, choosing to

concentrate on philology instead. He eventually reached the Collège de France in 1863, but Catholics were so scandalized by his inaugural lecture, in which he called Jesus "an incomparable man," that they prevailed upon the Emperor to suspend Renan's course. He did not get his chair back until after the Empire fell. See Peyre, *Renan,* 15.

110. See Ernest Renan, "La métaphysique et son avenir," in *Oeuvres complètes,* ed. Henriette Psichari (Paris: Calmann-Lévy, n.d.), 1:680–714.

111. Renan, "M. Cousin," in *Essais de morale et de critique* (Paris: Calmann-Lévy, n.d.), 70.

112. Canivez, *Jules Lagneau,* 1:185–90.

113. This put him outside the reach of academic philosophy (Cousin had retired by then), but it did not put him beyond the criticism of Catholics and other conservatives. His appointment "scandalized conservative academicians" (Leo Weinstein, *Hippolyte Taine* [New York: Twayne Publishers, 1972], 20). See also Charlton, *Positivist Thought in France,* 127–57.

114. Taine, *Les philosophes classiques,* 82.

CHAPTER 2. ECLECTIC BUDDHIST: THÉODULE RIBOT

1. Alfred Binet noted in 1890 that Ribot "is not a physician or an observer; the pathological facts that he uses are always second hand . . ." (introduction to *On Double Consciousness,* reproduced in *Centenaire de Th. Ribot* [Paris: Agen, 1939], 111). On the centenary of Ribot's birth, a number of prominent psychologists reminisced about his impact upon the development of the discipline in France. All praised him highly, yet several admitted that his empirical studies left much to be desired. G. Poyer thought that Ribot remained "literary" and "synthesizing," "too attached to his books" (*Centenaire de Th. Ribot* , 129). Henri Piéron, who succeeded Alfred Binet at the Laboratoire de Psychologie Physiologique, claimed that Ribot left the direction of the laboratory to Beaunis because he did not want to become an experimenter (ibid., 185). Paul Fraisse, a prominent midcentury psychologist and student of Piéron, had to admit that "Ribot never experimented, no more in the manner of Wundt than in that of Galton" (Paul Fraisse, Jean Piaget, and Maurice Reuchlin, *Traité de psychologie expérimentale,* vol. 1, *Histoire et méthode* [Paris: Presses Universitaires de France, 1970], 33). This statement is incorrect, since one of Ribot's only empirical studies was a questionnaire to find out what people thought about when words designating general ideas were presented to them. Such questionnaires were pioneered by Galton. Nevertheless, it reflects the ambiguous position of Ribot in the self-understanding of experimental psychology. Another prominent twentieth-century French psychologist, Maurice Reuchlin, came to the same conclusion. See his "Historical Background for National Trends in Psychology: France," *Journal of the History of the Behavioral Sciences* 1 (1965): 115.

2. Joseph Ben-David and Randall Collins, "Social Factors in the Origins of a New Science: The Case of Psychology," *American Sociological Review* 31 (1966): 457.

3. "Discours de M. Georges Dumas," in *Centenaire,* 38–39.

4. Jacqueline Thirard, "La fondation de la 'Revue philosophique,'" *Revue philosophique* (1976): 412.

5. Information on Ribot's early life is derived from François Picavet, "M. Théodule Ribot," *Revue bleue* (1894): 590–95; the *Grande encyclopédie;* S. Krauss, *Théodule Ribots Psychologie* (Jena: Hermann Costenoble, 1905), 5–9; and Ribot's personnel file with the Ministry of Public Instruction, AN F¹⁷ 21608.

6. Although he disagreed strongly with the content of Ribot's dissertation (see the next section), Caro's review for the Académie des Sciences Morales et Politiques is very respectful of Ribot's talent and learning (Elme Caro, review of *L'hérédité psycho-*

logique, by Théodule Ribot, *Journal des Savants* [1874]: 51–66). Georges Dumas reports that Ribot "always respected Caro" ("Discours," 39.) The identities and tenures of most professors and students at the Ecole Normale during the nineteenth century can be obtained from *Le centenaire de l'Ecole Normale, 1795–1895* (Paris: Hachette, 1895). See Appendix A for sources of information about these individuals. Unless otherwise noted, information about students and professors can be assumed to come from these sources.

7. Théodule Ribot, "Philosophy in France," *Mind* 2 (1877): 372.

8. Excerpts from these letters have been published by Raymond Lenoir, to whom Espinas gave them, in the *Revue philosophique:* letters dated 1866 through 1875 are in *Revue philosophique* 147 (1957): 1–14; 1876 through 5 May 1877, *Revue philosophique* 152 (1962): 337–40; 28 September 1877 through 1878, *Revue philosophique* 154 (1964): 1879 through 19 January 1884, *Revue philosophique* 160 (1970): 165–73, 339–48; 16 March 1884 through 1893, *Revue philosophique* 165 (1975): 157–72. The correspondence ended in 1893, when Espinas moved to Paris. According to Philippe Besnard, former editor of the *Etudes durkheimiennes,* the originals were lost when Lenoir died without heirs (personal communication, November 1985). Letters will hereafter be referred to by the date Ribot placed on them. Readers may find the volume of the *Revue philosophique* in which they were published by matching the date of the letter with the references in this note.

9. Ribot to Espinas, 18 November 1866.

10. Ribot to Espinas, 22 November 1868.

11. Théodule Ribot, *English Psychology* (New York: D. Appleton, 1891), 34.

12. In the first edition, there were chapters on James Mill, John Stuart Mill, Herbert Spencer, Alexander Bain, George Lewes, Samuel Bailey, John Morell, and Robert Murphy. In later editions, Morell and Murphy were dropped, and David Hartley, the subject of Ribot's Latin dissertation, was added.

13. Ribot to Espinas, 17 June 1870. There are not many reviews of this work by French philosophers, partly because Ribot was a young unknown at the time of its publication, and partly because there were not many periodicals for academic philosophers. Ribot's work was presented to the conservative Académie des Sciences Morales et Politiques, the philosophical section of which was dominated by eclectic spiritualists. However, it was presented by one of the few members sympathetic to Ribot's approach, Etienne Vacherot. Vacherot praised the work highly, but this tells us little about how it was received by the eclectics. Etienne Vacherot, "Rapport verbal sur un ouvrage de M. Ribot intitulé *La psychologie anglaise contemporaine,*" *Séances de l'Académie des Sciences Morales et Politiques* 92 (1870): 467–72.

14. Théodule Ribot, *L'hérédité psychologique,* 2d ed. (Paris: G. Baillière, 1882), 1.

15. This theory was held by Prospère Lucas. See Ribot's chapter on "Exceptions to the Law of Heredity" (ibid., 226–61). Ribot considered the notion of a law that would predict unpredictability self-contradictory (ibid., 235).

16. He noted, however, a fundamental difference between Kant and Spencer: ". . . on Kant's hypothesis, it is the forms of the subject that fashion the object; on the associationist hypothesis, it is the object that fashions the subject: for the one, the world depends upon thought; for the others, thought depends on the world. . . . It is to transport mechanism into the intelligence itself . . ." (ibid., 295–96). In both cases, though, the forms of thought were innate to the individual. As Ribot had remarked earlier, however, spiritualists already implicitly admitted that the categories were inherited, since they held them to be innate a priori determinations of thought common to the entire human species and reproduced invariably in each individual (ibid., 159–60).

17. For example, "Historians ordinarily explain decadence by the state of morals, institutions, character, which is true in a sense; but these are reasons that are a little

vague; and there is . . . a more profound, ultimate cause, an organic cause that can act only through heredity and which they ignore completely" (ibid., 279–80).

18. Ibid., 379. Ribot borrowed explicitly from the eugenic ideas of Galton.

19. Ibid., 171. A scientific law implies a complete determination of the object, allowing the possibility of infallible prediction. Ribot admits that his laws have not reached this degree of precision.

20. Ribot to Espinas, 15 March 1873.

21. *Le Temps* (Paris), 8 December 1873.

22. In reality, what do all these facts, so laboriously gathered, represent in the face of the immense, inexhaustible reality that fills life and history. A few isolated, exceptional, extraordinary cases, that strike the imagination precisely because of their singularity. If heredity were the visible, incontestable law, would we notice, for example, the memory of the Porsons or the political faculty of the Medicis? One would notice, on the contrary, the cases that would be exceptions to the rule; it would be non-heredity that would be called to our attention. What would happen to these famous lists of M. Galton or M. Ribot, if that of the negative cases were drawn up? (Caro, review of *L'hérédité psycho-logique*, 58–59).

See also the summary of this presentation in Elme Caro, review of *L'hérédité, étude psychologique*, by Th. Ribot, *Séances de l'Académie des Sciences Morales et Politiques* 101 (1874, part 1): 536–38.

23. Jean Nourrisson et al., Observations on *L'hérédité, étude psychologique*, by Th. Ribot, *Séances de l'Académie des Sciences Morales et Politiques* 101 (1874, part 1): 538–39.

24. Ribot to Espinas, 9 December 1873.

25. Ribot had almost finished his translation of the first edition, when, to his dismay, Spencer came out with the second, completely revised and greatly expanded.

26. Théodule Ribot, *La philosophie de Schopenhauer* (Paris: G. Baillière, 1874), 173–74.

27. Ribot to Espinas, 20 April 1875.

28. At least two other journals of philosophy already existed in France. The *Philo-sophie positive*, edited by Emile Littré and begun in 1867, represented his notion of Comtean orthodoxy, the notion that accepted the early scientific philosophy but rejected the Religion of Humanity. The *Critique philosophique* of Charles Renouvier, the origins of which may also be traced to 1867, voiced his particular brand of neo-Kantianism (see chapter 4). The liberalization of the Second Empire during the latter half of the 1860s played no small role in the founding of these reviews. Their editors wanted to create pulpits from which they could address the public and the governing elites as freer political discussion and greater political participation were made possible. This educational impulse gained even more urgency after the fall of the Empire and the proclamation of the Third Republic. By contrast, the *Revue philosophique* had no such directly political motivation, although its openness implicitly endorsed the ideals of the new Republic.

29. Théodule Ribot, preface to *Revue philosophique* 1 (1876): 1.

30. "The *Revue* will only exclude articles outside the philosophical movement, that is, which being devoted to doctrines already known, rejuvenated only by a talent for literary exposition, would have nothing to teach the readers" (ibid., 2).

31. Jacqueline Thirard, citing Paul Janet's failed attempts to start such a publication, suggests that official spiritualism did not have enough energy to sustain a journal of its own ("La fondation," 412). There is an element of truth in this, but it should be remembered that if academic eclectics did not have their own journal, they had exclusive access to several others, such as the *Journal des savants,* the official publication of the Académie des Sciences Morales et Politiques, and the *Revue des deux mondes,* a general

review for the educated elite. These publications generally frowned upon radical new doctrines, but they welcomed articles by familiar academics on the philosophical topics that were, for the reasons given in chapter 1, a part of French culture. The existence of these outlets seems to have satisfied academic philosophers and made the need for a professional journal less pressing. If anything, positivists were more dependent on the *Revue philosophique,* since they still did not have access to many of the most prestigious journals. Nevertheless, between these two sources, spiritualists and positivists, Ribot crystallized an enormous potential demand.

32. Paul Janet praised the openness of the *Revue,* its spirit of "appeasement and conciliation" (*Le Temps,* 2 March 1876). Such praise was not without irony, since these were code words for eclecticism. See also the remarks of the spiritualist Emile Beaussire, quoted in Picavet, "M. Théodule Ribot," 594; and the obituary in the *Revue de métaphysique et de morale* 24 (1917): 131, a review that was founded in explicit opposition to the *Revue philosophique* (see chapter 4).

33. Meletti Bertolini, *Il pensiero e la memoria,* 442–43.

34. Janet's remarks are revealing in this regard (see previous note). *La philosophie positive* also accused the new review of eclecticism, to which Ribot responded that "eclecticism means choice; well, the *Revue* does not choose" (*Revue philosophique* [1876, part 1]: 632).

35. Thirard, "La fondation," 402.

36. Thirard's analysis of articles from 1876 to 1890 shows that over a third were related to psychology, far more than any other category (ibid., 405).

37. Ribot, preface, 2–3.

38. Ribot, "Philosophy in France," 367. *Mind* in many ways paralleled the *Revue philosophique:* both were general philosophical journals begun in 1876 by advocates of a more scientific psychology (*Mind* was begun by Alexander Bain).

39. Several chapters had previously appeared in the *Revue philosophique* and the *Revue scientifique,* beginning in 1875. Ribot was often accused by the spiritualists of having "invented Wundt."

40. Théodule Ribot, *German Psychology of Today,* 2d ed., trans. James Mark Baldwin (New York: Scribner, 1886), 1–2, 4.

41. It is significant that in discussing the logic of experimentation, Ribot invoked John Stuart Mill's four methods of induction (ibid., 10–11). Surprisingly, Ribot did not mention this aspect of Mill's thought in his discussion of Mill in *English Psychology.* This indicates the extent to which his contact with German experimental psychology refocussed his attention on the conditions of proof in science. No longer would Ribot call "the collecting an innumerable multitude of facts around some fundamental principles" a "truly experimental method."

42. Ibid., 63. Ribot's uneasiness with this concept would lead him to disapprove of Espinas's "conscience collective" (see chapter 3).

43. T.-V. Charpentier, review of *La psychologie allemande contemporaine,* by Théodule Ribot, *Revue philosophique* 9 (1880): 350–57. It was also, for reasons I have been unable to discover, the only work of Ribot to be reviewed in his journal during his lifetime. This dearth of reviews could not have been an attempt to avoid the appearance of self-aggrandizement; he freely published his own work in it, and he published articles that defended or attacked his work incidentally.

44. Ribot, *German Psychology,* 45.

45. Cf. the general review by Emile Beaussire, "La personnalité humaine d'après les théories récentes," *Revue des deux mondes* 3d series, 55 (1883): 316–51. He also held that the "psychology without soul" had a legitimate place, "even in the philosophical sciences," but nonetheless it inevitably implied a metaphysical position (ibid., 334).

46. The classic statement of this proposition is in Claude Bernard's *Introduction to*

the Study of Experimental Medicine, trans. Henry Copley Greene (New York: Dover Publications, 1957), 10. On the implications of this doctrine for the history of biology, see Georges Canguilhem, *Le normal et la pathologique* (Paris: Presses Universitaires de France, 1966).

47. On French psychiatry, see Jan E. Goldstein, *Console and Classify: The French Psychiatric Profession in the Nineteenth Century* (Cambridge: Cambridge University Press, 1987).

48. E.g., Georges Dwelshauvers, *La psychologie française contemporaine* (Paris: Alcan, 1920), 110; Fraisse, *Traité,* 1:33–34.

49. Reuchlin, "National Trends in Psychology," 117. He is wrong, however, when he claims that Ribot "hardly looked at physiology" (ibid., 116). It is true that Ribot did no original research in physiology, but he certainly participated in laboratory work as a student.

50. Théodule Ribot, *Diseases of Memory,* trans. William H. Smith (New York: Appleton, 1882), 69–70.

51. Ribot, *The Diseases of the Will,* 8th ed., trans. Merwin-Marie Snell (Chicago: Open Court, 1896), 115.

52. Ribot, *Diseases of Memory,* 9. "Contemporary authors, such as Huxley, Clifford, and Maudsley, in maintaining that consciousness is only an adjunct of certain nervous processes, as incapable of reacting upon them as is a shadow upon the steps of the traveler whom it accompanies, have opened the way for a new theory which we shall attempt to formulate here" (ibid., 11). The individuals he named were associated with the doctrine of epiphenomenalism in its most extreme form, which held that humans were "conscious automata."

53. Ibid., 10–11.

54. Ribot, *The Diseases of the Will,* 2.

55. "Consciousness in itself and by itself is a new factor. . . . [A state of consciousness] marks a sequence, and it is capable of being recommenced, or modified, or inhibited. Automatic acts unaccompanied by consciousness do not admit of anything like this" (Théodule Ribot, *The Diseases of Personality,* trans. J. Fitzgerald, in The Humboldt Library of Popular Science Literature, vol. 9 [New York: Humboldt, 1887], 5–6). In this way Ribot distanced himself from the "sincerest partisans" of epiphenomenalism (ibid.).

56. Ribot, *Diseases of Memory,* 136.

57. Ribot, *The Diseases of the Will,* 114.

58. Ribot, *The Diseases of Personality,* 51–52.

59. Ribot, *The Diseases of the Will,* 122.

60. The reviews of Ribot in the *Revue scientifique* are uniformly positive. See Charles Richet, review of *La psychologie allemande contemporaine, Revue scientifique* (1879, part 1): 970–74; idem, review of *Les maladies de la volonté, Revue scientifique* (1883, part 2): 687–88; idem, review of *Les maladies de la personnalité, Revue scientifique* (1885, part 1): 502–4. Richet was an eminent physiologist with an interest in psychology. He published often in the *Revue philosophique.* He was also editor of the *Revue scientifique.* The only review with a hint of scientific condescension is by a doctor who thought Ribot's work very good—for a philosopher (J.-M. Guardia, review of *Les maladies de la mémoire, Revue scientifique* [1881, part 1]: 738–47).

61. The reviews of the philosopher Charles Lévêque, professor at the Sorbonne and member of the Académie des Sciences Morale et Politiques, are instructive in this regard. He always praised Ribot's talent and scientific method before launching into extended analyses that attempted to separate the factual from the hypothetical, the acceptable from the controversial, and he never failed to grant some part of Ribot's thesis as a valid contribution to philosophy. See his reviews of *Les maladies de la mémoire, Journal*

des savants (1881): 680–702, and (1882): 42–55, 91–106, 204–18; and *Les maladies de la volonté, Journal des savants* (1883): 481–84.

62. Ribot to Espinas, 25 February 1880.

63. Ribot to Espinas, 5 December 1881.

64. On Soury, see the article in *La grande encyclopédie.*

65. Ribot to Espinas, 5 December 1881 and 12 December 1881.

66. Ribot to Espinas, 26 November 1883.

67. On Janet, see chapter 1.

68. See chapter 4.

69. Ribot reported that Janet raged at him about his positivism and was jealous because Ribot rather than Janet had started the *Revue philosophique* (Ribot to Espinas, 17 March 1879, 17 March 1880, 24 March 1880).

70. AN AJ[16] 4747. A *cours complémentaire* was a course on a topic considered slightly outside the mainstream of the faculty curriculum; hence, it was "complementary" to the instruction of the chaired professors. The topics of such courses had to be approved by the faculty, and they were subject to renewal each year. This was a common way of introducing new subjects into the curriculum, if the subject was considered novel or the funds were not available to fund a chair. Such courses were taught by *chargés de cours,* individuals hired to teach only those courses. These individuals could not be given tenure; they were subject to renewal each year, depending upon an evaluation of the suitability of their topic, the adequacy of their teaching, or the availability of funds (Louis Liard, "Faculté," *La grande encyclopédie,* n.d.). They were often retained for many years and were effectively part of the faculty, but their positions remained tenuous in the extreme and their remuneration well below that of the *professeurs titularisés.*

71. A *conférence* was a course that duplicated to some extent the courses of the chaired professors but offered more contact between students and professor as well as practical exercises designed to prepare students for the state examinations and pedagogical instruction to help them develop their teaching skills. *Conférences* were begun in 1877 as part of the reform of higher education. They were taught by *maîtres de conférence,* who like *chargés de cours* were appointed annually (Liard, "Faculté," and Albert Guigue, *La Faculté des Lettres de l'Université de Paris* [Paris: Alcan, 1935], 32–33).

72. AN AJ[16] 4747.

73. Boutroux would go onto a brilliant university career as one of the leading philosophers of his generation.

74. Janet, Caro, and Waddington held the three chairs of philosophy at the Faculty of Letters at the time. To these three one might add Carrau, who taught a *conférence* in the history of philosophy, and Henri Joly, who taught a *conférence* in philosophy. All were eclectic spiritualists.

75. AN AJ[16] 4747.

76. On Liard, see chapter 4.

77. Ribot to Espinas, 14 July 1885.

78. "They're giving me 3,000 ff. . . . I had hoped to have at least what Soury got (4,000 ff)" (Ribot to Espinas, 11 August 1885).

79. Théodule Ribot, "Leçon d'ouverture: La psychologie nouvelle," *Revue scientifique* (1885, part 2): 781.

80. See his letters to Espinas for the years 1885–88, and Dumas, "Discours," in *Centenaire,* 39–40.

81. Ribot to Espinas, 17 November 1887. Ribot later told his student Georges Dumas that Renan had staked his reputation on getting him into the Collège de France, telling him that "I will resign as administrator if you are not elected" ("Discours de M. Georges Dumas," in *Centenaire,* 44).

82. "College de France: Documents relatifs à la création des chaires," AN F[17] 13556.

Ribot wrote to Espinas that Lévêque had campaigned "violently" against him and for Joly (Ribot to Espinas, 9 December 1887).

83. Of 29 voting, 19 voted for the transformation, 9 against, and 1 abstained (AN F¹⁷ 13556). The Collège has 40 members.

84. Three others, the eclectic Victor Egger, one M. Alaux, and Ribot's friend Espinas, asked to be considered for second place only—*en seconde ligne*. The purpose of doing this was to make oneself better known to the academic elite and thereby improve one's future chances in the event of a vacancy. In this particular case, Renan also wanted a positivist like Espinas in the second spot to remove the possibility that the minister could choose a spiritualist if Joly were named second (Ribot to Espinas, 9 December 1887).

85. Ernest Renan, "Rapport sur les titres des candidats présentés par l'assemblée de MM. les Professeurs du Collège de France à la chaire de Psychologie Expérimentale et comparée," in AN F¹⁷ 13556. According to Renan's report, Lévêque praised Ribot highly and considered him fully qualified to occupy a chair at the Collège. This suggests that Ribot's depiction of a "violent" campaign against him may have been somewhat exaggerated.

86. Franck presented Joly to the Académie as the candidate preferred by the philosophy section, which consisted of Franck, Lévêque, Jules Barthélemy-Saint-Hilaire, Janet, Nourrisson, Ravaisson, and Vacherot.

87. Jules Simon to the Minister of Public Instruction, 11 February 1888, in AN F¹⁷ 13556. See also the minutes of the Académie, published in *Séances et travaux de l'Académie des Sciences Morales et Politiques* 29, no. 1 (1888): 651–53.

88. Ribot to Espinas, 17 November 1887 and 17 February 1888

89. Published as Théodule Ribot, "La psychologie contemporaine" in the *Revue scientifique* (1888, part 1): 449–55.

90. Paul Janet, "Une chaire de psychologie expérimentale et comparée au Collège de France," *Revue des deux mondes* 3d ser., 86 (1888): 518–49. More on Janet's support of human sciences in chapter 4.

91. According to Ribot's tally, 2 of 8 philosophers voted for him, whereas 3 of 7 in the "Morale" section, 3 of 8 in "Economie politique," and 3 of 8 in the History section voted for him (Ribot to Espinas, 17 February 1888). Only the section on "Legislation, public law, and jurisprudence" gave him less support than philosophy. He received no votes from them at all, perhaps because of the vehemence of the opposition of one of its members, Arthur Desjardins (ibid.).

92. Séance du 2 juin 1888, in AN AJ¹⁶ 4747.

93. Ribot to Espinas, 16 March 1884.

94. Henri Piéron, "Le laboratoire de psychologie à la Sorbonne," in *Centenaire*, 185–98.

95. On Beaunis, see "Chronique," *Année psychologique* 22 (1920–1921): 598.

Chapter 3. Incurable Metaphysician: Alfred Espinas

1. Ribot to Espinas, 15 July 1871.

2. Paul Janet, "Le mouvement philosophique," *Le Temps* (Paris), 16 August 1878.

3. Georges Davy, "L'oeuvre d'Espinas," *Revue philosophique* (1923, part 2): 215.

4. Roger Geiger, "The Development of French Sociology, 1871–1905" (Ph.D. diss., University of Michigan, 1972), 84, 91.

5. See chapter 1.

6. André Lalande, *Notice sur la vie et les travaux de M. Alfred Espinas* (Paris: Firmin-Didot, 1925), 8.

7. The Ecole Normale had its students examined, not by their professors, but by outsiders, most commonly from the Sorbonne.

8. See the evaluations by Janet in Espinas's personnel file, AN F[17] 22193.

9. "M. Espinas frequents the café a bit too much, and does not always seems sufficiently measured in his words," Report for the year 1868–69. See also the letter of the Recteur of the Academy of Dijon, Lycée Chaumont, 24 June 1871, in which Espinas's café life and political opinions lead the Recteur to recommend that his young professor be transferred to another lycée. Both in AN F[17] 22193.

10. Compte Rendu of 3 April 1870, in AN F[17] 22193.

11. Alfred Espinas, "Etre ou ne pas être, ou du postulat de la sociologie," *Revue philosophique* 51 (1901): 449.

12. Ribot to Espinas, 17 June 1870.

13. Alfred Espinas, *La philosophie expérimentale en Italie* (Paris: G. Baillière, 1880), 19.

14. Compare the following statement, which also serves as the epigraph for this chapter: "To prove that one should not philosophize, one has to philosophize; and some metaphysics is necessary even to those who try to prove the impotence of metaphysics. It is therefore best to take questions as they come, without worrying too much about the label they wear, except to bring them back as soon as possible . . . to definite terms, to experimental research" (Alfred Espinas, "La nature et l'immatérialisme," *Revue scientifique* [1879]: 1222).

15. "I see that you are an incurable metaphysician, and that's all there is to it. You obstinately refuse to understand that metaphysics will never be able to provide anything but possibilities, because it is not verifiable, will never do anything, from the scientific point of view, but spoil any science into which it enters" (Ribot to Espinas, 15 July 1871).

16. See the reports of the inspector general and the *recteur* in AN F[17] 22193.

17. Espinas to the director of Higher Education, 28 September 1876, in AN F[17] 22193.

18. Alfred Espinas, *Des sociétés animales*, 2d ed. (Paris: G. Baillière, 1878), 67.

19. ". . . our doctrine is not made to surprise either the spiritualist psychologists, who tend everywhere to explain the inferior by the superior and nature by its ends, nor the physiologists, who do not deny, as far as I know, the reality of psychic phenomena, even though they strive above all to discover their conditions" (ibid., 407–8). All of the early evolutionists believed in "mind," and if they sought to give a naturalistic explanation of it, they did not reduce it to matter. See Richards, *Darwin,* 6.

20. Espinas, *Des sociétés animales,* 527. The contrast between social and material is Espinas's.

21. On Darwin, see the chapters on the mental powers of animals in the *The Descent of Man* (Princeton: Princeton University Press, 1981), 34–106.

22. Alfred Binet, "La vie psychique des micro-organismes," in *Etudes de psychologie expérimentale* (Paris: Octave Doin, 1888), 87–237.

23. Henri Joly, *La psychologie comparée* (Paris: Hachette, 1878).

24. Espinas, *Des sociétés animales,* 531.

25. Janet to Espinas, 15 January 1877, cited in Lalande, *Notice,* 13.

26. The report of the dissertation committee to the dean of the Faculty of Letters reiterates this praise: "It is a remarkable work, by the novelty of the subject and by the extent of the zoological knowledge it reveals. Until now there had not yet been a complete study of a scientific nature on animal societies.

"The zoological science of the author is as exact as it is rich and interesting. The facts are classified methodically, in a clear order" (A. Mézières, report to the dean, 9 June 1877, in AN F[17] 22193).

27. Alfred Espinas, *De civitate apud Platonem, qui fiat una* (Paris: G. Baillière, 1877).

28. "You have not entirely understood my objections; . . . they have to do much less with the content and with your doctrines than with the impudence of putting forward and for all the world to hear two names as compromising as those of MM. Auguste Comte and Spencer, to which, rightly or wrongly, are attached many passions good or bad in both directions" (Janet to Espinas, 19 January 1877, cited in Lalande, *Notice*, 15).

29. Ibid.

30. Mézières, report.

31. Janet, "Le mouvement philosophique."

32. For more on this reorientation, see chapter 4.

33. Théodule Ribot, review of *Des sociétés animales*, by Alfred Espinas, *Revue philosophique* 4 (1877): 327–34.

34. Ribot to Espinas, 25 February 1878.

35. On Fouillée, see Logue, *From Philosophy to Sociology*, 129–50.

36. Alfred Fouillée, *La science sociale contemporaine*, 2d ed. (Paris: Hachette, 1885), 246, note 2.

37. J. Collier, review of *Des sociétés animales: Étude de psychologie comparée*, by Alfred Espinas, *Mind* 4 (1879): 105.

38. Wilhelm Wundt, "Ueber den gegenwärtigen Zustand der Thierpsychologie," *Vierteljahrschrift für wissenschaftliche Philosophie* 2 (1978): 139.

39. Ibid., 148.

40. Edmond Perrier, *Les colonies animales et la formation des organismes*, 2d ed. (Paris: Masson, 1898), 781. Ribot wrote Espinas in 1877 that he had recommended *Des sociétés animales* to Pérrier (Ribot to Espinas, 27 April 1877).

41. George Romanes, *Animal Intelligence* (1882; reprint, Westmead, England: Gregg International Publishers, 1970), 130.

42. Alfred Espinas, "Les études sociologiques en France," *Revue philosophique* 13 (1882): 594–95.

43. See Janet's letter of recommendation and Lachelier's inspection report, both highly favorable, in AN F[17] 22193.

44. See chapter 4.

45. Emile Boutroux, "L'agrégation de philosophie," *Revue internationale de l'enseignement* (1883, part 2): 865–78.

46. Alfred Espinas, "L'agrégation de philosophie," *Revue internationale de l'enseignement* 7 (1884, part 1): 589.

47. Alfred Espinas, *Cours de philosophie: Leçon d'ouverture* (Paris: Librairie Léopold Cerf, 1885), 20.

48. The new religion even featured its own conversion experiences.

> Thus one saw little by little adult minds, in full possession of themselves, in the state of moral deprivation in which they found themselves, embrace this doctrine, without recoiling before its ultimate conclusions. I remember with what stupefaction several of us were seized, when we learned from irrefutable manifestations (like baptisms) in our milieu, beside us, among our comrades, the apostles of the new faith had found believers, when we saw that they dreamed of conquering bit by bit all the minds up to and including the popular masses and that from the high regions of thought a religion was descending on the ignorant and the imaginative. (ibid., 16–17)

On neo-Kantianism see chapter 4.

49. Ribot to Espinas, 17 November and 9 December 1887.

50. Ribot to Espinas, 25 January 1888.

51. Ribot to Espinas, 24 March 1888.

52. To the almost ghoulish point of reporting potentially fatal illnesses among his philosophical colleagues (e.g., Ribot to Espinas, 21 May 1889).

53. Actes de la Faculté des Lettres de Paris, séance du 4 novembre 1893, in AN AJ[16] 4748. George Weisz has well documented the creation of this course in sociology in his article "The Republican Ideology and the Social Sciences; The Durkheimians and the History of Social Economy at the Sorbonne," in *The Sociological Domain*, 90–119. While agreeing with his interpretation of the process by which the position was created, I am highlighting the role of philosophers.

54. Actes de la Faculté des Lettres de Paris, séance du 22 novembre 1893, in AN AJ[16] 4748. Waddington had also opposed the creation of Ribot's *cours complémentaire* in 1885 partly on the grounds that it was too "new" (chapter 2). As was noted there, this reason is consistent with the view of higher education as a transmitter of established knowledge, not as a creator of new knowledge. It may also, of course, have been merely a pretext for the old spiritualist's opposition to a scientific approach he considered too materialistic, but it is significant that whereas in 1885 he had frankly made this objection, in 1893 there is no record of his having given this as a reason for his hesitation.

55. Paul Janet, "Rapport de M. Janet sur la création d'un cours de sociologie à la faculté des lettres," in Conseil général des facultés: Procès-verbaux des séances, 1894, AN AJ[16] 2653.

56. Weisz, "The Republican Ideology and the Social Sciences," 104.

57. Ibid.

58. Espinas to Octave Hamelin, 22 July 1894, in Correspondance d'Hamelin (Paris: Bibliothèque Victor Cousin, Mss. 356).

59. Ibid., and Espinas to Hamelin, 11 January 1896 and 6 April 1897.

60. Davy, "L'oeuvre d'Espinas," 263–68.

61. Espinas, "Les études sociologiques en France," 514.

62. Alfred Espinas, *Cours de pédagogie, leçon d'ouverture* (Paris: Alcan, 1884), 19.

63. Espinas, "Les études sociologiques en France," 364.

64. Ibid., 567.

65. Espinas, *Cours de pédagogie*, 5.

66. Alfred Espinas, "Les origines de la technologie," *Revue philosophique* 30 (1890): 116.

67. Ibid.

68. Ibid.

69. Espinas, *Cours de pédagogie*, 25–26.

70. "Give me the whole of science, if there is not in me an emotional and impulsive spring prepared without my knowledge to correspond to it, I will not draw from it enough to make me lift my little finger" (Alfred Espinas, "Leçon d'ouverture d'un cours d'histoire de l'économie sociale," *Revue internationale de sociologie* 2 [1894]: 332).

71. ... the most difficult thing for each of us [after the defeat of 1870–1871] was to bow ourselves before the rule of national morality that asks citizens to suffer without bitterness on the part of other citizens differences of philosophical or religious opinion. The French Université understands better and better that the various philosophical doctrines are social forces that can bring honor to a country in the eyes of foreigners and which, even within the country . . . can be called to a useful role. Such was always our point of view, and, even though foreign to the Kantian philosophy, we saw with satisfaction a moral teaching inspired above all by this philosophy take possession of our primary schools—why? Because those manuals drafted in the name of the categorical imperative advocated on each page love of the French motherland and service in the public interest. (ibid., 342–43)

Such was decidedly *not* always Espinas's view, and his acceptance of this view involved a tremendous change.

72. Ribot to Espinas, 13 February 1893.

73. Durkheim to Célestin Bouglé, 24 November 1897, in *Textes,* ed. Victor Karady (Paris: Editions de minuit, 1975), 2:414.

74. Paul Fauconnet, review of *Les origines de la technologie,* by Alfred Espinas, *Revue philosophique* 47 (1899): 439.

75. Review of *Les origines de la technologie,* by Alfred Espinas, *Revue de Métaphysique et de Morale* 6 (1898, May supplement): 2.

76. See, for example, Espinas's preface to the second edition of the *Histoire des doctrines économiques,* "Economie politique, économie sociale et sociologie," *Revue philosophique* 99 (1925): 161. This edition never appeared, but the preface was published posthumously.

77. Weisz, "The Republican Ideology and the Social Sciences," 105.

Chapter 4. Becoming Philosophy: The Transformation of the Academic Tradition

1. On the "crisis" of philosophy, see Fabiani, "Enjeux et usages."

2. The importance of education in Republican ideology has been well described by Zeldin, *Intellect and Pride,* 139.

3. Digeon, *La crise allemande de la pensée française,* 364–83.

4. Karady, "Les professeurs de la République," 91–93.

5. Ibid. On the prestige of Third Republic philosophy, see Fabiani, "Les programmes," 1. Examples of the genre praising philosophy at this period include Canivez, *Jules Lagneau;* and Alain, *Souvenirs concernant Jules Lagneau,* 4th ed. (Paris: NRF, 1925). Even those who vilified the institution paid it the compliment of attributing to it great power. See, for example, Albert Thibaudet, *La république des professeurs* (Paris: B. Grasset, 1927); Julien Benda, *La trahison des clercs* (Paris: B. Grasset, 1927); and Paul Nizan, *Les chiens de garde* (1929; reprint, Paris: F. Maspero, 1969).

6. Falcucci, *L'humanisme,* 371.

7. Michel Bréal, *Quelques mots sur l'Instruction publique en France* (Paris: Hachette, 1872).

8. Elie Rabier, "Rapport," cited in the *Revue internationale de l'enseignement* 12 (1886): 359.

9. See Prost, *Histoire de l'enseignement,* 250.

10. Alfred Fouillée, *Les études classiques et la démocratie* (Paris: A. Colin, 1898).

11. Rabier, who denounced the idea of "classical French education," thought that this appellation merely corrupted the idea of special education, which was precisely to be practical. In his later career as director of secondary education, he was an advocate of special and modern education, put in their proper place.

12. Prost, *Histoire de l'enseignement,* 247.

13. Fabiani has described this transformation in "Les programmes."

14. The syllabus of 1880 is reproduced in Henri Marion, "Le nouveau programme de philosophie," *Revue philosophique* 10 (1880): 426–28. It appears translated in Appendix B.

15. See Canivez, *Jules Lagneau,* 2:362–67. Rabier in fact referred to the Taine incident in his correspondence with Lagneau. See also Alain, *Souvenirs concernant Jules Lagneau,* 37–41.

16. Falcucci's thesis chronicles the gradual acceptance of science into the notion of *culture générale.* Significantly, he shows that this was only possible because science eventually came to be seen as essentially "humanistic," in the sense that it was disinterested rather than practical.

17. Marion, "Le nouveau programme," 415. See also Goldstein, "The Advent of Psychological Modernism," 202–3.

18. Marion, "Le nouveau programme," 422.

19. On the importance of moral education in the Third Republic, see Phyllis Stock-Morton, *Moral Education for a Secular Society: The Development of Morale Laïque in Nineteenth-Century France* (Albany: SUNY Press, 1988). Cf. Prost, *Histoire de l'enseignement*, 195–96.

20. Ibid., 196–97.

21. Marion, "Le nouveau programme," 415.

22. See Janet, *Victor Cousin*, 325–26.

23. Canivez, *Jules Lagneau*, 277–88.

24. Fernand Vandérem, ed., *Pour et contre l'enseignement philosophique* (Paris: Alcan, 1894). There were substantial responses by such academic philosophers as Paul Janet and Emile Boutroux, as well as affiliated sympathizers such as Alfred Fouillée and Henri Marion. On the controversy generated by this article, see Fabiani, "Enjeux et usages," 392.

25. See chapter 1.

26. Cousin, for example, always distinguished between secondary and higher education, despite the fact that his own philosophy was one of the factors that held the two together. He was an early advocate of the research university, but he could make no headway against the practical and professional orientation of the faculties.

27. See chapters 2 and 3.

28. The following section on the reform of higher education draws on George Weisz's excellent book, *The Emergence of Modern Universities in France, 1863–1914* (Princeton: Princeton University Press, 1983).

29. Karady, "Les professeurs de la République," 96.

30. Ibid., 95.

31. Christophe Charle, *Les professeurs de la faculté des lettres de Paris: Dictionnaire biographique*, 2 vols. (Paris: Editions du CNRS, 1985–86).

32. Karady has shown this very well in his article, "The Durkheimians in Academe," 78. See also George Weisz, "The Republican Ideology and the Social Sciences."

33. Chapter 3.

34. Chapters 2 and 3.

35. See chapter 1.

36. Janet, *Traité*, v.

37. Janet, *Principes de métaphysique et de psychologie*, 1:153–67.

38. Janet divided psychological laws into two types, mechanical and dynamical, and noted: "These two modes of explanation are so legitimate that, most of the time, so-called physiological explanations consist of transporting them purely and simply into the brain and the nerve cells, sometimes introducing a cerebral mechanism, sometimes a dynamism, very often the two mixed together, and which are only the objective and material translation of mental mechanism and dynamism" (ibid., 1:162).

39. Ribot to Espinas, 15 January 1880.

40. Janet, *Principes de métaphysique et de psychologie*, 1:40.

41. On Ravaisson, see chapter 1. On the new spiritualism, see the next section.

42. Rabier, *Leçons*, 1:392.

43. "The facts that give birth to the idea are not the sufficient premises from which the idea follows as a consequence, but only the base from which thought as it were takes flight in order to conceive it [the idea] in surpassing them [the facts]" (ibid., 2:230).

44. On Caro's critique of positivism, see Logue, *From Philosophy to Sociology*, 32–33. Spiritualist philosophers seized upon Claude Bernard's *Introduction to the Study of Experimental Medicine* (1865), which stressed the importance of ideas in science, as a rebuttal

to empiricist theories of science. See Henri Bergson, "Ce que la philosophie doit à Claude Bernard," in *Le Collège de France, 1530–1930* (Paris: Presses Universitaires de France, 1932), 237–42.

45. Rabier, *Leçons,* 2:155.

46. G. Fonsegrive, review of *Leçons de philosophie,* by Elie Rabier, *Revue philosophique* 23 (1887): 183. See also the review of the first volume by Victor Brochard, *Revue philosophique* 19 (1885): 565–73.

47. Rabier, *Leçons,* 1:67–68. Janet had left open the question of the status of unconscious phenomena.

48. See James Mark Baldwin, "Contemporary Philosophy in France," *New Princeton Review* 3 (1887): 142; and William James, *Principles of Psychology,* 2 vols. (New York: Henry Holt, 1890), 1:604.

49. See chapter 5.

50. Ribot to Espinas, 15 January 1880.

51. For this reason, I do not agree with Fabiani that structural changes alone can explain the appearance of diversity in university philosophy. The rhetoric of free inquiry was a powerful force in this transformation, wielded by opponents of eclecticism and acknowledged by the eclectics themselves. Otherwise it is impossible to explain the vilification of Cousin that became a ritual part of the discourse of academic philosophy during the early Third Republic.

52. Emile Boutroux cited Janet's lectures on Kant in 1867 as the beginning of renewed academic interest (Boutroux, "Paul Janet," 31–47). Boutroux and Jules Lachelier both taught Kant at the Ecole Normale (see the next section).

53. On Renouvier, see Gaston Milhaud, *La philosophie de Charles Renouvier* (Paris: J. Vrin, 1927); Octave Hamelin, *Le système de Renouvier* (Paris: J. Vrin, 1927); and Logue, *From Philosophy to Sociology,* 51–72.

54. Comments on the rise of neocriticism abound in French philosophy at this period. Ravaisson has been credited with discovering Renouvier by including him in his *Report* of 1868. Ribot discussed him in his article on "Philosophy in France" (1877), although even at this date he was forced to conclude that "in France his works have not been sufficiently read, and . . . they are far from obtaining the success they deserve" (Ribot, "Philosophy in France," 379). Alfred Fouillée on the other hand claimed in 1880 that "for several years there has been a neo-Kantian movement in France that is not without importance" ("Le neo-Kantisme en France," *Revue philosophique* 11 [1881]: 1). As we have seen (chapter 3), Alfred Espinas thought that converts to Kant were taking over the Université. By the 1890s, Janet could claim that critical philosophy "has reigned almost exclusively in philosophy for some years" (*Principes,* 1:v). Moreover, the sense of liberation from the determinism of positivism and the dogmatism of spiritualism is well captured in the remarks of William James, who claimed that Renouvier freed him from the oppressive burden of scientism (Gerald E. Myers, *William James: His Life and Thought* [New Haven: Yale University Press, 1986], 46–47). Espinas's characterization of the neo-Kantians as acolytes expresses the same sentiment, with a different valuation.

55. Milhaud, *La philosophie de Charles Renouvier,* 16.

56. Ibid., 21.

57. Charles Renouvier, *Essais de critique générale, premier essai: Traité de logique générale et de logique formelle,* 2 vols. (Paris: A. Colin, 1912), 1:9. First published in 1854.

58. Cf. Charles Renouvier, *Essais de critique générale, deuxième essai: Traité de psychologie rationnelle d'après les principes du criticisme,* 2 vols. (Paris: A. Colin, 1912), 2:148–50. First published in 1858.

59. Ibid., 2:171–72.

60. Ibid., 1:7–8.

61. Ibid., 1:133–38. Renouvier and his collaborator François Pillon (1830–1914) carried out a lifelong battle with associationist psychology in the pages of their weekly journal, *Critique philosophique*. See, for example, Pillon's review of *La psychologie anglaise contemporaine*, by Théodule Ribot, *Critique philosophique* (1872, part 1): 32. On Pillon, see *Discours prononcés aux obsèques par MM. Gout, Raoul Allier, et Lionel Dauriac* (Paris: Gout, Raoul Allier, and Lionel Dauriac, 1914).

62. Charles Renouvier, review of *La psychologie allemande contemporaine*, by Théodule Ribot, *Critique philosophique* (1879, part 2): 110.

63. The sciences are divided and establish their specific objects with the aid of distinctions, not of confusions. It is inadmissible that psychology, which has its particular, very well defined domain, should pretend to encompass the physiology of nerves. It is true that a superior science must lean on an inferior science, we would not contest this truth with M. Ribot; but where, how, and in what sense this support must be sought, that, in each case, is what must be known. Physics leans on mathematics; one will not say for that reason that physics studies a mathematical process with two faces. (ibid., 109)

64. See, for example, Renouvier's review of *Les maladies de la mémoire*, by Théodule Ribot, *Critique philosophique* (1881, part 2): 92–96.

65. François Pillon, "La formation des idées abstraites et générales," *Critique philosophique* (1885, part 1): 118–33, 178–215. This article began as a review of Durand Desormeaux's *Etudes philosophiques*, published posthumously by Espinas. Durand Desormeaux held among other things that association by resemblance could be reduced to association by contiguity, which Pillon claimed was "to remove from the mind all spontaneous participation and action in the formation of abstract ideas" (Pillon, "La formation des idées abstraites," 121). He then cited other recent arguments of the same position, including Rabier. See also Rabier's response, "A propos de l'association par ressemblance," *Critique philosophique* (1885, part 1): 460–66.

66. François Pillon, "Réponse aux observations de M. Rabier sur l'association par ressemblance," *Critique philosophique* (1885, part 2): 60.

67. Renouvier, *Deuxième Essai*, 1:191.

68. Ibid., 1:305.

69. Charles Renouvier, *Introduction à la philosophie analytique de l'histoire: Les idées, les religions, les systèmes*, new ed. (Paris: Ernest Leroux, 1896), 556. First published in 1864.

70. Ibid., 107–27, 139–42.

71. "Examen des principes de psychologie de Herbert Spencer, V: Principes de la logique," *Critique philosophique* (1877, part 2): 185–87.

72. See the review of Spencer's *Principles of Sociology*, *Critique philosophique* (1879, part 2): 407–15; also "La préparation à la science sociale par la biologie, selon M. Herbert Spencer," *Critique philosophique* (1876, part 1): 235–40.

73. Renouvier, *Introduction à la philosophie analytique de l'histoire*, 550.

74. Charles Renouvier, *La science de la morale* (Paris: Alcan, 1869), 2–3. On Renouvier's moral theory, see Logue, *From Philosophy to Sociology*, 51–72.

75. Louis Liard, *Des définitions géométriques et des définitions empiriques*, new ed. (Paris: Alcan, 1888), 174–75. First published in 1873.

76. He was replaced at Bordeaux by Espinas. For biographical information on Liard, see Ernest Lavisse, "Louis Liard," *Revue internationale de l'enseignement* 72 (1918): 81–99.

77. Louis Liard, *La science positive et la métaphysique*, 2d ed. (Paris: Baillière, 1883), 474. First published in 1879.

78. See his chapters on "La doctrine de l'évolution," ibid., part 1, chapters 9–11.

79. Liard's argument was similar to the arguments developed by Catholic philoso-

phers of science around the turn of the century, notably Pierre Duhem. See Harry Paul, *The Edge of Contingency* (Gainesville: University Presses of Florida, 1979). As Paul and others have noted, however, these arguments were not necessarily tied to a religious point of view. Liard's formulation was perfectly secular.

80. Digeon, *La crise allemande de la pensée française*, 174.

81. Louis Liard, *L'enseignement supérieur en France, 1789–1889*, 2 vols. (Paris, 1888–94), 2:344–45.

82. Ibid., 2:342. See also idem, *Universités et facultés* (Paris: Armand Colin, 1890), 151–52.

83. Liard correlated the change of opinion from favoring professionally oriented faculties to research-oriented universities with a change in the view of science. "It happened that precisely at that moment [when the debate between professionally and research-oriented higher education was raging], practice, so long separated from theory, repudiated this divorce.

"Practice without science is empiricism, that is the brute fact, without the reason for the fact. Of this legacy from the past the insufficiency was now known, and they wanted no more of it. For some time already the alliance of theory and practice, of idea and fact, had been established and each day tended to become closer" (*L'enseignement supérieur en France*, 2:350). Liard cited the example of Claude Bernard, who made of medicine an experimental science rather than a purely empirical enterprise.

84. Liard, *Universités et facultés*, 153–56.

85. Ibid., 156.

86. Ibid., 157.

87. As we have seen (chapter 2), Liard provided Ribot his position at the Sorbonne and supported his candidacy for the Collège de France against considerable opposition. On Liard's support for Durkheim, see Clark, *Prophets and Patrons*, 163.

88. Paul Janet, "Une nouvelle phase de la philosophie spiritualiste," *Revue des deux mondes*, 2d ser., 108 (1873): 363–65.

89. On the revival of French spiritualism, see Benrubi, *La philosophie contemporaine en France*, 2:545; and Bréhier, *Histoire de la philosophie*, 3:870–71, 889–91. Lest one overestimate the demise of eclecticism, it is important to note that during the first twenty years of the Third Republic, 80 percent of spiritualists appointed to elite academic positions were still eclectic.

90. On the influence of Lachelier, see Canivez, *Jules Lagneau*, 1:255–61. Fabiani cites Lachelier as one source of the type of philosophy professor who thought much and wrote little ("Les programmes," 14). Janet was impressed by the severity of Lachelier's style, commenting that "those who have accused university philosophy of being a literary and superficial philosophy will no longer have anything to complain about" ("Une nouvelle phase," 375).

91. Jules Lachelier, *The Foundations of Induction*, in *The Philosophy of Jules Lachelier*, trans. Edward G. Ballard (The Hague: Martinus Nijhoff, 1960), 38–40.

92. Ibid., p. 54. By "materialist idealism" Lachelier means the attempt to characterize reality in terms of space, time, matter, movement, and efficient causes, ideas that make reality abstract and "ideal" at the same time that they reduce it to mechanical processes.

93. Ribot announced it to Espinas in the following, rather sarcastic, terms: "I will publish on May 1st the famous article . . . on which Father Lachelier has been working for four years. The candidates for the *Agrégation* are going to learn it by heart" (Ribot to Espinas, 6 March 1885).

94. Lachelier, "Psychology and Metaphysics," in *The Philosophy of Jules Lachelier*, 87.

95. I am departing from Janet's article to look at Boutroux rather than Fouillée. Fouillée left the university in 1875, after only three years at the Ecole Normale, and

his thought, which centered around what he called the *idée-force,* became more eclectic than neospiritualist. By contrast, Boutroux remained within the university throughout his career and developed the essential themes of neospiritualism.

96. Emile Boutroux, *De la contingence des lois de la nature,* 4th ed. (Paris: Alcan, 1902), 5. First published in 1874.

97. Ibid., 136.

98. Ibid., 145.

99. Emile Boutroux, *De l'idée de loi naturelle dans la science et la philosophie contemporaines,* new ed. (Paris: J. Vrin, 1949), 9–10.

100. Mary Jo Nye has shown the social and intellectual connection between Boutroux and Henri Poincaré. See her article, "The Boutroux Circle and Poincaré's Conventionalism," *Journal of the History of Ideas* 40 (1979): 107–20.

101. Boutroux, *De la contingence,* 121.

102. Boutroux, *De l'idée de loi naturelle,* 132–33.

103. On the founding of this journal, see the special centenary number of the *Revue de métaphysique et de morale* 98, nos. 1–2 (1993), especially Perrine Simon-Nahum, "Xavier Léon/Elie Halévy: Correspondance (1891–1898)," 3–58; and Dominique Merllié, "Les rapports entre la *Revue de métaphysique* et la *Revue philosophique:* Xavier Léon et Théodule Ribot; Xavier Léon et Lucien Lévy-Bruhl," 59–108.

104. Preface, *Revue de métaphysique et de morale* 1 (1893): 1.

105. Rabier cut the number of subscriptions by lycées to the *Revue philosophique* in half and used the money saved to buy subscriptions to the new *Revue de métaphysique et de morale* (Ribot to Espinas, 8 February 1893).

106. Merllié, "Les rapports," 69.

107. Paul, *The Edge of Contingency,* 9.

108. For example, Célestin Bouglé (1870–1940), a philosopher and collaborator with Durkheim, was classified as both positivist and neocriticist by different sources. Gabriel Séailles (1852–1917), who worked with Paul Janet, was considered by some an eclectic spiritualist, by others a neocriticist. Louis Rodier (1864–1913) was classified as neocriticist and neospiritualist. André Lalande (1867–1954) and Frédéric Rauh (1861–1909) were placed by different authors in all three of the major categories—positivist, neocriticist, and neospiritualist. This increased ambiguity is consistent with Fabiani's hypothesis about the transformation of the philosophical field (Fabiani, *Les philosophes de la République*).

CHAPTER 5. THE SYNTHESIS OF PHILOSOPHICAL AND MEDICAL PSYCHOLOGY: PIERRE JANET

1. A. R. G. Owen, *Hysteria, Hypnosis, and Healing: The Work of J.-M. Charcot* (London: Dennis Dobson, 1971), 62; Ellenberger, *Discovery,* 91.

2. The term *psychological automatism* can be traced at least as far back as the French psychiatrist Prosper Despine, who used it in 1868 (Ellenberger, *Discovery,* 359).

3. Ian Hacking has mentioned Pierre Janet as one of those who helped "invent" split personality in the nineteenth century (Ian Hacking, "The Invention of Split Personalities," in *Human Nature and Natural Knowledge: Essays Presented to Marjorie Grene on the Occasion of Her Seventy-Fifth Birthday,* eds. A. Donagan, A. N. Perovich Jr., and M. V. Wedin [Dordrecht: D. Reidel, 1986], 63–85). Without trying to determine the extent to which Janet imposed his categories upon his data, I am in fact assuming that his theory was not simply a transparent reflection of what he observed. Janet's theory may or may not be accurate or fruitful, but it was certainly the result of his conceptual as much as his empirical environment. Janet himself recognized that his contribution to

psychology lay as much in his interpretation as in his description of psychological phenomena.

4. Sjövall, *Psychology of Tension;* Ellenberger, *Discovery,* 401–2; Prévost, *La psycho-philosophie de Pierre Janet.*

5. Jan Goldstein recognizes and briefly discusses this relationship in "The Advent of Psychologial Modernism."

6. Fraisse, Piaget, and Reuchlin, *Traité de psychologie expérimentale,* 1:35.

7. See chapter 4. For biographical information on Pierre Janet, see Ellenberger, *Discovery,* 331–47.

8. Janet, "Autobiography," 1:123–24.

9. Charpentier taught at the Lycée Louis-le-Grand, where Janet and Durkheim both did their *khâgne.* Charpentier's review of Ribot's *Contemporary German Psychology* was examined in chapter 2.

10. Janet, "Autobiography," 1:123.

11. Rabier is cited several times, always approvingly, in *Psychological Automatism,* Boutroux only once, and Ollé-Laprune not at all. Despite Rabier's brief tenure at the Ecole Normale, Janet would remember his teaching many years later, in the obituary he wrote for Edmond Goblot (Pierre Janet and André Lalande, "Edmond Goblot," *Annuaire de l'Association amicale des anciens élèves de l'ENS* [1936], 36–41).

12. Ellenberger, *Discovery,* 335. This was, however, not a very advanced degree, and does not indicate a particularly intensive course of study.

13. Although Pierre Janet's merits were beyond question, his uncle had a hand in the selection of the *jury de l'agrégation* that evaluated him, according to Ribot. The elder Janet first nominated himself for the jury, "because of his nephew," and then recused himself to avoid the appearance of favoritism, "while reserving the right to name his replacement" (Ribot to Espinas, 19 May 1882). This kind of subtle influence could not rescue an incompetent, but it could smooth the path for a talented young philosopher like Pierre.

14. Ellenberger, *Discovery,* 337–38. Cf. Janet, "Autobiography," 1:124–25.

15. Dominique Barrucand, *Histoire de l'hypnose en France* (Paris: Presses Universitaires de France, 1967), 40–42.

16. Ellenberger, *Discovery,* 338. Pierre Janet, "Note sur quelques phénomènes de somnambulisme," *Revue philosophique* 21 (1886): 190–98.

17. One may speculate that Pierre's career may have been one of the motivating factors behind Paul's decision to join the society, which consisted mainly of physiologists. Ribot, at any rate, described Paul Janet as hesitating when originally asked to join (Ribot to Espinas, 6 March 1885).

18. Janet, "Note," 198.

19. Charles Richet, "La suggestion mentale et le calcul des probabilités," *Revue philosophique* 18 (1884): 609–71. This was one of only two works to which Janet referred in his first paper.

20. Later published in Alfred Binet and Charles Féré, *Le magnétisme animal* (Paris: Alcan, 1887).

21. Charles Richet, "Un fait de somnambulisme à distance," *Revue philosophique* 21 (1886): 199–200, and Henri Beaunis, "Un fait de suggestion mentale," *Revue philosophique* 21 (1886): 204.

22. The Society for Psychical Research, which counted as members most of the best-known names in early experimental psychology, had been established in 1882 in part to investigate paranormal phenomena. This is another indication of the growing interest in the subject and of the close relationship of experimental psychology and parapsychology at this period.

23. Ellenberger, *Discovery,* 338.

24. Ibid., 85–102; Barrucand, *Histoire de l'hypnose,* 101–82.

25. Pierre Janet, "Les phases intermédiaires de l'hypnotisme," *Revue scientifique* 19 (1886, part 1): 586–87.

26. Pierre Janet, "Les actes inconscients et le dédoublement de la personnalité pendant le somnambulisme provoqué," *Revue philosophique* 22 (1886): 577–92.

27. Ellenberger, *Discovery,* 339. Séailles and Janet coauthored a *History of the Problems of Philosophy.*

28. Paul Janet's objections, which are actually quite penetrating, may be found in the review of *Psychological Automatism* that appeared in an appendix to his *Principes de métaphysique et de psychologie* (Paul Janet, "L'automatisme psychologique," in *Principes* 2: 556–72). For an analysis of Paul Janet's reading of his nephew's work, see Goldstein, "The Advent of Psychological Modernism," 203–4.

29. "The most probable solution seems this: Brochard *chargé de cours* at the Sorbonne, *suppléé* by G. Lyon (to please Berthelot), *suppléé* by Pierre Janet (to please [Paul] Janet). . . ." (Ribot to Espinas, 12 April 1889). This would have placed Janet at the Ecole Normale, where Lyon taught. In the event, this particular arrangement failed to materialize, but it indicates the extent to which people were aware that Paul Janet was working for his nephew's career. Cf. another rumor involving Paul Janet, Lyon, and Pierre Janet Ribot reported to Espinas in his letter of 17 June 1891.

30. Ellenberger, *Discovery,* 351; Theodore Zeldin, *France, 1848–1945: Anxiety and Hypocrisy* (Oxford: Oxford University Press, 1981), 88–89. As Ellenberger notes, it is impossible to name many of Janet's clients, in part because of the lack of testimony on the part of former patients, and in part because, in accordance with his wishes, his more than five thousand case files were destroyed after his death.

31. Séance du 4 novembre 1893, Actes de la Faculté des Lettres de Paris, 1888–99, AN AJ¹⁶ 4748. The discussion was continued at the next meeting on 22 November.

32. This was the response of the dean of the Faculty of Letters when the issue came up at the meeting of the committee of the General Council of Faculties that considered the new request. The Faculty of Letters formally withdrew its request for a course in logic in favor of the new courses in psychology and sociology (Séance du 19 février 1894, in AN AJ¹⁶ 2653).

33. Paris Faculty of Letters, meeting of 22 November 1893, AN AJ¹⁶ 4748.

34. Ibid.

35. Paul Janet, "Rapport de M. Janet sur la création d'un cours de psychologie expérimentale ou objective," in AN AJ¹⁶ 2653.

36. Ibid.

37. See the minutes of the committee meeting of 19 February 1894, and the plenary session of 26 February in AN AJ¹⁶ 2653. The issue generated a prolonged discussion on both occasions.

38. To which Milne-Edwards responded that this proved that the Faculty of Letters was not competent to teach the course (ibid.).

39. Léon Beudant, representative of the Faculty of Law, lamented Janet's mention of Auguste Comte, asked that the report not be published, and decried the slide toward socialism. Louis-Auguste Sabatier, from the Faculty of Protestant Theology, raised the practical question of who would be qualified to teach such a course, since it seemed to require a range of knowledge few could be expected to have. Despite such objections, the council approved the request unanimously in committee and with only one negative vote in plenary session. The title of the course proved more controversial: the committee approved "objective" rather than "experimental" by a vote of five to four, and the council by seven to two (ibid.).

40. Séance du 30 octobre 1897, AN AJ¹⁶ 4748.

41. Gaston Paris to the Minister of Public Instruction, 12 November 1901, in AN F[17] 13556. Paris was the administrator of the Collège de France at the time, and his letter indicates that the members voted unanimously to retain the chair.

42. Henri Bergson, "Rapport sur le maintien d'une chaire de psychologie expérimentale et comparée," *Mélanges* (Paris: Presses Universitaires de France, 1972), 507–9.

43. Collège de France, séance du 19 janvier 1902, in AN F[17] 13556. For a summary of Bergson's remarks to the Collège de France, see Ellenberger, *Discovery,* 343.

44. Georges Picot to the Minister of Public Instruction, 15 February 1902, AN F[17] 13556. Picot was the permanent secretary of the Académie. Another indication of the changes that had overcome this body is the fact that Ribot was elected to it in 1899. In an irony of fate, he replaced Jean Nourrisson, an eclectic who had opposed his nomination to the Collège de France. He was elected by a wide margin over the neocriticist philosopher Victor Brochard (Séance du 23 décembre 1899, *Séances et travaux de l'Académie des Sciences Morales et Politiques* 53 [1900, part 1]: 265). By his election, Ribot was placed in the awkward position of having to pronounce an *éloge* of his predecessor. The resulting piece tries very hard to find good things to say without disguising Ribot's antipathy toward Nourrisson's work (Théodule Ribot, *Notice sur la vie et les travaux de M. F. Nourrisson, lue dans la séance du 6 juillet 1901* [Paris: F. Didot, 1901]).

45. Chapter 2.

46. Séance du 15 mars 1902, Actes de la Faculté des Lettres de Paris, AN AJ[16] 4749.

47. Ellenberger, *Discovery,* 343; Etienne Trillat, *Histoire de l'hystérie* (Paris: Editions Seghers, 1986), 181–212; Martha N. Evans, *Fits and Starts: A Genealogy of Hysteria in Modern France* (Ithaca: Cornell University Press, 1991), 51–76.

48. Prévost, *La psycho-philosophie de Pierre Janet,* 14–15.

49. This incident, alluded to but not explained by Ellenberger and Prévost, had many motivations. Jules Dejerine, who threw Janet out, had been defeated in his bid to succeed Charcot by Fulgence Raymond, who was not as respected a neurologist as Dejerine but who had been a student of Charcot; Dejerine had not (E. Gauckler, *Le professeur J. Dejerine, 1849–1917* [Paris: Masson, 1922], p. 97). Raymond sponsored Janet's research at the Salpêtrière after Charcot's death, and Dejerine probably wanted to get rid of reminders of his earlier humiliation. In addition to this personal vendetta, Dejerine also had principled objections to Janet's work. Dejerine did not approve of experimentation on psychiatric patients unless it was directly related to their treatment, because he did not think it was in the best interest of the patients to treat them like guinea pigs. He also did not approve of the use of hypnosis in the treatment of neuroses (Dejerine and Gauckler, *Les manifestations fonctionnelles des psychonévroses* [Paris: Masson, 1911], vii, 400–405).

50. Prévost, *La psycho-philosophie,* 14–15.

51. Frank Wesley and Michel Hurtig, "Masters and Pupils among French Psychologists," *Journal of the History of the Behavioral Sciences* 5 (1969): 320–25.

52. Catalog of Books Borrowed, 1881–82, Bibliothèque de l'Ecole Normale Supérieure, Paris.

53. Pierre Janet, "La psychologie expérimentale et comparée," in *Le Collège de France, 1530–1930* (Paris: Presses Universitaires de France, 1932), 230.

54. "It is a question of a single thing that is known and studied in two different manners. A phenomenon that I consider on the outside, using my organs of sensation, and that I interpret by the rules and the habits of my thought, cannot have the same aspect as if I consider it in myself through consciousness" (Pierre Janet, *L'automatisme psychologique,* 4th ed. [Paris: Alcan, 1894; reprint, Paris: Centre National de la Recherche Scientifique, 1973], 451).

55. Janet called Leibniz's thought "this profound philosophy, to which all the physi-

cal and moral sciences today seem to bring us back. . . ." (ibid., 55). This was in opposition to the Cartesian doctrine that consciousness was an all-or-nothing phenomenon.

56. Ibid., 450.

57. Ibid., 26.

58. Janet, *Manuel de baccalauréat de l'enseignement secondaire classique Philosophie* (Paris: Nony, 1894), 60.

59. Janet, *Névroses et idées fixes*, 2d ed., 2 vols. (Paris: Alcan, 1904), 1:6–7.

60. Janet, *L'automatisme psychologique*, 26; Cf. idem, "L'anesthésie systématisée et la dissociation des phénomènes psychologiques," *Revue philosophique* 23 (1887): 452.

61. Janet paraphrasing Bernard in *L'automatisme psychologique*, 27. Janet read Bernard while still at the Ecole Normale, and this principle stayed with him throughout his career.

62. Janet, *Névroses et idées fixes*, 1:xiii.

63. Janet, *The Mental State of Hystericals*, trans. Caroline Rolin Corson (New York: Putnam, 1901), 102; idem, *Névroses et idées fixes*, 1:135.

64. Janet, *L'automatisme psychologique*, 58; cf. idem, *Mental State of Hystericals*, 35.

65. Janet, *Les obsessions et la psychasthénie*, 2 vols. (Paris: Alcan, 1903), 1:478.

66. Ibid.

67. Ribot, *Les maladies de la mémoire*, 135.

68. Janet, "L'anesthésie systématisée," 461.

69. "What convinced us to place at the top of this hierarchy voluntary action that really modifies the given world is that we have seen many patients in whom this action is continually disturbed from the beginning" (Janet, *Les obsessions et la psychasthénie*, 1:479).

70. Ibid., 1:476.

71. Janet, "La psychologie expérimentale et comparée," 227–28.

72. Janet, *L'automatisme psychologique*, 329, 432.

73. Janet, *The Mental State of Hystericals*, v.

74. Janet, *Névroses et idées fixes*, 1:xiii; Janet, *Les obsessions et la psychasthénie*, 1:vii.

75. Pierre Janet, "Le troisième Congrès international de psychologie," *Revue générale des sciences* 8 (1897): 23.

76. Janet, *The Mental State of Hystericals*, 238.

77. Ibid., xiv–xv. Reproduced almost verbatim in Janet, *Névroses et idées fixes*, 1:67.

78. Janet, *L'automatisme psychologique*, 33–34.

79. Janet, "Les actes inconscients," 588–89.

80. In the 1890s and 1900s Ribot become less optimistic about the potential of psychophysiological methods. See his contribution to *De la méthode dans les sciences* (1909).

81. He tried retrospectively to show the same of Ribot, underlining the passages in *Psychological Heredity* in which Ribot expressed his reservations about measuring psychological phenomena (Janet, "La psychologie expérimentale et comparée," 228).

82. Janet, *Névroses et idées fixes*, 1:xiii.

83. Janet, *The Mental State of Hystericals*, xiv.

84. Janet, *Névroses et idées fixes*, 1:67.

85. Janet, "The Measure of Attention and the Graph of Reaction Times," in *Névroses et idées fixes*, 1:69–108.

86. "Some day, perhaps, these physiological modifications, which accompany cerebral insufficiencies, will be determined in a manner precise enough to enable us to show a fundamental physiological phenomenon, to which all the details of the delirium of persecution may be related, and another by which all the phenomena of hysteria may be explained with precision. We shall then have a physiological definition of hysteria"

(Janet, *The Mental State of Hystericals*, 514). Cf. a similar statement in *Névroses et idées fixes*, 1:134.

87. Ibid., 1:134.

88. Ibid., 1:347. Similar statements may be found throughout Janet's works: cf. *L'automatisme psychologique*, 171; Janet, *The Mental State of Hystericals*, 514–15; *Névroses et idées fixes*, 1:68; and Janet, *Les obsessions et la psychasthénie*, 1:445, 605, 496–97.

89. See chapter 4.

90. Janet, *L'utomatisme psychologique*, 49–50.

91. Ibid., 433–35.

92. Ibid., 431–32.

93. Ibid., 49.

94. Ibid., 73–77.

95. Janet, *Névroses et idées fixes*, 1:36.

96. Ibid., 1:48.

97. Janet, *Les obsessions et la psychasthénie*, 1:43, 264–66, 544–51.

98. Janet, *L'automatisme psychologique*, 57.

99. Ibid., 454.

100. Ibid., 452.

101. Janet, *Névroses et idées fixes*, 1:56.

102. Cf. Janet, *L'automatisme psychologique*, 12, 127, 454; idem, *The Mental State of Hystericals*, 36, 410.

103. Janet, *Les obsessions et la psychasthénie*, 1:474–77.

104. The expression "la misère psychologique" first appears in *L'automatisme psychologique* (426), "psychasthenia" in *The Mental State of Hystericals* (519).

105. Ibid., 40.

106. Ibid., 401.

107. Ibid., 528.

108. Janet, *Les obsessions et la psychasthénie*, 1:333.

109. Ibid., 1:735.

110. Ellenberger, *Discovery*, 403.

111. Sjövall, *Psychology of Tension*, 33 and note. Sjövall bases his estimation of Janet's debt to Ribot on essays Janet wrote on the occasion of Ribot's death and on the quatercentenary of the Collège de France. These were occasions on which Janet was not likely to downplay his debt to Ribot. The evidence from his early works seems more conclusive to me, although I was also struck by the fact that Janet mentions Ribot only once, very briefly, in his autobiographical essay. Both Charcot and Paul Janet receive more attention.

112. Goldstein, *Console and Classify*, 257–63.

113. Pierre Janet, "J.-M. Charcot: Son oeuvre psychologique," *Revue philosophique* 39 (1895): 569–604. Scholars have differed on the extent to which Charcot's later theory of hysteria was "psychological," with some (such as Sjövall and Micale) holding that he remained wedded to a physiological model and others, such as Owen, arguing that his theory was purely psychological (Sjövall, *Psychology of Tension;* Mark S. Micale, "Charcot and the Idea of Hysteria in the Male—Gender, Mental Science, and Medical Diagnosis in Late Nineteenth-Century France," *Medical History* 34 [1990]: 363–411; Owen, *Hysteria, Hypnosis, and Healing*). Whatever the "essential" Charcot might turn out to be, there is ample evidence to support both points of view, which is why disagreement has lasted so long. The relevant point in the present context is that Janet construed Charcot as a model of the combination of psychology and medicine. Others, like Charcot's neurological students Babinski and Sollier, thought that their own more neurologically oriented studies continued the "real" thought of Charcot, and contested Janet's

claim to the legacy. On the successors of Charcot, see Evans, *Fits and Starts,* and Trillat, *Histoire de l'hystérie.*

114. Goldstein, *Console and Classify,* 262–63.

115. Charles Richet, "Du somnambulisme provoqué," *Revue philosophique* 10 (1880): 477.

116. Ibid., 480.

117. Paul Janet called logical relations "synthetic" and said of association: "The laws of association and conflict we have just studied constitute what can be called intellectual *mechanics:* it is the part played by *automatism* in our intellectual life; if this mechanics were alone, one can say that there would be no intelligence properly speaking" (*Traité,* 72, 76; emphasis in the original).

118. Janet, *L'automatisme psychologique,* 443.

119. Ibid., 442–43. He cites Rabier as one source of this theory. "I am only repeating the conclusions brilliantly sustained by several authors and in particular by M. Rabier" (ibid.).

120. Ibid.

121. "We have here not an association, that is to say a pure possibility persisting in a latent state, but genuine psychological phenomena, remarks, counting, in a word judgments persisting for thirteen days in the head of an individual without her being conscious of it; an unconscious judgment is an entirely different thing from a latent association" (Janet, "Les actes inconscients," 583–84; reproduced in *L'automatisme psychologique,* 253).

122. Alfred Binet, *La psychologie du raisonnement,* 5th ed. (Paris: Alcan, 1911), 9–10.

123. Pierre Janet, review of *La psychologie du raisonnement,* by Alfred Binet, *Revue philosophique* 22 (1886): 191–92.

124. Janet, *L'automatisme psychologique,* 432, 433.

125. Ibid., 49.

126. Ibid., 59–61; cf. idem, *The Mental State of Hystericals,* 36.

127. Janet, *L'automatisme psychologique,* 10, 60.

128. Ibid., 60.

129. His text was originally intended for the philosophy of the "modern" curriculum as well as of the "classics" or traditional curriculum. In the more streamlined syllabus for nontraditional curricula, philosophy was usually shortened to "moral" and "scientific" philosophy.

130. Janet, "L'anesthésie systématisée," 452.

131. Janet, *L'automatisme psychologique,* 33–34.

132. "Synthesis must verify analysis, and we should now take as our point of departure the most profound psychological disturbance we have found, and show how it brings in its wake all the characteristic phenomena of the delirium of doubt, abulia, and obsessions" (Janet, *Névroses et idées fixes,* 49).

133. Janet, *L'automatisme psychologique,* 448; cf. idem, *Névroses et idées fixes,* 138: ". . . this hypothesis, which is only a pure representation of the facts."

134. Janet, *The Mental State of Hystericals,* 485.

135. Janet, *Les obsessions et la psychasthénie,* 539.

136. Janet, *The Mental State of Hystericals,* 485.

137. Janet, *Manuel de baccalauréat,* 38; emphasis in the original.

138. Ibid., 39.

139. Ibid., 45.

140. Janet, *L'automatisme psychologique,* 50: idem, *Manuel du baccalauréat,* 60.

141. Janet, *L'automatisme psychologique,* 50.

142. Ibid., 31; idem, *Névroses et idées fixes,* xi

143. Janet, *L'automatisme psychologique,* 9–10.

144. Bernheim wrote a letter to the editor of the *Revue générale des sciences* to protest Janet's report of his words at the International Congress of Psychology in London, which Janet had included in a summary of the Congress for the *Revue*. Janet quoted Bernheim as saying, "Hypnotism is nothing, nothing at all; suggestion is completely innocuous, a piece of sound advice, that's all; hallucination is a dream, a little daydream; does hallucination exist? No, of course not, it is nothing, nothing at all" (Pierre Janet, "Le congrès international de psychologie expérimental," *Revue générale des sciences* 3 [1892]: 615). Bernheim protested, "M. Janet has misrepresented and made a travesty of the ideas that, at the request of the president—M. Sidgwick—I expounded at the *Congress* on suggestion and hypnotism. *I never uttered the sentence he has me saying in quotation marks*" (Hippolyte Bernheim, Letter to the editor, *Revue générale des sciences* 3 [1892]: 691). Janet responded that he had been so struck by the exaggerated statements at the Congress that "I believed I had to write them as dictated by M. Bernheim. I reproduced these terms from my notes, precisely so that I could not be accused of misinterpreting his words" (Pierre Janet, response, *Revue générale des sciences* 3 [1892]: 691). It is difficult to determine which of the two psychologists is correct, and it may well be that Janet's notes were accurate. Nevertheless, Bernheim felt that Janet's theoretical orientation had caused him to misconstrue his words, and one cannot exclude the possibility that Janet's record of his patients was equally selective.

145. Janet, *Névroses et idées fixes,* v–viii.

146. Even the mention of the "laws of social hierarchy," obviously inspired by the rise of sociology, was not entirely new. As we have seen, Paul Janet had also introduced a section on "l'homme social" into his *Traité* (chapter 4).

147. Janet, *Névroses et idées fixes,* 36.

Chapter 6. Metaphysician: Emile Durkheim's Science of the Moral

1. Except for Roger Geiger, who recognized that without Espinas, Durkheim would have been impossible (Geiger, "The Development of French Sociology," 91).

2. Jones, "Robertson Smith, Durkheim, and Sacrifice," 199–200. See also Schmaus, *Durkheim's Philosophy of Science,* 12; and Fabiani, "Métaphysique, morale, sociologie," 176.

3. See the next section for references to scholars who have characterized Durkheim in different ways. Among the few who have recognized this difficulty in pigeonholing Durkheim are Fabiani and Schmaus. Both note that Durkheim himself recognized only the appellation "rationalist" (Fabiani, "Métaphysique, morale, sociologie," 175–79; Schmaus, *Durkheim's Philosophy of Science.*) Rather than take Durkheim's self-characterization at face value, or, in the case of Schmaus, painstakingly work out a coherent philosophy from Durkheim's statements, I hope to show that Durkheim's philosophy and practice of sociology cannot be reduced to anything as tidy as "rationalism." Rationalism for Durkheim is a cover for a complex set of ideas taken from a variety of sources.

4. Philippe Besnard, "The 'Année sociologique' Team," in *The Sociological Domain,* 11–39; Schmaus, *Durkheim's Philosophy of Science,* 32.

5. Those who think that Durkheim's position remained fundamentally the same include Anthony Giddens, *Durkheim* (London: Fontana Press, 1978), 82; Robert A. Strikwerda, "Emile Durkheim's Philosophy of Science: Framework for a New Social Science" (Ph.D diss., University of Notre Dame, 1982), 167–69, 175–76; and Schmaus, *Durkheim's Philosophy of Science,* 12–17. Those who find a substantial shift between the early work, as exemplified by *The Division of Labor,* and the later work, represented by

The Elementary Forms, include Parsons, *The Structure of Social Action,* 356–68; Lukes, *Durkheim,* 229–36; and Dénes Némedi, "Collective Consciousness, Morphology, and Collective Representations: Durkheim's Sociology of Knowledge, 1894–1900," *Sociological Perspectives* 38 (1995): 41–56.

6. Most of the following information on Durkheim's education and career is taken from Lukes, *Durkheim.*

7. Georges Davy, in "Centenaire de la naissance de Durkheim," *Annales de l'Université de Paris* 1 (1960): 16.

8. See Emile Durkheim, *De la division du travail social* (Paris: Alcan, 1893), 261 note (hereafter referred to as *The Division of Labor;* all references are to this edition unless otherwise noted); idem, "Représentations individuelles et représentations collectives," *Revue de Métaphysique et de Morale* 6 (1898): 277–79 (hereafter referred to as "Individual and Collective Representations").

9. He appears to have stayed in Boutroux's apartment briefly upon his arrival in Paris in 1902 (Durkheim to Octave Hamelin, 21 October 1902, in *Textes,* 2:456).

10. Lukes, *Durkheim,* 52.

11. Ibid., 64.

12. Ibid., 64–65.

13. Emile Durkheim, Cours de philosophie faite au Lycée de Sens, Papers of André Lalande, ms. 2351, Bibliothèque de la Sorbonne, Paris. On these notes, see Neil Gross, "A Note on the Sociological Eye and the Discovery of a New Durkheim Text," *Journal of the History of the Behavioral Sciences* 32 (1996): 408–23.

14. S. G. Mestrovic, "The Social World as Will and Idea: Schopenhauer's Influence on Durkheim," *Sociological Review* 36 (1988): 674.

15. R. Lacroze, "Emile Durkheim à Bordeaux," *Actes de l'Académie Nationale des Sciences, Belles-Lettres et Arts de Bordeaux,* 4th ser., 17 (1960): 1.

16. Cited in Lukes, *Durkheim,* 100–1.

17. Emile Durkheim, *Les règles de la méthode sociologique,* 21st ed. (Paris: Presses Universitaires de France, 1983), 2. This text, hereafter referred to as *The Rules,* is based on the second edition (1901). Durkheim first published this work in the *Revue philosophique* in 1894 and in book form the following year.

18. Lukes, *Durkheim,* 407.

19. On Durkheim's friendships at Bordeaux, see Lukes, *Durkheim,* 103–4; and Némedi and Pickering, "Durkheim's Friendship," 107–25.

20. With the exception of Brochard, the committee was the same as the one that constituted Pierre Janet's *jury.* Durkheim wrote the required Latin thesis on Montesquieu.

21. See the summary by L. Muhlfield for the *Revue universitaire,* in *Textes,* 2:288–91.

22. Ibid., 289–90.

23. Janet's demonstrative gesture is reported by Durkheim's protégé Célestin Bouglé, in "L'oeuvre sociologique d'Emile Durkheim," *Europe* 23 (1930): 281. Lukes cites this as evidence of the obstacles Durkheim faced in advancing his career (*Durkheim,* 296).

24. "Celui-là sera un maître" (Muhlfield, *Revue universitaire,* in *Textes,* 2:289.) Cf. other reports summarized in Lukes, *Durkheim,* 296–99.

25. Of course, this was not always couched in terms favorable to Durkheim. Bergson later recalled of Durkheim at the École Normale that "his conversation was already nothing but polysyllogisms and sorites. Having [subsequently] taken as premises the Totem and the Taboo, I am not surprised that he has been able to deduce the whole world from them" (reported in Lukes, *Durkheim,* 52).

26. Léon Brunschvicg and Elie Halévy, "L'année philosophique 1893," *Revue de métaphysique et de morale* 2 (1894): 565–66.

27. On Durkheim's relations with the *Revue de métaphysique et de morale*, see Merllié, "Les rapports," 64–68; and Louis Pinto, "Le détail et la nuance: La sociologie vue par les philosophes dans la *Revue de métaphysique et de morale*, 1893–99," *Revue de métaphysique et de morale* 98 (1993): 141–74.

28. Gabriel Tarde, "Les deux éléments de la sociologie," in *Etudes de psychologie sociale* (Paris: V. Giard & E. Brière, 1898), 74. This work was originally a lecture given at the First International Congress of Sociology in October 1894.

29. Marcel Bernès, "Sur la méthode de la sociologie," *Revue philosophique* 39 (1895): 235.

30. See, for example, Durkheim's correspondence with Bouglé and François Simiand, in *Textes*, 2:389–451. Both were only in their twenties and just out of school when these letters were written. Former students in philosophy made up half the contributors to the *Année sociologique*, more than three times the number of any other discipline (Besnard, "The 'Année sociologique' Team," in *The Sociological Domain*, 32–37).

31. Paul Lapie, who later wrote for the *Année sociologique*, wrote the "Année sociologique" column in the *Revue* during 1895 and 1896. The column then passed to François Simiand until 1898, when he abandoned it to assist Durkheim as coeditor of the *Année sociologique*. Paul Fauconnet, another close associate of Durkheim, wrote the review of *Suicide* (*Revue de métaphysique et de morale* 5 [1897], November Supplement: 2–3). See Besnard, "The 'Année sociologique' Team," 12–13. It is true that Simiand's replacement was less receptive to Durkheimian sociology, to the point that Durkheim wrote a letter to the editor, Xavier Léon, to complain about the reviews he was getting (Durkheim to Léon, 21 September 1902, in *Textes*, 2:465–66). But he remained on extremely good terms with Léon, as the rest of his letters attest (ibid., 2:466–81).

32. As indicated in chapter 3, it is unclear for whom this position was intended, but since Durkheim had defended his thesis earlier that year, and since it was Durkheim's principal advisor, Emile Boutroux, who made the proposal, one may hazard a guess that Durkheim was at least a possible candidate.

33. Espinas to Hamelin, 16 March 1899.

34. "I add that the course founded by the generosity of M. le Comte de Chambrun does not duplicate the course in sociology that we propose and request. The one is exclusively historical, the other dogmatic. The one does the history of science, the other will do the science itself" (Janet, "Rapport de M. Janet," AN AJ[16] 2653).

35. Séance du 16 avril 1894, AN AJ[16] 2653.

36. On Tarde, see J. Milet, *Gabriel Tarde et la philosophie de l'histoire* (Paris: Vrin, 1970); and Ian Lubek, "Histoire des psychologies sociales perdues: Le cas de Gabriel Tarde," *Revue française de sociologie* 22 (1981): 361–95.

37. Bernès, "Sur la méthode de la sociologie." See also idem, "La sociologie: ses conditions d'existence, son importance scientifique et philosophique," *Revue de métaphysique et de morale* 3 (1895): 149–83. As far as I can tell, Bernès was never able to make his course or himself a permanent part of the faculty at Montpellier. He did publish an extended summary of his course, however. See Marcel Bernès, "Programme d'un cours de sociologie générale," *Revue internationale de sociologie* 3 (1895): 981–1015; and 4 (1896): 1–32, 497–511, 589–603, 688–715, 770–99.

38. Espinas to Hamelin, 16 March 1899.

39. Ibid.

40. Ibid.

41. On Izoulet, see Logue, *From Philosophy to Sociology*, 111–17. On Worms, see Terry Clark, "Marginality, Eclecticism, and Innovation: René Worms and the *Revue*

internationale de sociologie from 1893 to 1914," *Revue internationale de sociologie*, 2d ser., 3 (1967): 12–27; and Roger Geiger, "René Worms, l'organicisme et l'organisation de la sociologie," *Revue française de sociologie* 22 (1981): 345–60. If these candidates were problematic in some ways—Izoulet because of his mixture of science, philosophy, and religion, Worms because of his youth—as graduates of the Ecole Normale and holders of the doctorate they presented formidable academic credentials.

42. Letourneau held a chair in the history of civilizations at the Ecole d'Anthropologie, a private institution associated with the Société d'Anthropologie in Paris. On Letourneau, see "Nécrologie: Le Dr. Charles Letourneau," *L'anthropologie* 13 (1902): 295–97; and René Worms, "Charles Letourneau," *Revue internationale de sociologie* 10 (1902): 89–91. Lacassagne was a medical doctor who taught in Lyons. On Lacassagne, see Robert A. Nye, *Crime, Madness, and Politics in Modern France: The Medical Concept of National Decline* (Princeton: Princeton University Press, 1984), 103–19.

43. When Espinas mentioned Durkheim's name to Mlle. Weill, a patroness of the social sciences, she responded, "But he is not known in Paris! *He is not seen!* Why doesn't he come give lectures at the Collège des Sciences Sociales [an organization supported in part by Weill]. Write him to tell him that he should come make himself heard!" (quoted in Espinas to Hamelin, 16 March 1899).

44. Victor Karady, "Stratégies de réussite et modes de faire-valoir de la sociologie chez les durkheimiens," *Revue française de sociologie* 20 (1979): 51–63.

45. Laurent Mucchielli, "Pourquoi réglementer la sociologie? Les interlocuteurs de Durkheim," in *La sociologie et sa méthode: Les* Règles *de Durkheim un siècle après,* eds. Massimo Borlandi and Laurent Mucchielli (Paris: Editions L'Harmattan, 1995), 29–32.

46. Lukes, *Durkheim,* 406; Dominick LaCapra, *Emile Durkheim: Sociologist and Philosopher* (Ithaca, N.Y.: Cornell University Press, 1972), 3–4.

47. On Durkheim's wait for an appointment in Paris, see Lukes, *Durkheim,* 301; Logue, *From Philosophy to Sociology,* 153; and Giddens, *Durkheim,* 16.

48. On Durkheim's use of his credentials and connections as pedagogue, see Vogt, "Political Connections."

49. Emile Durkheim, "Cours de science sociale: Leçon d'ouverture," in *La science sociale et l'action,* ed. Jean-Claude Filloux (Paris: Presses Universitaires de France, 1970), 96. Of course, it is not surprising that this lesson, given at Bordeaux while Espinas was dean of the Faculty of Letters, should contain a laudatory reference to someone who had made his position possible. But upon closer examination, Durkheim's account of the history of sociology in this lecture echoes some of the terms of Espinas's historical preface to *Animal Societies.* For example, Durkheim referred to an opposition between an "organic" and a "mechanical" conception of society that went back to the ancient Greeks, with the mechanical view that society is a machine contructed by individuals dominating for most of that time (ibid., 79).

50. Emile Durkheim, "L'état actuel des études sociologiques en France," in *Textes,* 1:90–91. Originally published in the Italian journal *La riforma sociale* in 1895.

51. Letter to the editor of the *Revue néo-scholastique* (1907), in *Textes,* 1:401.

52. Espinas complained of his mistreatment by Durkheim in his letters to Hamelin (16 March 1899, 12 January 1902, 6 July 1902). In these letters, he sounds bitter, but he consoled himself by remarking that "One always opposes one's immediate precursors; it is a law" (Espinas to Hamelin, 16 March 1899).

53. On Espinas's exclusion from the *Année sociologique,* see Espinas to Hamelin, 16 March 1899. On Durkheim's criticism of Espinas, see his review of *Les castes et la sociologie biologique,* by Novicow, and "'Etre ou ne pas être' ou du postulat de la sociologie," by Espinas, *Année sociologique* 5 (1902): 127–29; cf. Espinas, "Etre ou ne pas être." Even in this review Durkheim distinguished Espinas's organicism from "the analogies and metaphors with which certain organicists are satisfied" (Durkheim, review of "Etre

ou ne pas être," 128). And in private Durkheim defended Espinas against those of his younger colleagues who considered him *nul*. See, for example, Durkheim's letter to Bouglé, 13 August 1901, in which Durkheim defended Espinas against his critics, saying that Espinas was not prepared for the kind of course that was expected of him at the Sorbonne (*Revue française de sociologie* 17, no. 2 [1976]: 178).

54. Simiand's column on "L'année sociologique" for the year 1896 did not mention Espinas directly, but it contained an incisive attack on the organic analogy (*Revue de métaphysique et de morale* 5 [1897]: 489–519). Bouglé's dissertation on the *Caste Regime in India* was also directed against the use of biological explanation in sociology, and he criticized Espinas directly in an exchange over the latter's article "To Be or Not to Be" (Célestin Bouglé, "Le procès de la sociologie biologique," *Revue philosophique* 52 [1901]: 121–46).

55. Espinas is the first author mentioned in the review (Emile Durkheim, review of *Bau und Leben des sozialen Körpers*, by Albert Schaeffle, in *Textes* 1:355). The review was originally published in the *Revue philosophique*.

56. Ribot to Espinas, 30 September 1884. Espinas's article is discussed in chapter 3.

57. Lukes, *Durkheim*, 48, 52–3.

58. Emile Durkheim, "La philosophie dans les universités allemandes," in *Textes*, 3:483. Originally published in the *Revue internationale de l'enseignement* in 1887.

59. Durkheim, *The Division of Labor*, 451–52.

60. Emile Durkheim, *L'évolution pédagogique en France*, 2d ed. (Paris: Presses Universitaires de France, 1969), especially the chapters on Erasmian humanism, 220–60. These lectures were originally given in 1904–5 at the Sorbonne. They were first published by Maurice Halbwachs in 1938.

61. Emile Durkheim, "L'enseignement philosophique et l'agrégation de philosophie," *Revue philosophique* 39 (1895): 121–47.

62. Durkheim, *The Division of Labor*, i.

63. For example: "The old political economy also claimed the right to abstraction, and as a principle, it could not be denied, but the use it made of it was vitiated, because it proposed as a basis for all its deductions an abstraction it did not have the right to use—that is, the notion of a man who would be exclusively guided in his actions by his own personal interest" (Emile Durkheim, "La sociologie et son domaine scientifique," in *Textes*, 1:16; originally published in 1900). Cf. "Sociologie et sciences sociales" (1903), in *Textes*, 1:125.

64. A number of secondary sources discuss Durkheim's use of biological metaphors, as well as the issue of the extent to which he outgrew them. Roger Geiger sees Durkheim's early work as part of the organic paradigm that dominated French sociology from around 1870 to the end of the century, although Durkheim managed to work his way out of it ("The Development of French Sociology," 131–32). Strikwerda also believes biological analogies played an essential role in Durkheim's early work ("Emile Durkheim's Philosophy of Science," 155–71). See also M. J. Hawkins, "Traditionalism and Organicism in Durkheim's Early Writings, 1885–1893," *Journal of the History of the Behavioral Sciences* 16 (1980): 31–44.

65. For example, Durkheim cited Perrier to show why individual types could remain heterogeneous even as collective types became more similar (*The Division of Labor*, 136–37). Durkheim also used Perrier as the source of his principle of classification of types of societies by the nature and mode of connection of the social units (ibid., 190–93). This would be expanded upon in *The Rules*, and it became the basis of his theory of classification.

66. For an analysis of the origins of Durkheim's use of the normal and the pathological in analogies of social health, see Robert A. Nye, "Heredity, Pathology, and Psycho-

neurosis in Durkheim's Early Work," *Knowledge and Society: Studies in the Sociology of Culture* 4 (1982): 103–42.

67. Durkheim, *The Rules*, 104.

68. Cited by Lukes, *Durkheim*, 53–54.

69. Catalog of books borrowed, 1881–82, Bibliothèque de l'Ecole Normale Supérieure, Paris.

70. Durkheim, *The Division of Labor*, 448.

71. Ibid., 33.

72. Ibid., 105. This basic description of the fusion of individual into collective representations repeats Espinas's account almost verbatim. Moreover, it was to play a major role in Durkheim's thought. He invoked it in his discussion of imitation in *Suicide*, and again in his description of collective effervescence in *Elementary Forms*. See Emile Durkheim, *Suicide*, new ed. (Paris: Presses Universitaires de France, 1960), 108–12; idem, *The Elementary Forms of the Religious Life*, trans. Joseph Ward Swain (New York: Free Press, 1915), 246–48.

73. Emile Durkheim, "La science positive de la morale en Allemagne," in *Textes*, 1:336 (originally published in 1887 in the *Revue philosophique*). Cf. Durkheim, *The Rules*, 141.

74. Durkheim, "L'agrégation de philosophie," 132, 138–39.

75. Durkheim, *The Rules*, 2.

76. Durkheim, *The Division of Labor*, i.

77. Durkheim, *The Rules*, 15.

78. "One proves nothing when, as so often happens, one is content to show by more or less numerous examples that, in scattered cases, the facts have varied as the hypothesis demands. From these sporadic and fragmentary agreements one can draw no general conclusions. To illustrate the idea is not to demonstrate it" (ibid., 134). Cf. the preface to the first edition of *Suicide:* "An illustration does not constitute a demonstration" (in *Textes*, 1:44).

79. Durkheim, *The Rules*, 30.

80. Durkheim, "Individual and Collective Representations," 300.

81. E.g., Durkheim, *The Division of Labor*, 99.

82. E.g., Durkheim, *The Rules*, 102; idem, "Individual and Collective Representations," 295.

83. Durkheim, *The Division of Labor in Society*, trans. George Simpson (New York: Free Press, 1933), 1–31. This is the only reference in this chapter to this edition; all others are to the first edition.

84. Durkheim, *The Division of Labor*, 218–51.

85. Durkheim, *The Elementary Forms*, 245–55.

86. Durkheim, *The Rules*, 140.

87. Ibid.

88. Durkheim, letter to the editor of the *Revue néo-scolastique*, in *Textes*, 1:403.

89. Durkheim, *The Division of Labor*, 389.

90. Durkheim, "Individual and Collective Representations," 302.

91. Durkheim to Célestin Bouglé, circa 1898–99, in *Textes*, 2:420.

92. Durkheim, *The Rules*, xi.

93. Brunschvicg and Halévy, "L'année philosophique 1893," 565.

94. Tarde, "Les deux éléments de la sociologie," 67, 69.

95. Charles Andler, "Sociologie et démocratie," *Revue de Métaphysique et de Morale* 4 (1896): 245.

96. Lukes, *Durkheim*, 58.

97. Durkheim, "L'agrégation de philosophie," 132.

98. "Everything takes place mechanically" (Durkheim, *The Division of Labor,* 299). Cf. the discussion of causes in *The Rules,* 89–97.

99. See, for example, the discussions of Ravaisson in chapter 1 and of Lachelier and Boutroux in chapter 4. Bernès faults Durkheim's failure to allow for the contingency of human action ("Sur la méthode de la sociologie," 393–94). Cf. the discussion of Bernès in Pinto, "Le détail et la nuance," 164–66.

100. Cited in Durkheim, "L'agrégation de philosophie," 133–34.

101. Ibid., 134. The emphasis is added, but the French is quite emphatic too: "Est-ce là ce qu'on appelle philosopher? Est-ce là l'idée qu'on se fait dans nos classes de la science, de la vérité objective et impersonelle?"

102. Lukes, *Durkheim,* 54.

103. Catalog of books borrowed, 1879–82, Bibliothèque de l'Ecole Normale Supérieure, Paris.

104. LaCapra refers to Durkheim's epistemology as "Cartesianized neo-Kantianism" (LaCapra, *Durkheim,* 8); Strikwerda calls it a modified Kantian rationalism ("Emile Durkheim's Philosophy of Science," 27). A more recent assessment of Renouvier's influence, which argues that "his system is the starting point for Durkheim's sociology," may be found in S. G. Stedman Jones, "Charles Renouvier and Emile Durkheim: Les Règles de la méthode sociologique," *Sociological Perspectives* 38 (1995): 27–40.

105. Durkheim, *The Division of Labor,* 5–7.

106. François Pillon, review of *De la division du travail social,* by Emile Durkheim, *L'année philosophique* 4 (1893), 275–77; idem, review of *Les règles de la méthode sociologique* by Emile Durkheim, *L'année philosophique* 5 (1894), 271–73; idem, review of *Suicide,* by Emile Durkheim, *L'année philosophique* 8 (1897), 252–54.

107. In parts of the introduction to *The Division of Labor* that Durkheim suppressed after the first edition, he quoted Janet approvingly to support the view that the first principles of philosophers are always taken from some observation (Durkheim, *The Division of Labor,* 5). He also approved of Janet's placing practical before theoretical morality: "So far as we know, M. Janet is the only French moralist who has assigned more importance to the improperly called 'practical' ethics than to the so-called 'theoretical.' We believe this innovation important. But to be truly fruitful, it would be necessary for this examination of duties not to be reduced to a purely descriptive and very general analysis. Each would have to be established in all its complexity. . . . Only by these particular researches can we little by little extricate notions of the whole and a philosophical generalization" (ibid., 19, note). Nor was this the only time Durkheim invoked Janet. He cited the same arguments about the priority of observation over speculation and practical over theoretical ethics in his 1888–89 course on the sociology of the family (in *Textes,* 3:28).

108. Durkheim, "L'agrégation de philosophie," 133, n. 1.

109. Ibid., 140.

110. Durkheim, *The Rules,* 2.

111. Cf. Janet's *Traité de philosophie* and Rabier's *Leçons de philosophie,* discussed in chapter 4. In this sense, it seems to me that the articles in the special issue of *Sociological Perspectives* (vol. 38, no. 1 [1995]), devoted to *The Rules,* miss the point. A variety of influences are discussed, but the humble tradition of textbook philosophy of science is nowhere mentioned.

112. Cf. Rabier, *Leçons,* 2:199–202; Janet, *Traité,* 497–500.

113. Durkheim, *The Rules,* 35; cf. Rabier, *Leçons,* 2:201–2.

114. Durkheim, *The Rules,* xx.

115. See, for example, Bernès, "Sur la méthode de la sociologie," 240–41. For more on the influence of textbook philosophy of science on Durkheim's use of external indices, see Brooks, "The Definition of Sociology."

116. Schmaus, *Durkheim's Philosophy of Science,* 58–62.

117. Rabier, for example, argues that a cause involves a necessary connection, not simply an invariable antecedent (*Leçons,* 2:117–19).

118. Rabier, *Leçons,* 2:128–38.

119. Some of those who view Durkheim as a Kantian are listed in the previous section. Among those who view him as a positivist are Parsons, *The Structure of Social Action;* and Thomas F. Gieryn, "Durkheim's Sociology of Scientific Knowledge," *Journal of the History of the Behavioral Sciences* 18 (1982): 107.

120. Brunschvicg and Halévy, "L'année philosophique 1893," 568. Cf. the analysis of Brunschvicg and Halévy's critique in Pinto, "Le détail et la nuance," 159–60.

121. His own younger colleague Simiand thought it unnecessary for Durkheim to maintain that social forces are "real" (François Simiand, "L'année sociologique 1897," *Revue de métaphysique et de morale* 6 [1898]: 649–50). Bernès also criticized Durkheim for not recognizing the "conventional" nature of scientific concepts and hypotheses (Bernès, "Sur la méthode de la sociologie," 243). This was part of the neospiritualist critique of science discussed in chapter 4.

122. Bernès, "Sur la méthode de la sociologie," 397. For overviews of responses to Durkheim's method as presented in *The Rules,* see Brooks, "The Definition of Sociology," and Giovanni Paoletti, "La réception des *Règles* en France, du vivant de Durkheim," in *La sociologie et sa méthode,* 247–83.

123. Durkheim, letter to the editor of the *Revue néo-scolastique,* in *Textes,* 1:403. Cf. "Sociologie et sciences sociales," in *Textes,* 1:122–34. Also "La sociologie," in *Textes,* 1:110–12 (originally published in 1915).

124. For Durkheim's contention that Comte remained essentially philosophical, see *Textes,* 1:95, 1:112–13, 1:122–27, 1:160. For his suggestion that this was necessary, given the stage of development of the discipline, see "Sociologie et sciences sociales," where he explained that "Not only did nascent sociology present this character [philosophical], but indeed it was necessary that it present it. It could only be born of a philosophy, because it was philosophical traditions that were opposed to its constitution" (*Textes,* 1:123).

125. In *Textes,* 1:111.

126. For a recent assessment of Durkheim's ambivalent relationship with Comte, see Annie Petit, "De Comte à Durkheim: Un héritage ambivalent," in *La sociologie et sa méthode,* 49–70.

127. Durkheim, *The Rules,* 140.

128. Janet's monism has been discussed in chapter 5. See Gabriel Tarde, "Monadologie et sociologie," in *Essais et mélanges sociologiques* (Paris: Masson, 1895), 309–89; and Charles Renouvier and L. Prat, *La nouvelle monadologie* (Paris: A. Colin, 1899). For Durkheim's monism, which he characterizes as a "spiritualist realism," see his Cours de philosophie, 52a–53a, 251–51a (an "a" represents a page intercalated between consecutively numbered pages in the manuscript). The question of when Durkheim developed his emergent realism has been rendered problematic by the discovery of these lectures by Neil Gross ("A Note on the Sociological Eye").

129. Durkheim, "Sociologie et sciences sociales," in *Textes,* 1:124.

130. Emile Durkheim, "La philosophie dans les universités allemandes," in *Textes,* 3:478. Originally published in 1887 in the *Revue internationale de l'enseignement.*

131. Durkheim, *The Rules,* 29–30. S. G. Stedman Jones suggests that this passage refers to the publication of Renouvier's *Traité de psychologie rationelle* in 1859 ("Charles Renouvier and Emile Durkheim," 32). It is far more likely that it refers to German psychophysics, and in particular to G. T. Fechner's *Elemente der Psychophysik,* published in 1860.

132. Durkheim, *Suicide,* 351. In this passage he was trying to refute Tarde's claim

that we have by introspection clear access to the causes of social phenomena. Cf. Emile Durkheim and Marcel Mauss, "De quelques formes primitives de classification," in *Journal sociologique,* ed. Jean Duvignand (Paris: Presses Universitaires de France, 1969), 396 (originally published in 1903 in the *Année sociologique;* hereafter referred to by its usual English translation, "Primitive Classification").

133. Durkheim, "Individual and Collective Representations," 301.

134. Durkheim, *The Rules,* 103. For an examination of the interaction between Durkheimian sociology and contemporary French psychology over the issue of collective psychology, see Laurent Mucchielli, "Sociologie et psychologie en France, l'appel à un territoire commun: Vers une psychologie collective (1890–1940)," *Revue de synthèse* 115 (1994): 445–83.

135. Durkheim, "Individual and Collective Representations," 299

136. Durkheim, *The Division of Labor,* 257–59. On Wundt's influence, see Jan Jacob de Wolf, "Wundt and Durkheim: A Reconsideration of a Relationship," *Anthropos* 82 (1987): 1–23.

137. Durkheim, *The Division of Labor,* 370–71. Durkheim mentions the same phenomenon of substitution, without reference, in Durkheim, "Individual and Collective Representations," 296.

138. Durkheim, *The Division of Labor,* 372.

139. Emile Durkheim, *Moral Education: A Study in the Theory and Application of the Sociology of Education,* trans. Everett K. Wilson and Herman Schnurer (New York: Free Press, 1961), 139–43. Although Durkheim does not give specific references—the book consists of lectures Durkheim gave at Bordeaux and the Sorbonne that he never worked up for publication—it would seem likely that the reference is to Guyau's *Education and Heredity: A Study in Sociology,* 2d ed., trans. W. J. Greenstreet (New York: Scribner, 1895), first published in 1889. Guyau himself draws on research into hypnotism. The other reference is undoubtedly to Binet and Henry, "De la suggestibilité naturelle chez les enfants," *Revue philosophique* 38 (1894), 337–47. See also the article "Enfance," which Durkheim wrote with Ferdinand Buisson for the *Nouveau dictionnaire de pédagogie and d'éducation primaire* (1911), in *Textes,* 3 : 363–69.

140. Durkheim, *Suicide,* 115.

141. Durkheim, "Individual and Collective Representations," 290.

142. Durkheim, *The Division of Labor,* 338–66; idem, *Suicide,* chapter 2, "Suicide and Normal Psychological States: Race and Heredity," 54–81.

143. Durkheim and Mauss, "Primitive Classification," 396.

144. We now know from what multiplicity of elements the mechanism by which we construct, project outwards, and localize in space our representations of the sensible world has been formed. But this work of dissociation has only rarely been applied to properly logical operations. . . . Psychologists think that the simple play of association of ideas, of the laws of contiguity and resemblance between mental states, suffice to explain the agglutination of images, their organization into concepts, and into concepts classed with respect to each other. (ibid., 395–96).

145. Brooks, "Analogy and Argumentation."

146. See, for example, the response to Durkheim's article by Edmond Goblot, a philosopher and former classmate of Durkheim at the Ecole Normale: "Sur la théorie physiologique de l'association," *Revue philosophique* 45 (1898): 487–503.

147. Durkheim, "Individual and Collective Representations," 297–98.

148. Charles Elmer Gehlke, *Emile Durkheim's Contributions to Sociological Theory,* Columbia University Studies in Political Science 63 (New York: Columbia University Press, 1915), 104; Brooks, "Analogy and Argumentation," 239–40. Schmaus discusses this issue nicely in Durkheim's *Philosophy of Science,* 241–43.

149. Durkheim, chapter 1, "Suicide and Psychopathic States," *Suicide,* 19–53.

150. Ibid., 22–26. On the decline of the monomania diagnosis, see Goldstein, *Console and Classify,* 189–96.

151. Durkheim, *Suicide,* 33–35. See Nye, "Heredity, Pathology, and Psychoneurosis," 128–33, where he compares Durkheim's description of neurasthenia with that of a medical text on the subject published the same year. Again, Durkheim gave no references for his description.

152. There is a substantial literature on the history of crowd psychology: Susanna Barrows, *Distorting Mirrors: Visions of the Crowd in Late Nineteenth-Century France* (New Haven, Conn.: Yale University Press, 1981); Robert A. Nye, *The Origins of Crowd Psychology: Gustave LeBon and the Crisis of Mass Democracy in the Third Republic* (Beverly Hills, Calif.: Sage Publications, 1975); Serge Moscovici, *The Age of the Crowd: A Historical Treatise on Mass Psychology* (Cambridge: Cambridge University Press, 1985); Jaap van Ginnekin, *Crowds, Psychology, and Politics, 1871–1899* (Cambridge University Press, 1992); Carl F. Graumann and Serge Moscovici, eds., *Changing Conceptions of Crowd Mind and Behavior* (New York: Springer-Verlag, 1986).

153. Gustave LeBon, *Psychologie des foules,* 2d. ed. (Paris: Alcan, 1896). First published in 1895, it has since appeared in numerous editions and translations. Tarde's articles on "Les crimes des foules" (first given as a paper in 1892) and "Foules et sectes du point de vue criminel" (1893) were republished in *Essais et mélanges sociologiques* in 1895. Tarde continued to develop his ideas on crowd psychology and later published *L'opinion et la foule* (Paris: Alcan, 1901).

154. Erika Apfelbaum and Gregory R. McGuire, "Models of Suggestive Influence and the Disqualification of the Social Crowd," in *Changing Conceptions of Crowd Behavior,* 27–50.

155. Durkheim and Mauss, "Primitive Classification," 461.

156. See Lukes, *Durkheim,* 58–63.

157. Durkheim, "Sociologie et sciences sociales," in *Textes,* 1:146–47.

158. Durkheim, *The Rules,* 76. Durkheim argued that history was only scientific when it studied general causes rather than particular events; see, for example, his review articles in the *Année sociologique,* reproduced in *Textes,* 1:195–99.

159. Durkheim, *The Rules,* 24–28.

160. Durkheim, "L'état actuel des études sociologiques," in *Textes,* 1:105.

161. Durkheim, *The Division of Labor,* 71–72, 117, 138–41.

162. Ibid., 71.

163. Durkheim, "L'état actuel des études sociologiques," in *Textes,* 1:105.

164. Durkheim, "La sociologie et son domaine scientifique," in *Textes,* 1:32.

165. Durkheim, "La science positive de la morale," in *Textes,* 1:335.

166. The Catholic apologist and anti-Durkheimian Simon Deploige was perhaps the first to make an issue of the matter. In an article on "La Genèse du système de M. Durkheim" in the *Revue néo-scolastique* in 1906, he traced Durkheim's social realism to his reading of German sources. The article was obviously intended to play on French chauvinism and Germanophobia in order to discredit Durkheim's ideas by association. It drew a reply from Durkheim in which, while acknowledging his debt to German thought, he drew attention to his French and English antecedents as well. See Durkheim, letters to the editor of the *Revue néo-scolastique,* in *Textes,* 1:401–3.

167. Ibid., and Emile Durkheim, "Note sur l'influence allemande dans la sociologie française," in *Textes,* 1:400 (originally published in the *Mercure de France,* 1902). Lukes (*Durkheim,* 92–93) tends to take Durkheim at his word, while Chamboredon, emphasizing the extent to which Durkheim tried to distance himself from his German roots, thinks they were deeper than he cared to admit (Jean-Claude Chamboredon, "Emile Durkheim: Le social, objet de science; Du moral au politique?" *Critique* 40 [1984]: 490).

168. Durkheim, "L'état actuel des études sociologiques," in *Textes*, 1:106.
169. See his discussions of Letourneau and Lacassagne in "L'état actuel des études sociologiques," in *Textes*, 1:76–82. Durkheim's reservations about anthroposociology may be found in his introduction to the section of the *Année sociologique* by that name: "L'anthroposociologie," *Année sociologique* 1 (1898): 519. Despite his opposition to such doctrines, Durkheim explained that they were included in his journal because he felt obligated to cover all aspects of sociology; besides, one could always glean useful information, even from misguided research.
170. "L'état actuel des études sociologiques," in *Textes*, 1:86–89.
171. Lukes, *Durkheim*, 302–13. For an account that proclaims Tarde the victor, see Milet, *Gabriel Tarde*, 256. For an account that makes Durkheim the aggressor, see Philippe Besnard, "Durkheim critique de Tarde: Des *Règles* au *Suicide*," in *La sociologie et sa méthode*, 221–43.
172. Compare, for example, Durkheim's argument in "Individual and Collective Representations" with Tarde's description of the mind as a society of nerve cells in *La logique social* (Paris: Alcan, 1895), 117–18.

173. And if I dared . . . push this idea to the limit, . . . perhaps I too would be led into arcana such as the Leibnizian region of monads, upon which scientific thought seems to be converging today from so many different paths. Perhaps then I would be led to say that between the ontological fantasmagoria of M. Durkheim and our neo-monadological hypothesis, one must choose: if the latter is rejected, the former must be accepted. *But I do not want to embark upon such metaphysical flights of fancy. Let us remain attached to the shore of facts*" (Tarde, "Les deux éléments de la sociologie. 76; emphasis added)

Cf. the following passage from "Individual and Collective Representations":

In short, individualistic sociology is only applying to social life the principle of the old materialistic metaphysics: it pretends in effect to explain the complex by the simple, the superior by the inferior, the whole by the part, which is a contradiction in terms. Of course, the contrary principle seems to us no less untenable; one should not, with idealist and teleological metaphysics, derive the part from the whole. . . . It remains then to explain the phenomena that are produced in the whole by the properties characteristic of the whole, the complex by the complex, social facts by society. . . . *This is the only path that science can follow*" (298; emphasis added)

A few pages later, after characterizing his social realism as "hyper-spirituality," he adds, "Despite its metaphysical aspect, the word designates nothing except an ensemble of natural facts that must be explained by natural causes" (302).
174. Durkheim, "Individual and Collective Representations," 302.
175. Durkheim, "L'agrégation de philosophie," 136.
176. Ibid., 138.
177. Marcel Bernès, "La philosophie au lycée et à l'agrégation," *Revue philosophique* 39 (1895): 617–18.
178. Ibid., 617.
179. Ibid., 625.
180. Durkheim, "L'agrégation de philosophie," 125.
181. See, for example, Durkheim's characterization of the criminological school as "a kind of eclecticism"(Durkheim, "L'état actuel des études sociologiques," in *Textes*, 1:81). He criticized the members of this school for confusing biology and sociology to the point of introducing ambiguities in their fundamental concepts. In the conclusion to this piece, Durkheim expressed a desire to avoid "a vain and fluctuating eclecticism" (ibid., 108).
182. "La philosophie dans les universités allemandes," in *Textes*, 3:453–55; "La morale positive en Allemagne," in *Textes*, 1:299.

183. Durkheim, "Contribution à une enquête sur l'oeuvre de Taine" (1897), in *Textes*, 1:175.

CONCLUSION

1. On Janet, see Ellenberger, *Discovery*, 407–9; on Durkheim, see Karady, "The Durkheimians in Academe," 89.

2. Besnard, "The 'Année sociologique' Team," 32–37.

3. On the filiation of early psychologists in France, see Wesley and Hurtig, "Masters and Pupils," 320–25. An analysis of the contributors to the collective and multivoloume *Traité de psychologie*, inspired by Ribot and published in 1923, shows that fully two-thirds of the contributors—among them Pierre Janet—had backgrounds in academic philosophy (Georges Dumas, ed., *Traité de psychologie*, 2 vols. [Paris: Alcan, 1923–24]).

4. Harry Paul has analyzed the emergence of the new philosophy of science primarily in relation to Catholicism. He recognizes, however, that academic philosophy also contributed to the new view of science (Paul, *The Edge of Contingency*, 10).

5. Fabiani has also emphasized this divergence between the new philosophy of science and the epistemological assumptions of Durkheimian sociology (Fabiani, "La crise du champ philosophique," 72).

6. On the crisis in philosophy, see Descombes, *Modern French Philosophy*, 3–32; on the history of Freudianism in France, see Elizabeth Roudinesco, *La bataille de cent ans: Histoire de la psychanalyse en France*, 2 vols. (Paris: Seuil, 1986); on Marx in France, see Michael Kelly, *Modern French Marxism* (Baltimore: Johns Hopkins, 1982).

APPENDIX A. ACADEMIC PHILOSOPHERS, NOTABLE PHILOSOPHERS, AND SOCIAL SCIENTISTS IN NINETEENTH-CENTURY FRANCE

1. Christophe Charle, *Les professeurs de la faculté des lettres de Paris*.

2. *Le Collège de France, 1530–1930*.

3. *Le centenaire de l'Ecole Normale, 1795–1895*. I supplemented this volume with Charle and other sources to complete the list up to 1910.

4. Havelange et al., *Les inspecteurs généraux*.

5. W. Paul Vogt, "Identifying Scholarly and Intellectual Communities: A Note on French Philosophy, 1900–1939," *History and Theory* 21 (1982): 267–78.

6. Fabiani, "La crise du champ philosophique" and *Les philosophes de la République*. Fabiani's list is good for identifying notable philosophers but less adequate as a measure of academic philosophy. Any history of philosophy will generally select for notability rather than for institutional affiliation, and even though Benrubi includes a large number of academic philosophers, he also includes nonacademic philosophers and excludes academic philosophers who were not notable. Fabiani's reliance on this source raises some questions, since his thesis focuses on academic philosophy.

7. Charles Adam, *La philosophie en France* (1894); J. Benrubi, *Les sources et les courants de la philosophie contemporaine en France* (1933); George Boas, *French Philosophies of the Romantic Period* (1964); Emile Bréhier, *Histoire de la philosophie* (1932); Lucien Lévy-Bruhl, *History of Modern Philosophy in France*, (1899; reprint, Chicago: Open Court Publishing, 1924); Dominique Parodi, *La philosophie contemporaine en France* (Paris: Alcan, 1919); Eugène Poitou, *Les philosophes français contemporains* (Paris: Charpentier, 1864); Félix Ravaisson, *La philosophie en France au XIXe siècle* (1868).

8. John I. Brooks III, "Academic Philosophy and the Human Sciences in Nineteenth-Century France" (Ph.D. diss., University of Chicago, 1990), Appendix 2.

APPENDIX B. THE *Programmes* of 1832, 1874, 1880, and 1902

1. Source: Cousin, *Défense de l'Université*, 359–63.
2. Source: *Bulletin Administratif du Ministère de l'Instruction Publique* 17 (1874): 490–93.
3. Source: Marion, "Le nouveau programme," 426–28.
4. Source: *Bulletin Administratif du Ministère de l'Instruction Publique* (1902, part 2): 760–62.

Works Cited

MANUSCRIPT AND ARCHIVAL MATERIAL

Archives Nationales, Paris (AN). AJ[16] 2653. Conseil Géneral des Facultés: Procès-verbaux des séances, 1894.

AN AJ[16] 4747. Actes et délibérations de la Faculté des Lettres de Paris, no. 3, 1881 à 1888.

AN AJ[16] 4748. Actes de la Faculté des Lettres de Paris, 1888–99.

AN AJ[16] 4749. Actes de la Faculté des Lettres de Paris, 1900–1906.

AN F[17] 13556. College de France: Documents relatifs à la création des chaires.

AN F[17] 21608. Théodule Ribot: Dossier personnel.

AN F[17] 22193. Alfred Espinas: Dossier personnel.

Bibliothèque de la Sorbonne, Paris. Ms. 2351. Papers of André Lalande. Emile Durkheim. Cours de philosophie faite au Lycée de Sens.

Bibliothèque Victor Cousin. Paris. Mss. 356. Correspondance d'Hamelin.

Ecole Normale Supérieure, Bibliothèque des Lettres, Paris (ENS). Catalog of books borrowed, 1879–82.

PRINTED MATERIAL

Adam, Charles. *La philosophie en France*. Paris: Alcan, 1894.

Alain. *Souvenirs concernant Jules Lagneau*. 4th ed. Paris: NRF, 1925.

Anderson, Roger. *Education in France, 1848–1870*. Oxford: Clarendon Press, 1975.

Andler, Charles. "Sociologie et démocratie." *Revue de métaphysique et de morale* 4 (1896): 243–56.

Aron, Raymond. *Les étapes de la pensée sociologique*. Paris: Gallimard, 1967.

"Arrêté du 5 octobre 1869." *Bulletin administratif du Ministère de l'Instruction Publique* (1869, part 2): 215.

Bachelard, Gaston. *La formation de l'esprit scientifique*. 12th ed. Paris: Vrin, 1983.

Baker, Donald N., and Patrick J. Harrigan. *The Making of Frenchmen: Current Directions in the History of Education in France, 1679–1979*. Waterloo, Ontario: Historical Reflections Press, 1980.

Baldwin, James Mark. "Contemporary Philosophy in France." *New Princeton Review* 3 (1887): 137–44.

Barrows, Susanna. *Distorting Mirrors: Visions of the Crowd in Late Nineteenth-Century France*. New Haven, Conn.: Yale University Press, 1981.

Barrucand, Dominique. *Histoire de l'hypnose en France*. Paris: Presses Universitaires de France, 1967.

Beaunis, Henri. "Un fait de suggestion mentale." *Revue philosophique* 21 (1886): 204.

Beaussire, Emile. "La personnalité humaine d'après les théories récentes." *Revue des deux mondes,* 3d ser., 55 (1883): 316–51.

Ben-David, J., and R. Collins. "Social Factors in the Origins of a New Science: The Case of Psychology." *American Sociological Review* 31 (1966): 451–65.

Benda, Julien. *La trahison des clercs.* Paris: B. Grasset, 1927.

Benrubi, J. *Les sources et les courants de la philosophie contemporaine en France.* 2 vols. Paris: Alcan, 1933.

Bergson, Henri. "Notice sur la vie et les oeuvres de M. Félix Ravaisson-Mollien." In *La pensée et le mouvant.* Paris, 1934.

———. "Rapport sur le maintien d'une chaire de psychologie expérimentale et comparée." In *Mélanges,* 507–9. Paris: Presses Universitaires de France, 1972.

Bernard, Claude. *Introduction to the Study of Experimental Medicine.* Translated by Henry Copley Greene. New York: Dover Publications, 1957.

Bernès, Marcel. "La philosophie au lycée et à l'agrégation." *Revue philosophique* 39 (1895): 605–25.

———. "Programme d'un cours de sociologie générale." *Revue internationale de sociologie* 3 (1895): 981–1015; and 4 (1896): 1–32, 497–511, 589–603, 688–715, 770–99.

———. "La sociologie: ses conditions d'existence, son importance scientifique et philosophique." *Revue de métaphysique et de morale* 3 (1895): 149–83.

———. "Sur la méthode de la sociologie." *Revue philosophique* 39 (1895): 233–57, 372–99.

Bernheim, Hippolyte. Letter to the editor. *Revue générale des sciences* 3 (1892): 691.

Besnard, Philippe, ed. *The Sociological Domain: The Durkheimians and the Founding of French Sociology.* Cambridge: Cambridge University Press, 1983.

Binet, Alfred. *Etudes de psychologie expérimentale.* Paris: Octave Doin, 1888.

———. *La psychologie du raisonnement: Recherches expérimentales par l'hypnotisme.* 5th ed. Paris: Alcan, 1911.

Binet, Alfred, and Charles Féré. *Le magnétisme animal.* Paris: Alcan, 1887.

Binet, Alfred, and Victor Henry. "De la suggestibilité naturelle chez les enfants." *Revue philosophique* 38 (1894): 337–47.

Bloor, David. *Knowledge and Social Imagery.* London: Routledge and Kegan Paul, 1976.

Boas, George. *French Philosophies of the Romantic Period.* New York: Russell and Russell, 1964.

Borlandi, Massimo, and Laurent Mucchielli, eds. *La sociologie et sa méthode: Les Règles de Durkheim un siècle après.* Paris: Editions L'Harmattan, 1995.

Bottomore, Tom, and Robert Nisbet, eds. *A History of Sociological Analysis.* New York: Basic, 1978.

Bouasse, H. et al. *De la méthode dans les sciences.* 2d ed. Paris: Alcan, 1910.

Bouglé, Célestin. "Le procès de la sociologie biologique." *Revue philosophique* 52 (1901): 121–46.

Bouglé, Célestin et al. "L'oeuvre sociologique d'Emile Durkheim." *Europe* 23 (1930): 281–304.

Boutroux, Emile. "L'agrégation de philosophie." *Revue internationale de l'enseignement* (1883, part 2): 865–78.

———. *De l'idée de loi naturelle dans la science et la philosophie contemporaines.* New ed. Paris: J. Vrin, 1949.

———. *De la contingence des lois de la nature.* 4th ed. Paris: Alcan, 1902.

———. "Paul Janet." *Annuaire de l'Association Amicale des Anciens Elèves de l'Ecole Normale* (1900): 31–47.

Bréal, Michel. *Quelques mots sur l'Instruction publique en France.* Paris: Hachette, 1872.

Bréhier, Emile. *Histoire de la philosophie.* 3 vols. Paris: Presses Universitaires de France, 1932.

Brochard, Victor. Review of *Leçons de philosophie.* Vol. 1, *Psychologie,* by Elie Rabier. *Revue philosophique* 19 (1885): 565–73.

Brooks, John I. III. "Academic Philosophy and the Human Sciences in Nineteenth-Century France." Ph.D. diss., University of Chicago, 1990.

———. "Analogy and Argumentation in an Interdisciplinary Context: Durkheim's 'Individual and Collective Representations'." *History of the Human Sciences* 4 (1991): 223–59.

———. "The Definition of Sociology and the Sociology of Definition: Durkheim's *Rules of Sociological Method* and High-School Philosophy in France." *Journal of the History of the Behavioral Sciences* 32 (1996): 379–407.

———. "Philosophy and Psychology at the Sorbonne, 1885–1913." *Journal of the History of the Behavioral Sciences* 29 (1993): 123–45.

Brown, Richard H. *A Poetic for Sociology: Toward a Logic of Discovery for the Human Sciences.* Cambridge: Cambridge University Press, 1977.

———, ed. *Writing the Social Text: Poetics and Politics in Social Science Discourse.* New York: Aldine de Gruyter, 1992.

Brunschvicg, Léon, and Elie Halévy. "L'année philosophique 1893." *Revue de métaphysique et de morale* 2 (1894): 473–96, 563–90.

Canguilhem, Georges. *Le normal et la pathologique.* Paris: Presses Universitaires de France, 1966.

Canivez, André. *Jules Lagneau, professeur de philosophie.* 2 vols. Paris: Les Belles Lettres, 1965.

Caro, Elme. Review of *L'hérédité, étude psychologique,* by Th. Ribot. *Séances de l'Académie des Sciences Morales et Politiques* 101 (1874, part 1): 536–38.

———. Review of *L'hérédité psychologique,* by Théodule Ribot. *Journal des Savants* (1874): 51–66.

Le centenaire de l'Ecole Normale, 1795–1895. Paris: Hachette, 1895.

Centenaire de Théodule Ribot: Jubilé de la psychologie scientifique française, 1839–1889–1939. Paris: Agen, 1939.

Chamboredon, Jean-Claude. "Emile Durkheim: le social, objet de science. Du moral au politique?" *Critique* 40, nos. 445–46 (1984): 460–531.

Charle, Christophe. *Les professeurs de la faculté des lettres de Paris: Dictionnaire biographique.* 2 vols. Paris: Editions du CNRS, 1985–86.

Charlton, D. G. *Positivist Thought in France During the Second Empire, 1852–1870.* Oxford: Oxford University Press, 1959.

Charpentier, Thomas-Victor. Review of *La psychologie allemande contemporaine,* by Théodule Ribot. *Revue philosophique* 9 (1880): 350–57.

"Chronique." *Année psychologique* 22 (1920–21): 598.

Clark, Terry N. "Marginality, Eclecticism, and Innovation: René Worms and the *Revue internationale de sociologie* from 1893 to 1914." *Revue internationale de sociologie,* 2d ser., 3 (1967): 12–27.

———. *Prophets and Patrons: The French University and the Emergence of the Social Sciences.* Cambridge, Mass.: Harvard University Press, 1973.

Le Collège de France, 1530–1930. Paris: Presses Universitaires de France, 1932.

Collier, J. Review of *Des sociétés animales: Etude de psychologie comparée*, by Alfred Espinas. *Mind* 4 (1879): 105–12.

Comte, Auguste. *Cours de philosophie positive*. 2 vols. Edited by Michel Serres, François Dagognet, and Allal Sinaceur. Paris: Hermann, 1975.

Cousin, Victor. *Défense de l'Université et de la philosophie*. 3d ed. Paris: Joubert, 1844.

———. *Du vrai, du beau, et du bien*. 20th ed. Paris: Didier, 1878.

[Darlu, Alphonse]. Preface. *Revue de métaphysique et de morale* 1 (1893): 1–3.

Darwin, Charles. *The Descent of Man*. Princeton: Princeton University Press, 1981.

Davy, Georges. "Centenaire de la naissance de Durkheim." *Annales de l'Université de Paris* 1 (1960): 15–19.

———. "L'oeuvre d'Espinas." *Revue philosophique* (1923, part 2): 214–70.

Dejerine, J., and E. Gauckler. *Les manifestations fonctionnelles des psychonévroses: leur traitement par la psychothérapie*. Paris: Masson, 1911.

Derrida, Jacques. *Writing and Difference*. Translated by Alan Bass. Chicago: University of Chicago Press, 1978.

Descombes, Vincent. *Modern French Philosophy*. Translated by L. Scott-Fox and J. M. Harding. Cambridge: Cambridge University Press, 1979.

Digeon, Claude. *La crise allemande de la pensée française, 1870–1914*. Paris: Presses Universitaires de France, 1959.

Discours prononcés aux obsèques [de François Pillon] *par MM. Gout, Raoul Allier, et Lionel Dauriac, le 21 décembre 1914*. Paris: Gout, Raoul Allier, and Lionel Daurice, 1914.

Dumas, Georges, ed. *Traité de psychologie*. 2 vols. Paris: Alcan, 1923.

Durand Désormeaux, F. *Etudes philosophiques*. Edited by Alfred Espinas. 2 vols. Paris: Alcan, 1884.

Durkheim, Emile. "L'anthroposociologie." *Année sociologique* 1 (1898): 519.

———. Review of *Les castes et la sociologie biologique*, by Novicow, and "'Etre ou ne pas être' ou du postulat de la sociologie," by Espinas. *Année sociologique* 5 (1902): 127–29.

———. *De la division du travail social*. Paris: Alcan, 1893.

———. *The Division of Labor in Society*. Translated by George Simpson. New York: Free Press, 1933.

———. *The Elementary Forms of the Religious Life*. Translated by Joseph Ward Swain. New York: Free Press, 1915.

———. "L'enseignement philosophique et l'agrégation de philosophie." *Revue philosophique* 39 (1895): 121–47.

———. *L'évolution pédagogique en France*. 2d ed. Paris: Presses Universitaires de France, 1969.

———. *Moral Education: A Study in the Theory and Application of the Sociology of Education*. Translated by Everett K. Wilson and Herman Schnurer. New York: Free Press, 1961.

———. *Les règles de la méthode sociologique*. 21st ed. Paris: Presses Universitaires de France, 1983.

———. "Représentations individuelles et représentations collectives." *Revue de métaphysique et de morale* 6 (1898): 273–302.

———. *La science sociale et l'action*. Edited by Jean-Claude Filloux. Paris: Presses Universitaires de France, 1970.

———. *Le suicide: Etude de sociologie*. New ed. Paris: Presses Universitaire de France, 1960.

————. *Textes*. Edited by Victor Karady. 3 vols. Paris: Editions de minuit, 1975.

————. to Célestin Bouglé, 13 August 1901. *Revue française de sociologie* 17, no. 2 (1976): 178.

————, and Marcel Mauss. "De quelques formes primitives de classification." In *Journal sociologique*, edited by Jean Duvignand, 395–461. Paris: Presses Universitaires de France, 1969.

Dwelshauvers, Georges. *La psychologie française contemporaine*. Paris: Alcan, 1920.

Ellenberger, Henri. *The Discovery of the Unconscious: The History and Evolution of Dynamic Psychiatry*. New York: Basic, 1970.

Espinas, Alfred. "L'agrégation de philosophie." *Revue internationale de l'enseignement* 7 (1884, part 1): 585–607.

————. *De Civitate apud Platonem qui fiat una*. Paris: G. Baillière, 1877.

————. *Des sociétés animales*. 2d ed. Paris: Baillière, 1878.

————. "Economie politique, économie sociale et sociologie." *Revue philosophique* 99 (1925): 161–78.

————. "Etre ou ne pas être: Ou, du postulat de la sociologie." *Revue philosophique* 51 (1901): 449–80.

————. "Les études sociologiques en France." *Revue philosophique* 13 (1882): 565–607; 14 (1882): 337–67, 509–28.

————. *Faculté des lettres de Bordeaux. Cours de pédagogie, leçon d'ouverture, par M. A. Espinas, professeur de philosophie (24 avril 1884): Idée générale de pédagogie ou art de l'éducation*. Paris: Felix Alcan, 1884.

————. *Faculté des lettres de Bordeaux. Cours de philosophie. Leçon d'ouverture faite le 21 novembre 1884*. Paris: Librairie Léopold Cerf, 1885.

————. *Histoire des doctrines économiques*. Paris: A. Colin, 1891.

————. "Leçon d'ouverture d'un cours d'histoire de l'économie sociale." *Revue internationale de sociologie* 2 (1894): 321–51.

————. "La nature et l'immatérialisme." *Revue scientifique* (1879): 1222–28.

————. *Les origines de la technologie*. Paris: Alcan, 1897.

————. "Les origines de la technologie." *Revue philosophique* 30 (1890): 113–35, 295–314.

————. *La philosophie expérimentale en Italie*. Paris: G. Baillière, 1880.

————. *La philosophie sociale du XVIIIe siècle et la Révolution*. Paris: Alcan, 1898.

Evans, Martha N. *Fits and Starts: A Genealogy of Hysteria in Modern France*. Ithaca: Cornell University Press, 1991.

"Examen des principes de psychologie de Herbert Spencer; V: Principes de la logique." *Critique philosophique* 12 (1877, part 2): 180–88.

Fabiani, Jean-Louis. "La crise du champ philosophique, 1880–1914." Thèse de 3e cycle, Ecole des Hautes Etudes en Sciences Sociales, 1980.

————. "Enjeux et usages de la 'crise' dans la philosophie universitaire en France au tournant du siècle." *Annales: Economies, sociétés, civilisations* 2 (1985): 377–409.

————. "Métaphysique, morale, sociologie: Durkheim et le retour à la philosophie." *Revue de métaphysique et de morale* 98 (1993):175–91.

————. *Les philosophes de la République*. Paris: Les Editions de Minuit, 1988.

————. "Les programmes, les hommes, et les oeuvres: Professeurs de philosophie en classe et en ville au tournant du siècle." *Actes de la recherche en sciences sociales* 47–48 (June 1983): 3–20.

Falcucci, Clermont. *L'humanisme dans l'enseignement secondaire en France au XIXe siècle.* Toulouse: Edouard Privat, 1939.

Fauconnet, Paul. Review of *Les origines de la technologie,* by Alfred Espinas. *Revue philosophique* 47(1899): 430–41.

———. Review of *Suicide,* by Emile Durkheim. *Revue de métaphysique et de morale* 5 (1897, November Supplement): 2–3.

Ferry, Luc, and Alain Renaut. *French Philosophy of the Sixties: An Essay on Anti-Humanism.* Translated by Mary H. S. Cattani. Amherst: University of Massachusetts Press, 1990.

Fonsegrive, G. Review of *Leçons de philosophie: Logique,* by Elie Rabier. *Revue philosophique* 23 (1887): 172–83.

Foucault, Michel. *The Archaeology of Knowledge.* Translated by A. M. Sheridan Smith. New York: Harper, 1972.

———. "What Is an Author?" In *Language, Counter-Memory, Practice,* translated by Donald F. Bouchard and Sherry Simon, 113–38. Ithaca: Cornell University Press, 1977.

Fouillée, Alfred. *Les études classiques et la démocratie.* Paris: A. Colin, 1898.

———. "Le neo-Kantisme en France." *Revue philosophique* 11 (1881): 1–45, 337–69, 598–625.

———. *La science sociale contemporaine.* 2d ed. Paris: Hachette, 1885.

Fraisse, Paul, Jean Piaget, and Maurice Reuchlin. *Traité de psychologie expérimentale.* Vol. 1, *Histoire et méthode.* Paris: Presses Universitaires de France, 1970.

Franck, Adolphe, ed. *Dictionnaire des sciences philosophiques.* 2d ed. Paris: Hachette, 1875.

Fuller, Steve. *Philosophy, Rhetoric, and the End of Knowledge: The Coming of Science and Technology Studies.* Madison: University of Wisconsin Press, 1993.

Gauckler, E. *Le professeur J. Dejerine, 1849–1917.* Paris: Masson, 1922.

Gehlke, Charles Elmer. *Emile Durkheim's Contributions to Sociological Theory.* Columbia University Studies in Political Science 63. New York: Columbia University Press, 1915.

Geiger, Roger Lewis. "The Development of French Sociology, 1871–1905." Ph.D. diss., University of Michigan, 1972.

———. "René Worms, l'organicisme et l'organisation de la sociologie." *Revue française de sociologie* 22 (1981): 345–60.

Giddens, Anthony. *Durkheim.* London: Fontana Press, 1978.

Gieryn, Thomas F. "Durkheim's Sociology of Scientific Knowledge." *Journal of the History of the Behavioral Sciences* 18 (1982): 107–29.

Ginnekin, Jaap van. *Crowds, Psychology, and Politics, 1871–1899.* Cambridge: Cambridge University Press, 1992.

Goblot, Edmond. "Sur la théorie physiologique de l'association." *Revue philosophique* 45 (1898): 487–503.

Goldstein, Doris. "Official Philosophy in Modern France: The Example of Victor Cousin." *Journal of Social History* 1 (spring 1968): 259–79.

Goldstein, Jan E. "The Advent of Psychological Modernism in France: An Alternate Narrative." In *Modernist Impulses in the Human Sciences, 1870–1930,* edited by Dorothy Ross, 190–209. Baltimore: Johns Hopkins, 1994.

———. *Console and Classify: The French Psychiatric Profession in the Nineteenth Century.* Cambridge: Cambridge University Press, 1987.

———, ed. *Foucault and the Writing of History.* Cambridge, Mass.: Blackwell, 1994.

Graumann, Carl F., and Serge Moscovici, eds. *Changing Conceptions of Crowd Mind and Behavior.* New York: Springer-Verlag, 1986.

Gross, Neil. "A Note on the Sociological Eye and the Discovery of a New Durkheim Text." *Journal of the History of the Behavioral Sciences* 32 (1996): 408–23.

Guardia, J.-M. Review of *Les maladies de la mémoire,* by Théodule Ribot. *Revue scientifique* (1881, part 1): 738–47.

Guigue, Albert. *La Faculté des Lettres de l'Université de Paris.* Paris: Alcan, 1935.

Gusfield, J. "The Literary Rhetoric of Science." *American Sociological Review* 41 (1976): 16–34.

Guyau, Jean-Marie. *Education and Heredity: A Study in Sociology.* Translated by W. J. Greenstreet. 2d ed. New York: Scribner, 1895.

Hacking, Ian. "The Invention of Split Personalities." In *Human Nature and Natural Knowledge: Essays Presented to Marjorie Grene on the Occasion of her Seventy-Fifth Birthday,* edited by A. Donagan, A. N. Perovich Jr., and M. V. Wedin, 63–85. Dordrecht: D. Reidel, 1986.

Hamelin, Octave. *Le système de Renouvier.* Paris: J. Vrin, 1927.

Havelange, Isabelle, Françoise Huguet, and Bernadette Lebedeff. *Les inspecteurs généraux de l'Instruction publique: Dictionnaire biographique, 1802–1914.* Paris: Institut National de Recherche Pédagogique, 1986.

Hawkins, M. J. "Traditionalism and Organicism in Durkheim's Early Writings 1885–1893." *Journal of the History of the Behavioral Sciences* 16 (1980): 31–44.

James, William. *Principles of Psychology.* 2 vols. New York: Henry Holt, 1890.

Janet, Paul. "Une chaire de psychologie expérimentale et comparée au Collège de France." *Revue des deux mondes* 3d ser., 86 (1888): 518–49.

———. *La crise philosophique: MM. Taine, Renan, Littré, Vacherot.* Paris: G. Baillière, 1865.

———. "Le mouvement philosophique: II." *Le Temps,* 16 August 1878.

———. "Une nouvelle phase de la philosophie spiritualiste." *Revue des deux mondes* 2d series 108 (1873): 363–88.

———. *Principes de métaphysique et de psychologie.* 2 vols. Paris: Delagrave, 1897.

———. *Le Temps* (Paris). 2 March 1876.

———. *Traité élémentaire de philosophie, à l'usage des classes.* Paris: Delagrave, 1879.

———. *Victor Cousin et son oeuvre.* Paris: Calmann Lévy, 1885.

Janet, Pierre. "Les actes inconscients et le dédoublement de la personnalité pendant le somnambulisme provoqué." *Revue philosophique* 22 (1886): 577–92.

———. "L'anesthésie systématisée et la dissociation des phénomènes psychologiques." *Revue philosophique* 23 (1887): 449–72.

———. "Autobiography." In *A History of Psychology in Autobiography,* edited by Carl Murchison, 123–33. Worcester, Mass.: Clark University Press, 1930–.

———. *L'automatisme psychologique: Essai de psychologie expérimentale sur les formes inférieures de l'activité humaine.* 4th ed. Paris: Alcan, 1894. Reprint. Paris: Centre National de la Recherche Scientifique, 1973.

———. "Le congrès international de psychologie expérimentale." *Revue générale des sciences* 3 (1892): 609–16.

———. *De l'enseignement de la philosophie.* Le Havre, 1884.

———. "J.-M. Charcot: Son oeuvre psychologique." *Revue philosophique* 39 (1895): 569–604.

————. *Manuel de baccalauréat de l'enseignement secondaire classique Philosophie.* Paris: Nony, 1894.

————. *The Mental State of Hystericals.* Translated by Caroline Rolin Corson. New York: Putnam, 1901.

————. *Névroses et idées fixes.* 2d ed. 2 vols. Paris: Alcan, 1904.

————. "Note sur quelques phénomènes de somnambulisme." *Revue philosophique* 21 (1886): 190–98.

————. *Les obsessions et la psychasthénie.* 2 vols. Paris: Alcan, 1903.

————. "Les phases intermédiaires de l'hypnotisme." *Revue scientifique* 19 (1886, part 1): 577–87.

————. Review of *La psychologie du raisonnement,* by Alfred Binet. *Revue philosophique* 22 (1886):188–93.

————. Response. *Revue générale des sciences* 3 (1892): 691.

————. "Le troisième Congrès international de psychologie." *Revue générale des sciences* 8 (1897): 22–27.

————, and André Lalande. "Edmond Goblot." *Annuaire de l'Association amicale des anciens élèves de l'ENS* (1936), 36–41.

Janet, Pierre, Henri Piéron, and Charles Lalo. *Manuel de baccalauréat.* Paris: Vuibert, 1925.

Joly, Henri. *La psychologie comparée.* Paris: Hachette, 1878.

Jones, Robert Alun. "Robertson Smith, Durkheim and Sacrifice: An Historical Context for *The Elementary Forms of the Religious Life.*" *Journal of the History of the Behavioral Sciences* 17 (1981): 184–205.

Jourdain, Charles. *Notions de philosophie.* 9th ed. Paris: Hachette, 1865.

Karady, Victor. "Les professeurs de la République." *Actes de la recherche en sciences sociales* 47–48 (June 1983): 90–112.

————. "Stratégies de réussite et modes de faire-valoir de la sociologie chez les durkheimiens." *Revue française de sociologie* 20 (1979): 49–82.

Kelly, Michael. *Modern French Marxism.* Baltimore: Johns Hopkins, 1982.

Keylor, William R. *Academy and Community: The Foundation of the French Historical Profession.* Cambridge, Mass.: Harvard University Press, 1975.

Koyré, Alexandre. *Etudes d'histoire de la pensée scientifique.* Paris: Gallimard, 1973.

Krauss, S. *Théodule Ribots Psychologie.* Jena: Hermann Costenoble, 1905.

Kuhn, Thomas S. *The Structure of Scientific Revolutions.* 2d ed. Chicago: University of Chicago Press, 1970.

LaCapra, Dominick. *Emile Durkheim: Sociologist and Philosopher.* Ithaca, N.Y.: Cornell University Press, 1972.

Lachelier, Jules. *The Philosophy of Jules Lachelier.* Translated by Edward G. Ballard. The Hague: Martinus Nijhoff, 1960.

Lacroze, R. "Emile Durkheim à Bordeaux." *Actes de l'Académie Nationale des Sciences, Belles-Lettres et Arts de Bordeaux,* 4th ser., 17 (1960): 1–6.

Lalande, André. *Institut de France. Académie des Sciences Morales et Politiques. Notice sur la vie et les travaux de M. Alfred Espinas, lue dans la séance du 24 janv. 1925.* Paris: Firmin-Didot, 1925.

Lapie, Paul. "L'année sociologique 1894." *Revue de métaphysique et de morale* 3 (1895): 308–39.

Latour, Bruno, and Paolo Fabbri. "La rhétorique de la science: Pouvoir et devoir dans un article de science exacte." *Actes de la Recherche en Sciences Sociales* 13 (February 1977): 81–95.

Lavisse, Ernest. "Louis Liard." *Revue internationale de l'enseignement* 72 (1918): 81–99.

LeBon, Gustave. *Psychologie des foules*. 2d ed. Paris: Alcan, 1896.

Leroux, Pierre. *Oeuvres, 1825–1850*. 2 vols. Geneva: Slatkine Reprints, 1978.

Lévêque, Charles. Review of *Les maladies de la mémoire*, by Théodule Ribot. *Journal des savants* (1881): 680–702, and (1882): 42–55, 91–106, 204–18.

———. Review of *Les maladies de la volonté*, by Théodule Ribot. *Journal des savants* (1883): 481–84.

Lévy-Bruhl, Lucien. *History of Modern Philosophy in France*. Reprint. Chicago: Open Court Publishing Co., 1924.

———. *La philosophie d'Auguste Comte*. 4th ed. Paris: Alcan, 1921.

Liard, Louis. *Des définitions géométriques et des définitions empiriques*. New ed. Paris: Alcan, 1888.

———. *L'enseignement supérieur en France, 1789–1889*. 2 vols. Paris, 1888–94.

———. "Faculté." *La grande encyclopédie*. n.d.

———. *La science positive et la métaphysique*. 2d ed. Paris: Germer Baillière et Cie., 1883.

———. *Universités et facultés*. Paris: Armand Colin, 1890.

Lilla, Mark, ed. *New French Thought: Political Philosophy*. Princeton: Princeton University Press, 1994.

Logue, William. *From Philosophy to Sociology: The Evolution of French Liberalism, 1870–1914*. Dekalb: Northern Illinois University Press, 1983.

Lubek, Ian. "Histoire des psychologies sociales perdues: Le cas de Gabriel Tarde." *Revue française de sociologie* 22 (1981): 361–95.

Lukes, Steven. *Emile Durkheim, His Life and Work: A Historical and Critical Study*. New York: Harper, 1972.

Review of *Les maladies de la mémoire*, by Théodule Ribot. *Critique philosophique* (1881, part 2): 92–96.

Marion, Henri. "Le nouveau programme de philosophie." *Revue philosophique* 10 (1880): 414–28.

McCloskey, Donald. *The Rhetoric of Economics*. Madison: University of Wisconsin Press, 1985.

Meletti Bertolini, Mara. *Il pensiero e la memoria: Filosofia e psicologia nella "Revue philosophique" di Théodule Ribot (1876–1916)*. Milan: Franco Angeli, 1991.

Merllié, Dominique. "Les rapports entre la *Revue de métaphysique* et la *Revue philosophique*: Xavier Léon et Théodule Ribot; Xavier Léon et Lucien Lévy-Bruhl." *Revue de métaphysique et de morale* 98 (1993): 59–108.

Mestrovic, Stjepan G. *Emile Durkheim and the Reformation of Sociology*. Totowa, N. J.: Rowman and Littlefield, 1988.

———. "The Social World as Will and Idea: Schopenhauer's Influence on Durkheim." *Sociological Review* 36 (1988): 674–705.

Micale, Mark S. "Charcot and the Idea of Hysteria in the Male—Gender, Mental Science, and Medical Diagnosis in Late Nineteenth-Century France." *Medical History* 34 (1990): 363–411.

Milet, J. *Gabriel Tarde et la philosophie de l'histoire*. Paris: J. Vrin, 1970.

Milhaud, Gaston. *La philosophie de Charles Renouvier*. Paris: J. Vrin, 1927.

Moscovici, Serge. *The Age of the Crowd: A Historical Treatise on Mass Psychology*. Translated by J. C. Whitehouse. Cambridge: Cambridge University Press, 1985.

Mucchielli, Laurent. "Sociologie et psychologie en France, l'appel à un territoire commun: Vers une psychologie collective (1890–1940)." *Revue de synthèse* 115 (1994): 445–83.

Münsterberg, Hugo et al. *Subconscious Phenomena*. Boston: Richard G. Badger, 1910.

Myers, Gerald E. *William James: His Life and Thought*. New Haven: Yale University Press, 1986.

"Nécrologie: Le Dr. Charles Letourneau." *L'anthropologie* 13 (1902): 295–97.

"Nécrologie: Théodule Ribot, 1839–1916." *Revue de métaphysique et de morale* 24 (1917): 129–32.

Nelson, John S., Allan Megill, and Donald N. McCloskey, eds. *The Rhetoric of the Human Sciences*. Madison: University of Wisconsin Press, 1987.

Némedi, Dénes. "Collective Consciousness, Morphology, and Collective Representations: Durkheim's Sociology of Knowledge, 1894–1900." *Sociological Perspectives* 38 (1995): 41–56.

———, and W. S. F. Pickering. "Durkheim's Friendship with the Philosopher Octave Hamelin: Together with Translations of Two Items by Durkheim." *British Journal of Sociology* 46 (1995): 107–25.

Nizan, Paul. *Les chiens de garde*. Paris: F. Maspero, 1969.

Nourrisson, Jean et al. Observations on *L'hérédité, étude psychologique*, by Th. Ribot. *Séances de l'Académie des Sciences Morales et Politiques* 101 (1874, part 1): 538–40.

Nye, Mary Jo. "The Boutroux Circle and Poincaré's Conventionalism." *Journal of the History of Ideas* 40 (1979): 107–20.

Nye, Robert A. *Crime, Madness, and Politics in Modern France: The Medical Concept of National Decline*. Princeton: Princeton University Press, 1984.

———. "Heredity, Pathology, and Psychoneurosis in Durkheim's Early Work." *Knowledge and Society: Studies in the Sociology of Culture* 4 (1982): 103–42.

———. *The Origins of Crowd Psychology: Gustave LeBon and the Crisis of Mass Democracy in the Third Republic*. Beverly Hills, Calif.: Sage Publications, 1975.

Review of *Les origines de la technologie*, by Alfred Espinas. *Revue de métaphysique et de morale* 6 (1898, May supplement): 2.

Ostrowski, Jean J. *Alfred Espinas, précurseur de la praxéologie (ses antécédents et ses successeurs)*. Paris: R. Pichon and R. Durand-Anzias, 1973.

Overington, Michael A. "The Scientific Community as Audience: Toward a Rhetorical Analysis of Science." *Philosophy and Rhetoric* 10 (summer 1977): 143–64.

Owen, A. R. G. *Hysteria, Hypnosis, and Healing: The Work of J.-M. Charcot*. London: Dennis Dobson, 1971.

Parodi, Dominique. *La philosophie contemporaine en France*. Paris: Alcan, 1919.

Parsons, Talcott. *The Structure of Social Action: A Study in Social Theory with Special Reference to a Group of Recent European Writers*. 2d ed. New York: Free Press, 1949.

Paul, Harry W. *The Edge of Contingency: French Catholic Reaction to Scientific Change from Darwin to Duhem*. Gainesville: University Presses of Florida, 1979.

Perelman, Chaim, and Olbrechts-Tyteca, L. *The New Rhetoric: A Treatise on Argumentation*. Translated by John Wilkinson and Purcell Weaver. Notre Dame, Indiana: University of Notre Dame Press, 1969.

Perrier, Edmond. *Les colonies animales et la formation des organismes*. 2d ed. Paris: Masson. 1898.

Peyre, Henri. *Renan*. Paris: Presses Universitaires de France, 1969.

Picavet, F. "M. Théodule Ribot." *Revue bleue* (1894): 590–95.

Pickering, Mary. *Auguste Comte: An Intellectual Biography.* Vol. 1. Cambridge: Cambridge University Press, 1993.

Pillon, François. Review of *De la division du travail social,* by Emile Durkheim. *L'année philosophique* 4 (1893): 275–77.

———. "La formation des idées abstraites et générales." *Critique philosophique* (1885, part 1): 118–33, 178–215.

———. Review of *La psychologie anglaise contemporaine,* by Théodule Ribot. *Critique philosophique* (1872, part 1): 28–32.

———. Review of *Les règles de la méthode sociologique,* by Emile Durkheim. *L'année philosophique* 5 (1894): 271–73.

———. "Réponse aux observations de M. Rabier sur l'association par ressemblance." *Critique philosophique* (1885, part 2): 55–66.

———. Review of *Suicide,* by Emile Durkheim. *L'année philosophique* 8 (1897): 252–54.

Pinto, Louis. "Le détail et la nuance: La sociologie vue par les philosophes dans la *Revue de métaphysique et de morale,* 1893–99." *Revue de métaphysique et de morale* 98 (1993): 141–74.

Poitou, Eugène. *Les philosophes français contemporains.* Paris: Charpentier, 1864.

"La préparation à la science sociale par la biologie, selon M. Herbert Spencer." *Critique philosophique* 9 (1876, part 1): 235–40.

Prévost, Claude M. *La psycho-philosophie de Pierre Janet: Economies mentales et progrès humain.* Paris: Payot, 1973.

Review of *Principes de sociologie,* vol. 2, by Herbert Spencer. *Critique philosophique* 16 (1879, part 2): 407–15.

"Programme officiel du cours: Classe de philosophie." *Bulletin Administratif du Ministère de l'Instruction Publique* 17 (1874): 490–93.

"Programme officiel du cours: Classe de philosophie." *Bulletin Administratif du Ministère de l'Instruction Publique* (1902, part 2): 760–62.

Prost, Antoine. *Histoire de l'enseignement in France, 1800–1967.* Paris: Librairie Armand Colin, 1968.

Rabier, Elie. "A propos de l'association par ressemblance." *Critique philosophique* (1885, part 1): 460–66.

———. *Leçons de philosophie.* 2 vols. Paris: Hachette, 1884–86.

———. "Rapport." *Revue internationale de l'enseignement* 12 (1886): 353–60.

Ravaisson, Félix. *La philosophie en France au XIXe siècle.* Paris: Imprimerie Impériale, 1868.

Renan, Ernest. *Essais de morale et de critique.* Paris: Calmann-Lévy, n.d.

———. *Etudes philosophiques: De l'Ecosse à V. Cousin.* Edited by Jean Pommier. Paris: Editions A.-G. Nizet, 1972.

———. "La métaphysique et son avenir." In *Oeuvres complètes,* edited by Henriette Psichari, 1:680–714. Paris: Calmann-Lévy, n.d.

Renouvier, Charles. *Essais de critique générale, deuxième essai: Traité de psychologie rationnelle d'après les principes du criticisme.* 2 vols. Paris: Armand Colin, 1912.

———. *Essais de critique générale; Premier essai: Traité de logique générale et de logique formelle.* 2 vols. Paris: Armand Colin, 1912.

———. *Introduction à la philosophie analytique de l'histoire: Les idées, les religions, les systèmes.* New ed. Paris: Ernest Leroux, 1896.

———. Review of *Les maladies de la mémoire,* by Théodule Ribot. *Critique philosophique* (1881, part 2): 92–96.

―――. Review of *La psychologie allemande contemporaine*, by Théodule Ribot. *Critique philosophique* (1879, part 2): 103–12.

―――. *La science de la morale*. 2d ed. 2 vols. Paris: Alcan, 1908.

―――, and L. Prat. *La nouvelle monadologie*. Paris: A. Colin, 1899.

Reuchlin, Maurice. "The Historical Background for National Trends in Psychology: France." *Journal of the History of the Behavioral Sciences* 1 (1965): 115–23.

Ribot, Théodule. Review of *Des sociétés animales*, by Alfred Espinas. *Revue philosophique* 4 (1877): 327–34.

―――. *Diseases of Memory*. Translated by William H. Smith. New York: Appleton, 1882.

―――. *The Diseases of Personality*. Translated by J. Fitzgerald. Humboldt Library of Popular Science Literature, vol. 9. New York: Humboldt, 1887.

―――. *The Diseases of the Will*. 2d ed. Translated from the 8th French ed. by Merwin-Marie Snell. Chicago: Open Court, 1896.

―――. *English Psychology*. New York: D. Appleton, 1891.

―――. *German Psychology of Today*. 2d ed. Translated by James Mark Baldwin. New York: Scribner, 1886.

―――. *L'hérédité psychologique*. 2d ed. Paris: G. Baillière, 1882.

―――. "Leçon d'ouverture: La psychologie nouvelle." *Revue scientifique* (1885, part 2): 780–87.

―――. Letters to Alfred Espinas. Edited by Raymond Lenoir. *Revue philosophique* 147 (1957): 1–14; 152 (1962): 337–40; 154 (1964): 79–84; 160 (1970): 165–73, 339–48; 165 (1975): 157–72.

―――. *Notice sur la vie et les travaux de M. F. Nourrisson. Lue dans la séance du 6 juillet 1901*. Paris: F. Didot, 1901.

―――. *La philosophie de Schopenhauer*. Paris: G. Baillière, 1874.

―――. "Philosophy in France." *Mind* 2 (1877): 366–86.

―――. Preface. *Revue philosophique* 1 (1876): 1–4.

―――. "La psychologie contemporaine." *Revue scientifique* (1888, part 1): 449–55.

―――. *Revue philosophique* 1 (1876): 632.

Richards, Robert J. *Darwin and the Emergence of Evolutionary Theories of Mind and Behavior*. Chicago: University of Chicago Press, 1987.

Richet, Charles. "Du somnambulisme provoqué." *Revue philosophique* 10 (1880): 336–74, 462–84.

―――. "Un fait de somnambulisme à distance." *Revue philosophique* 21 (1886): 199–200.

―――. Review of *Les maladies de la personnalité*, by Théodule Ribot. *Revue scientifique* (1885, part 1): 502–4.

―――. Review of *Les maladies de la volonté*, by Théodule Ribot. *Revue scientifique* (1883, part 2): 687–88.

―――. Review of *La psychologie allemande contemporaine*, by Théodule Ribot. *Revue scientifique* (1879, part 1): 970–74.

―――. "La suggestion mentale et le calcul des probabilités." *Revue philosophique* 18 (1884): 609–71.

Ringer, Fritz. *Fields of Knowledge: French Academic Culture in Comparative Perspective, 1890–1920*. Cambridge: Cambridge University Press, 1992.

Roblet, Nicole, ed. *Annuaire de l'Association Amicale des Anciens Elèves de l'Ecole Normale Supérieure. Table des notices nécrologiques*. Paris: Presses de l'ENS, 1975.

Romanes, George. *Animal Intelligence*. 1882. Reprint. Westmead, England: Gregg International Publishers, 1970.

Rosenau, Pauline Marie. *Post-Modernism and the Social Sciences: Insights, Inroads, and Intrusions*. Princeton: Princeton University Press, 1992.

Roudinesco, Elizabeth. *La bataille de cent ans: Histoire de la psychanalyse en France*. 2 vols. Paris: Seuil, 1986.

Sarton, George. *Sarton on the History of Science*. Cambridge, Mass.: Harvard University Press, 1962.

Schmaus, Warren. *Durkheim's Philosophy of Science and the Sociology of Knowledge: Creating an Intellectual Niche*. Chicago: University of Chicago Press, 1994.

"Séance du 4 février 1888." *Séances et travaux de l'Académie des Sciences Morales et Politiques* 29 (1888, part 1): 651–53.

"Séance du 23 décembre 1899." *Séances et travaux de l'Académie des Sciences Morales et Politiques* 53 (1900, part 1): 265.

Shapin, Steven. "Discipline and Bounding: The History and Sociology of Science as Seen Through the Externalism-Internalism Debate." *History of Science* 30 (1992): 333–69.

Simiand, François. "L'année sociologique 1897." *Revue de métaphysique et de morale* 6 (1898): 608–53.

———. "L'année sociologique française 1896." *Revue de métaphysique et de morale* 5 (1897): 489–519.

Simon, Jules. *Victor Cousin*. Translated by Melville B. Anderson and Edward Anderson. Chicago: McClurg, 1888.

Simon, Walter M. *European Positivism in the Nineteenth Century: An Essay in Intellectual History*. Port Washington, N.Y.: Kennikat Press, 1963.

———. "Two Cultures in Nineteenth-Century France: Victor Cousin and Auguste Comte." *Journal of the History of Ideas* 26 (1965): 45–68.

Simon-Nahum, Perrine. "Xavier Léon/Elie Halévy: Correspondance (1891–1898)." *Revue de métaphysique et de morale* 98 (1993): 3–58.

Sjövall, Björn. *Psychology of Tension*. Stockholm: Norstedt, 1967.

Skinner, Quentin. "Social Meaning and the Explanation of Social Action." In *The Philosophy of History*, edited by Patrick Gardiner, 106–26. Oxford: Oxford University Press, 1974.

Smith, Roger. *The Ecole Normale Supérieure and the Third Republic*. Albany, N.Y.: SUNY Press, 1982.

Spitzer, Alan B. *The French Generation of 1820*. Princeton, N.J.: Princeton University Press, 1987.

Stedman Jones, S. G. "Charles Renouvier and Emile Durkheim: *Les règles de la méthode sociologique*." *Sociological Perspectives* 38 (1995): 27–40.

Stock-Morton, Phyllis. *Moral Education for a Secular Society: The Development of Morale Laïque in Nineteenth-Century France*. Albany: SUNY Press, 1988.

Strikwerda, Robert A. "Emile Durkheim's Philosophy of Science: Framework for a New Social Science." Ph.D diss., University of Notre Dame, 1982.

Taine, Hippolyte. *De l'intelligence*. 2 vols. 11th ed. Paris: Hachette, 1906.

———. *Les philosophes classiques du XIXe siècle en France*. 6th ed. Paris: Hachette, 1888. Reprint. Paris: Slatkine, 1979.

Tarde, Gabriel. *Essais et mélanges sociologiques*. Paris: G. Masson, 1895.

———. *Etudes de psychologie sociale*. Paris: V. Giard & E. Brière, 1898.

————. *La logique sociale*. Paris: Alcan, 1895.

————. *L'opinion et la foule*. Paris: Alcan, 1901.

Telzrow, Thomas M. "The Watchdogs: French Academic Philosophy in the Nineteenth Century—The Case of Paul Janet." Ph.D. diss., University of Wisconsin, 1973.

Le Temps (Paris). 8 December 1873.

"Théodule Ribot." *La grande encyclopédie,* n.d.

Thibaudet, Albert. *La république des professeurs*. Paris: B. Grasset, 1927.

Thirard, Jacqueline. "La fondation de la 'Revue philosophique.'" *Revue philosophique* (1976): 401–13.

Toulmin, Stephen. *Human Understanding: The Collective Use and Evolution of Concepts*. Princeton: Princeton University Press, 1972.

Trillat, Etienne. *Histoire de l'hystérie*. Paris: Editions Seghers, 1986.

Vacherot, Etienne. "Rapport verbal sur un ouvrage de M. Ribot intitulé: *La psychologie anglaise contemporaine.*" *Séances de l'Académie des Sciences Morales et Politiques* 92 (1870): 467–72.

Vandérem, Fernand, ed. *Pour et contre l'enseignement philosophique*. Paris: Alcan, 1894.

Vogt, W. Paul. "Identifying Scholarly and Intellectual Communities: A Note on French Philosophy." *History and Theory* 21 (1982): 267–78.

————. "Political Connections, Professional Advancement, and Moral Education in Durkheimian Sociology." *Journal of the History of the Behavioral Sciences* 27 (1991): 56–75.

Weaver, Richard M. "The Rhetoric of Social Science." In *The Ethics of Rhetoric*, 187–91. Chicago: Henry Regney, 1968.

Weinstein, Leo. *Hippolyte Taine*. New York: Twayne Publishers, 1972.

Weisz, George. *The Emergence of Modern Universities in France, 1863–1914*. Princeton: Princeton University Press, 1983.

Wesley, Frank, and Michel Hurtig. "Masters and Pupils among French Psychologists." *Journal of the History of the Behavioral Sciences* 5 (1969): 320–25.

Wolf, Jan Jacob de. "Wundt and Durkheim: A Reconsideration of a Relationship." *Anthropos* 82 (1987): 1–23.

Worms, René. "Charles Letourneau." *Revue internationale de sociologie* 10 (1902): 89–91.

Wundt, Wilhelm. "Ueber den gegenwärtigen Zustand der Thierpsychologie." *Vierteljahrschrift für wissenschaftliche Philosophie* 2 (1978): 137–49.

Yvert, Benoît, ed. *Dictionnaire des ministres, 1789–1989*. Paris: Perrin, 1990.

Zeldin, Theodore. *France, 1848–1945: Anxiety and Hypocrisy*. Oxford: Oxford University Press, 1981.

————. *France, 1848–1945: Intellect and Pride*. Oxford: Oxford University Press, 1980.

————. *France, 1848–1945: Taste and Corruption*. Oxford: Oxford University Press, 1980.

Index